Tree Physiology

VOLUME 3

Series Editor:
Professor Heinz Rennenberg, *University of Freiburg, Germany*

Aims and scope

The series **"Tree Physiology"** is aimed to cover recent advances in all aspects of the physiology of woody plants, i.e.: environmental physiology including plant-atmosphere, plant-pedosphere/hydrosphere, and organismic interactions; whole tree physiology including storage and mobilization as well as maturation and senescence; woody plant tissue culture and molecular physiology for micropropagation, transformation of chemicals, preservation of gen-pools, membrane transport, transformation of tree species, etc. In addition, technological advances in tree physiology will be covered together with the achievements obtained with these techniques.

Although the main emphasis will be on basic research, applied questions will also be addressed. It is the scope of the series to provide advanced students in forestry and plant biology as well as researchers working in the various fields of tree physiology and woody plant physiology. Since also applied questions will be addressed, part of the series will also be of interest for environmental and applied biologists.

The titles published in this series are listed at the end of this volume.

TRACE GAS EXCHANGE IN FOREST ECOSYSTEMS

TRACE GAS EXCHANGE IN FOREST ECOSYSTEMS

Edited by

R. GASCHE
Institut Atmosphärishe Umweltforschung,
Garmisch-Partenkirchen, Germany

H. PAPEN
Institut Atmosphärishe Umweltforschung,
Garmisch-Partenkirchen, Germany

and

H. RENNENBERG
Albert-Ludwig Universität,
Institut für Forstbotanik und Baumphysiologie, Freiburg, Germany

KLUWER ACADEMIC PUBLISHERS
DORDRECHT / BOSTON / LONDON

A C.I.P. Catalogue record for this book is available from the Library of Congress.

ISBN 1-4020-1113-X

Published by Kluwer Academic Publishers,
P.O. Box 17, 3300 AA Dordrecht, The Netherlands.

Sold and distributed in North, Central and South America
by Kluwer Academic Publishers,
101 Philip Drive, Norwell, MA 02061, U.S.A.

In all other countries, sold and distributed
by Kluwer Academic Publishers,
P.O. Box 322, 3300 AH Dordrecht, The Netherlands.

Printed on acid-free paper

Printed in the Netherlands.

Dedicated to Prof. Dr. Wolfgang Seiler for his outstanding contributions to Biosphere-Atmosphere Exchange Research

Contents

Contributors

Baldocchi, Dennis D. (229–242)
 Department of Environmental Science, Policy and Management, 151 Hilgard Hall,
 University of California, Berkeley,Berkeley, CA 94720, United States of America
Butterbach-Bahl, Klaus (141–156)
 Fraunhofer Institut für Atmosphärische Umweltforschung, Abteilung für
 Bodenmikrobiologie, Kreuzeckbahnstrasse 19, D-82467 Garmisch-Partenkirchen,
 Germany
Cape, J. Neil (245–255)
 Centre for Ecology and Hydrology, Bush Estate, Penicuik, Midlothian, EH26 0QB, United
 Kingdom
Clough, Elisabeth C. M. (53–77)
 Department of Biology, University College London, Gower Street, London, WCIE 6BT,
 United Kingdom
Conrad, Ralf (3–33)
 Max-Planck-Institut für terrestrische Mikrobiologie, Karl-von-Frisch-Str., D-35043
 Marburg, Germany
Erisman, Jan W. (159–173)
 Energy Research Centre of the Netherlands (ECN), Petten, The Netherlands
Ernst, Dieter (307–324)
 Institute of Biochemical Plant Pathology, GSF - National Research Center for
 Environment and Health, D-85764 Neuherberg, Germany
Forkel, Renate (257–276)
 Fraunhofer Institut für Atmosphärische Umweltforschung, Kreuzeckbahnstrasse 19, D-
 82467 Garmisch-Partenkirchen, Germany
Gasche, Rainer (117–140)
 Fraunhofer Institut für Atmosphärische Umweltforschung, Abteilung für
 Bodenmikrobiologie, Kreuzeckbahnstrasse 19, D-82467 Garmisch-Partenkirchen,
 Germany
Geunther, Alex (175–191)
 Atmospheric Chemistry Division, National Center for Atmospheric Research, Boulder CO
 80307, United States of America

Hall, Sharon J. (279–306)
   Environmental Science Program, The Colorado College, Colorado Springs, CO 80903,
   United States of America
Kreuzwieser, Jürgen (101–114; 193–209)
   Institut für Forstbotanik und Baumphysiologie, Professur für Baumphysiologie, Georges-
   Köhler-Allee Geb. 053/054, D-79110 Freiburg i. Br., Germany
Langebartels, Christian (307–324)
   Institute of Biochemical Plant Pathology, GSF - National Research Center for
   Environment and Health, D-85764 Neuherberg, Germany
Lichtenthaler, Hartmut K. (79–99)
   Bontany II, University of Karlsruhe, D-76128 Karlsruhe, Germany
Matson, Pamela A. (279–306)
   Department of Geological and Environmental Sciences, Standford University, Stanford,
   CA 94305, United States of America
Nielsen, Kent H. (53–77; 159–173))
   Plant Nutrition Laboratory, Department of Agricultural Sciences, Royal Veterinary and
   Agricultural University, Thorvaldsensvej 40, 8, 5., DK-1871 Frederiksberg C, Denmark
Papen, Hans (117–140)
   Fraunhofer Institut für Atmosphärische Umweltforschung, Abteilung für
   Bodenmikrobiologie, Kreuzeckbahnstrasse 19, D-82467 Garmisch-Partenkirchen,
   Germany
Pearson, John (53–77; 159–173)
   Department of Biology, University College London, Gower Street, London, WCIE 6BT,
   United Kingdom
Sandermann, Heinrich (307–324)
   Institute of Biochemical Plant Pathology, GSF - National Research Center for
   Environment and Health, D-85764 Neuherberg, Germany
Schjørring, Jan K (53–77; 159–173)
   Plant Nutrition Laboratory, Royal Veterinary and Agricultural University, Copenhagen
   Denmark
Seufert, Guenther (175–191)
   Environment Institute, Joint Research Centre, T.P. 051, 21020 Ispra, Italy
Steinbrecher, Rainer (175–191)
   Fraunhofer Institut für Atmosphärische Umweltforschung, Kreuzeckbahnstrasse 19, D-
   82467 Garmisch-Partenkirchen, Germany
Stockwell, William R. (257–276)
   Desert Research Institute (DRI), Reno, Nevada, United States of America
Thomas, Gabriele (307–324)
   Institute of Biochemical Plant Pathology, GSF - National Research Center for
   Environment and Health, D-85764 Neuherberg, Germany
Vogg, Gerd (307–324)
   Institute of Biochemical Plant Pathology, GSF - National Research Center for
   Environment and Health, D-85764 Neuherberg, Germany
Wellburn, Alan R. (35–52)
   Department of Biological Sciences, Lancaster University, Lancaster LA1 4YQ,, United
   Kingdom
Wieser, Gerhard (211–226)
   Forstliche Bundesversuchsanstalt, Abteilung Forstpflanzenphysiologie, Rennweg 1, A-
   6020 Innsbruck, Austria

Wildt, Jürgen (307–327)
   Institut of Chemistry and Dynamics of the Geosphere (ICG-III Phytosphere),
   Forschungszentrum Jülich, D-52425 Jülich, Germany
Wilson, Kell (229–242)
   Atmospheric Turbulence and Diffusion Division, NOAA/ATDD, PO Box 2456, Oak
   Ridge, TN, United States of America
Woodall, Janet (53–77)
   Department of Biology, University College London, Gower Street, London, WCIE 6BT,
   United Kongdom
Zeidler, Johannes G. (79–99)
   Bontany II, University of Karlsruhe, D-76128 Karlsruhe, Germany

1

BIOLOGICAL PROCESSES INVOLVED IN TRACE GAS EXCHANGE

Chapter 1.1

Microbiological and biochemical background of production and consumption of NO and N_2O in soil

Ralf Conrad
Max-Planck-Institut für terrestrische Mikrobiologie, Karl-von-Frisch-Str., D-35043 Marburg, Germany

1.   INTRODUCTION

Nitric oxide (NO) and nitrous oxide (N_2O) play a central role in the chemistry of the troposphere and stratosphere, respectively, and affect the habitability of Earth (Crutzen 1979; Logan 1983). Recent evaluations indicate that soils are a major source for NO and N_2O, with a source strength of ca. 20 Tg NO-N yr^{-1} (Davidson and Kingerlee 1997; Skiba et al. 1997) and ca. 7 Tg N_2O-N yr^{-1} (Bouwman et al. 1995; Cole et al. 1997), thus contributing as much as 20% and 70% to the total budget of atmospheric NO and N_2O, respectively (Conrad 1996b).

The flux of NO and N_2O between soil and atmosphere is the result of dynamic production and consumption processes in soil. The processes that may be involved in determining the flux between soil and atmosphere of NO (Johansson 1989; Conrad, 1990; Davidson 1991; Williams et al. 1992; Meixner 1994; Conrad 1995; 1996a, b; Ludwig et al. 2001) and N_2O (Knowles 1985; Sahrawat and Keeney 1986; Firestone and Davidson 1989; Granli and Boeckman 1994; Conrad 1995, 1996b; Davidson and Schimel 1995; Beauchamp 1997; Davidson and Verchot 2000) have been reviewed by several authors. There is agreement that chemical reactions in soil play no role in the turnover of N_2O (Bremner et al. 1980), but may be significant for the turnover of NO under certain circumstances (Conrad 1996a, b); for example, chemical decomposition of high nitrite concentrations (Blackmer and Cerrato 1986; Davidson 1992; Venterea and Rolston 2000), or chemical

3

R. Gasche et al. (eds.), Trace Gas Exchange in Forest Ecosystems, 3–33.

oxidation of NO with O_2 to NO_2 in the presence of acetylene (Bollmann and Conrad 1997b; McKenney et al. 1997). Under most soil conditions, however, the turnover of both NO and N_2O is caused by biochemical reactions taking place within microbial cells.

It is obvious that any flux of NO and N_2O between the soil and the atmosphere is ultimately caused by microbial processes. So far, most work used a macroscopic approach by measuring fluxes and turnover rates on various soil plots or soil samples under all kinds of natural or experimentally controlled conditions. Other investigations concentrated on studying microbial cultures, also looking at biochemical and genetic details. The knowledge obtained from the microbial studies is essential to interpret the observations from soil experiments. It is still completely unknown, however, which physiotypes and taxa of microorganisms are exactly involved in the NO and N_2O turnover within a particular soil sample, which biochemical background these microorganisms have and how they use it under which soil conditions. The functioning of soil processes on the microscopic scale of individual microorganisms has not even been tackled. Knowledge is missing, since the appropriate tools have not been available. During the last decades, however, microbiologists learned to affiliate individual microorganisms within a phylogenetic system (Woese 1987; Amann et al. 1995; Stackebrandt et al. 1999). It is now possible to exactly classify microorganisms based on their molecular structures. This possibility is presently revolutionizing the taxonomic classification of microorganisms. Our knowledge emphasizes that most (>99%) of the soil microorganisms are still unknown (Torsvik et al. 1990). Unambiguous taxonomy is the basis for learning and describing the biochemical machinery that is characteristic of each individual microorganism. Hence, it is just now that we may start thinking of investigating the functioning of NO and N_2O turnover on a microscopic scale in soil.

For this purpose, I will briefly summarize our knowledge about the principal NO and N_2O turnover processes in soil, which microbial physiotypes and taxa may be involved, and what the principal enzymatic reactions are that produce and consume NO and N_2O in soil.

2.   NO AND N_2O TURNOVER BY MICROBIAL OXIDATION AND REDUCTION PROCESSES

Production and consumption of NO and N_2O has been demonstrated in a large number of different organisms, and has been reviewed (Knowles 1985; Zumft 1993; Stamler and Feelisch 1996; Conrad 1995; 1996a, b). The two most important groups of soil organisms seem to be the denitrifiers and the nitrifiers. However, there are also other microorganisms that produce and consume NO and/or N_2O. All these microorganisms are phylogenetically diverse and differ in their metabolic pathways. Despite the large diversity of different microorganisms and metabolic reactions, one can basically distinguish two distinct types of microbial processes in soil, one involving the reduction of nitrate and the other the oxidation of ammonium. Please note, that the terms nitrification, nitrifiers and denitrification, denitrifiers are avoided in the next paragraph, since they need a careful biochemical definition.

2.1   Differentiation of production processes

In general, process-based studies of production of NO and N_2O in soil have attempted to distinguish between production by reduction of nitrate and production by oxidation of ammonia. For this purpose three different approaches were most widely used.

(1) The simplest approach is to stimulate NO and N_2O production by defined addition of substrates. For example, stimulation by addition of nitrate versus ammonium indicates that production by nitrate reduction is more important than by ammonium oxidation. Such stimulation experiments have also been applied in field experiments (Breitenbeck et al. 1980; Conrad and Seiler 1980; Slemr and Seiler 1984; Sanhueza et al. 1990). Although the conclusions drawn from such experiments seem to be plausible, it should be considered that they are only valid for the condition with the surplus nitrogen. Without the surplus in nitrogen, the pathway of NO or N_2O production may be completely different. Moreover, stimulation of NO or N_2O production by the addition of ammonium may be complex, since ammonium may first be oxidized to nitrate which can then be reduced to NO and N_2O. Such coupled oxidation-reduction pathways have often been reported for aquatic sediments where nitrogen is limiting, but they may also apply for soil sites with high C/N ratio (Rysgaard et al. 1993; Nielsen et al. 1996).

Furthermore, oxidative processes can be inhibited by incubation of soil under anoxic conditions, while reductive processes can be inhibited by incubation of soil at the presence of pure O_2. Both techniques have frequently been used in laboratory studies, mostly in combination with addition of nitrate or ammonium or of other inhibitors (see below) to differentiate and quantify NO and N_2O production by oxidative or reductive pathways (Robertson and Tiedje 1987; Remde and Conrad 1991b; Papen et al. 1993).

(2) The most straightforward approach to determine the production pathway of NO or N_2O is the application of isotopically labeled $^{15}NH_4^+$ and

$^{15}NO_3^-$ (Remde and Conrad 1991b; Stevens et al. 1997). The isotopic analysis of the NO and N_2O (also N_2) produced allows to distinguish between ammonium or nitrate as the primary source of the gaseous nitrogen compounds. This approach is suited for laboratory studies as well as for field work. Since the technique requires the addition of isotopically enriched nitrogen compounds, it has to be considered that this addition may result in stimulated microbial activities if the amount of added nitrogen is of considerable magnitude as compared to the endogenous nitrogen pool, i.e. at least at sites that are characterized by N-limitation. Even more problematic is the necessity to homogeneously mix the added isotopes with the natural nitrogen pool, thus the isotopes reach the same microsites where the microorganisms would produce NO and N_2O even in the absence of any addition. Even slight inhomogeneities may thus bias the result of isotopic studies (Davidson and Hackler 1994). A possible solution may be the application of isotopic nitrogen in gaseous form, e.g. application of NO_2 instead of addition of nitrite/nitrate (Stark and Firestone 1995) and addition of gaseous NH_3 instead of ammonium.

(3) The most widely used approach to distinguish between oxidative and reductive production pathways of NO and N_2O is the application of "specific" inhibitors against ammonium-oxidizing bacteria, e.g. nitrapyrin (N-serve), dicyandiamide, methyl fluoride, dimethyl ether, acetylene etc.. Acetylene is probably the most suitable inhibitor, since it is gaseous and thus easily distributed within the soil, is cheap, and its reaction mechanism, which is an irreversible inhibition of the enzyme ammonium monooxygenase, is well understood (Hyman and Wood 1985; Hyman and Arp 1992). A partial pressure of 1-10 Pa acetylene is sufficient to completely inhibit the oxidation of ammonium in soil (Berg et al. 1982; Klemedtsson et al. 1988). Application of higher partial pressures is problematic for two reasons: Since the reduction of N_2O by the enzyme nitrous oxide reductase is inhibited at elevated acetylene partial pressures (completely inhibited at <10 kPa acetylene), N_2O production rates may be overestimated due to inhibition of simultaneous N_2O consumption. Acetylene partial pressures >0.1 kPa in the presence of >20 Pa O_2 catalyze in the oxidation of NO to NO_2 and thus result in an underestimation of NO production (Bollmann and Conrad 1997a, b; McKenney et al. 1997). Low partial pressures of acetylene are safe in this respect, but complete removal of acetylene may result in a rather rapid recovery of ammonium oxidation probably due to *de-novo* synthesis of new enzyme (Bollmann and Conrad 1997c). Coupled oxidation-reduction pathways, as discussed above for the stimulation experiments, would be inhibited by acetylene even at low partial pressures. The choice of the correct partial pressure for avoiding artifacts makes the acetylene inhibition

technique very problematic for application under field conditions (Klemedtsson et al. 1990; Arah et al. 1991; Henault and Germon 1995).

Inhibitors of ammonium oxidation have also been used for differentiating between so-called autotrophic and heterotrophic nitrification. This test is based on the assumption that only autotrophic nitrifiers possess the inhibitor-sensitive ammonium monooxygenase which, however, is not necessarily true (see below). It should be noted that the term "heterotrophic nitrification" actually comprises two different activities. Strictly speaking, heterotrophic nitrifiers would use organic compounds as carbon source for cell biomass formation, but oxidize ammonium similar to autotrophic nitrifiers, e.g. *Alcaligenes faecalis* (Papen et al. 1989; Anderson et al. 1993), *Paracoccus denitrificans* (Moir et al. 1996), *Pseudomonas putida* (Daum et al. 1998). A completely different microbial process, which unfortunately is also called "heterotrophic nitrification", is the oxidation of organic nitrogen compounds, instead of ammonium, to nitrate (Schimel et al. 1984; Killham 1986; Pedersen et al. 1999). This process does not use the enzyme ammonium monooxygenase and thus, is not inhibited by acetylene, nitrapyrin etc. (see below).

It should be noted that specific inhibitors for the reduction of nitrate or nitrite are so far missing. Such inhibitors would be extremely helpful, since they would allow verification of the results obtained with the use of inhibitors of ammonium oxidation. The only option is to generally inhibit all anaerobic processes by incubation under pure O_2 (Robertson and Tiedje 1987; Papen et al. 1993). However, this option is problematic since it does not exclude the persistence of anoxic microniches due to microbial O_2 consumption. Chlorate and chlorite have occasionally been used for inhibition of nitrate/nitrite reduction (Remde and Conrad 1990; 1991a), but these compounds may cause artifacts due to inhibition of other processes (e.g. heterotrophic nitrification; Bauhus et al. 1996) or insensitivity of periplasmic nitrate reductases against chlorate (Zumft 1997).

2.2 Differentiation of consumption processes

Although soils are usually net sources for atmospheric NO and N₂O, they can also act as sinks, at least temporarily (Ryden 1981; Slemr and Seiler 1984). The flux of NO and N₂O obtained at the soil/atmosphere interface is the result of dynamic production and consumption processes in soil. Most studies concentrated on the production of NO and N₂O and neglected consumption processes. The reason for this is that production rates are usually larger than consumption rates and therefore, net production is the overall obtained result, which in turn means that consumption is masked. However, much of the N₂O that is produced within the soil column may be consumed and never reach the atmosphere (Seiler and Conrad 1981; Arah et al. 1991). In the case of NO, turnover is even more dynamic and results in NO lifetimes within the soil of just a few minutes and the establishment of a compensation concentration which is often comparable to the ambient NO concentration (Conrad 1994; Rudolph et al. 1996b, Conrad and Dentener 1999). It also should be realized that negligence of consumption prohibits the accurate determination of the factors which regulate the resulting net production. Since the technical effort for measuring both production and consumption of NO is not greater than measuring just the net result of the two processes, there is no reason to confine oneself to the latter and lose valuable information (Gödde and Conrad 1998; Bollmann et al. 1999).

Consumption of N_2O is to our knowledge only accomplished by reduction to N_2. Oxidation of N_2O by soil catalase and peroxidase has been mentioned only in two publications, but never been investigated again (Knowles 1985). On the other hand, NO can be both oxidized and reduced by soil microorganisms (Remde and Conrad 1991a; Rudolph et al. 1996a). Differentiation of these two consumption pathways is so far only possible by relatively cumbersome kinetic studies (Conrad 1995; Koschorreck and Conrad 1997; Dunfield and Knowles 1997; 1998) or by application of isotopically enriched ^{15}NO (Rudolph et al. 1996a). The latter technique is problematic under environmentally realistic conditions because of the low NO concentrations in the soil atmosphere and problems with the detection limit of GC-MS systems (Sich and Russow 1998).

Therefore, a quantification of the contribution of oxidative and reductive NO consumption processes to total NO consumption has so far not been achieved and thus, studies of the regulation of these two processes under real soil conditions are still missing. Whereas the reductive NO consumption is part of the general microbial nitrate reduction sequence and thus seems to be connected to bacteria reducing nitrate and/or nitrite, the oxidative NO consumption seems to be widespread among heterotrophic microorganisms (Koschorreck et al. 1996). It is noteworthy that the oxidative NO consumption has a much lower affinity for NO than the reductive consumption, but seems to be dominant in many soils (Koschorreck and Conrad 1997; Dunfield and Knowles 1998).

2.3   Regulation of NO and N_2O turnover by soil conditions

A crucial question is the regulation of NO and N_2O turnover by environmental conditions. This question has been studied on the level of net fluxes between soil and atmosphere both in the field and in laboratory incubations of intact soil columns or soil samples from various soil horizons, and on the level of oxidative versus reductive production processes (for literature see reviews cited in the Introduction). The observed fluxes are typically log-normally distributed and cover about 4 orders of magnitude. Hence, a highly dynamic regulation on the microbial process level is quite likely, which results in highly variable fluxes both temporarily and spatially. Quite conspicuous is a hot-spot phenomenon, i.e. that a macroscopic flux is sometimes caused by only a small fraction of the area investigated (Parkin 1987; Ambus and Christensen 1994). At the moment we understand very little about the regulation of NO and N₂O turnover in-situ. The problems are compounded by the fact that the effect of single soil variables cannot be separated from others since soil variables influence each other. A conceptual model has been proposed in which various environmental regulators affect ammonium oxidation and nitrate reduction in a hierarchical way (Tiedje 1988; Robertson 1989). For example, precipitation events cause an increase in soil water content which stimulates respiratory processes and impedes gaseous diffusion. This results in local consumption of O_2 thus enhancing anaerobic microbial metabolism and possibly stimulating reductive NO and N₂O production. On the other hand, however, diffusion of NO and N₂O is also impeded and thus consumption of these gases is stimulated. Soil moisture affects not only the metabolism of nitrogen but also of carbon, which in itself is another soil variable that regulates the turnover of NO and N₂O.

For mathematical models of NO and N₂O soil fluxes the following variables are useful (since data can relatively easily be obtained) and have in fact been incorporated into models (Li et al. 1996; Potter et al. 1996; Nevison et al. 1996; Parton et al. 1996; Wang et al. 1997; Frolking et al. 1998; Li et al. 2000; Stange et al. 2000): temperature, soil moisture, soil texture, vegetation (primary productivity), and fertilizer usage. In process models, the actual NO and N₂O production rates are obtained in submodels describing the oxidation of ammonium and the reduction of nitrate and assuming that a certain percentage of the converted nitrogen is lost as NO and N₂O depending on the environmental conditions.

In numerous studies, soil moisture has been found to be one of the most important regulators (Davidson 1991). Using various circumstantial evidence it has been suggested that NO and N₂O are the dominant products below and above field capacity, respectively, and that NO and N₂O are mainly produced by nitrification and denitrification, respectively (Anderson and Levine 1987; Davidson 1991; Papen and Butterbach-Bahl 1999; Gasche and Papen 1999). This pattern has largely been confirmed in a systematic study, in which NO and N₂O production by nitrification and denitrification was quantified in soil samples incubated under defined O_2 partial pressures and at different moisture regimes (Bollmann and Conrad 1998). It should also be noted that the theoretical basis of these empirically obtained patterns is unclear or at least imprecise. Although it is established that the microbial enzymes that are critical for NO and N_2O production are regulated by O_2 (see below), it is by no means clear why exactly such a pattern as observed in experiments with soil has to result.

Comparatively little is known about the regulation of NO and N_2O production by other soil variables. Decreasing soil pH seems to increase the proportion of NO and N_2O on the products of nitrogen conversion (Martikainen 1985; Nägele and Conrad 1990; Martikainen and DeBoer 1993; Stevens et al. 1998). This may partially be caused by enhancing the chemical decomposition of the nitrite accumulated during ammonium oxidation (Venterea and Rolston 2000). Increasing temperature results in increasing enzyme activities and thus in increasing turnover rates, but may differentially affect the activity of consecutive enzymatic steps resulting in accumulation of intermediate NO and N_2O (McKenney et al. 1984). In addition, soil microbial communities may adapt to environmental changes by selecting for populations with a different pH or temperature optimum (Parkin et al. 1985; Saad and Conrad 1993a; DeBoer et al. 1995; Gödde and Conrad 1999). Furthermore, production and consumption processes may be stimulated to varying extents thus making a prediction of the net effect problematic (Saad and Conrad 1993b). Multivariate statistical analysis of NO and N_2O production in various soils indicates that vegetation, pH, ammonium content and texture were major regulating variables besides soil moisture and temperature (Robertson 1994; Gödde and Conrad 2000).

Removing the relatively strong regulatory effects of temperature and moisture by incubating soil samples at constant 25°C and at constant 60% water holding capacity (WHC), the production rates of NO and N_2O in soils from different locations still range over 2 orders of magnitude (Fig. 1A). Similarly, the contribution of nitrate reduction to NO and N_2O production ranges from zero to 100%, although the relative soil moisture is the same (Fig. 1B). However, for the chosen conditions (25°C, 60% WHC) the following two trends are striking: (1) Rates of NO production are almost an order of magnitude higher than rates of N_2O production, stressing the relative importance of NO compared to N_2O as trace gas emitted from soil. Similar trends have been observed by Keller and Reiners (1994) and Butterbach et al. (1997). (2) The contribution of nitrate reduction to N_2O production is consistently larger than to NO production. NO is apparently mainly produced from ammonium oxidation. Other investigators have come to the same conclusion (Skiba et al. 1997). This could partially be explained by different co-ordination of NO production and NO consumption in nitrate-reducing and ammonium-oxidizing microorganisms (see below).

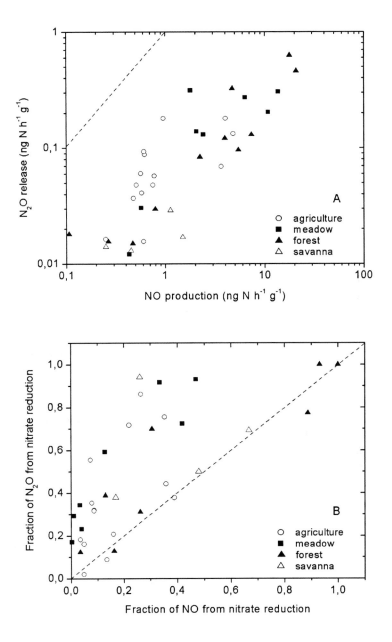

Figure 1. Correlation of (A) the rates of N₂O production with those of NO production, and (B) the fractions of N₂O production with those of NO production both derived from nitrate reduction, all assayed in soils from various sites at uniform temperature (25°C) and soil moisture (60% WHC); data are from Bollmann and Conrad (1997a) and Gödde and Conrad (2000); the dashed line indicates a 1:1 ratio of the correlated data.

In summary, little is known about regulation of NO and N_2O turnover in soil, and of what is known is based on empirical observations but is not understood theoretically on the basis of microbial metabolism. Possibly this lack of theoretical understanding is due to the unexpected high diversity of microbial physiotypes, taxa and regulatory patterns that are potentially involved in NO and N_2O turnover.

3.   METABOLISM OF NO AND N_2O BY MICROBIAL PHYSIOTYPES, TAXA, AND ENZYMES

The enzyme machinery which is ultimately responsible for production and consumption of NO and N_2O and for regulation of these processes has gradually evolved since the beginning of life. At present, we know almost nothing about this evolution, but realize that the enzyme machinery of extant "species" of microorganisms is more or less different when comparing one "species" with the other.

A biological species is defined as "groups of interbreeding or potentially interbreeding populations that are reproductively isolated from other such groups" (E. Mayr cited in Stackebrandt et al. 1999). The biological species concept cannot be applied to life forms which do not reproduce sexually. Soil microorganisms mainly consist of fungi and bacteria. Bacterial life forms which phylogenetically belong to the domains *Bacteria* or *Archaea* do not reproduce sexually and populations are mainly clonal in structure, although some genetic exchange exists not only within but also among different "species" (Cohan 1994). The classical species concept cannot be applied to *Bacteria* or *Archaea*. The microbe that is isolated from soil does not represent a species but a strain. However, the isolate can be phylogenetically affiliated within the tree of life by using the genetic sequence information of conservative macromolecules such as the ribosomal RNA (usually the 16S rRNA) which is one of the reliable phylogenetic markers (Stackebrandt et al. 1999). For pragmatic reasons, a bacterial "species" is defined by a >70% DNA similarity (identical base pair sequence). Usually, this is the case when the 16S rRNA have >97% similarity (Stackebrandt and Goebel 1994). This bacterial "species" concept has to be borne in mind when comparing the enzymatic machinery among different microorganisms.

3.1   Microbial physiotypes and taxa

Modern bacterial taxonomy takes the similarities and dissimilarities of phylogenetic markers into account. However, not all of the described bacterial "species" have yet been reclassified on the basis of phylogenetic markers. Most of the described bacterial "species" were isolated as physiotypes with a more or less conspicuous physiology: e.g. nitrifiers growing on ammonium, O_2 and CO_2 as sole sources for energy and carbon. Some of the microbial physiotypes, e.g. many nitrifiers, are phylogenetically related (monophyletic) and classified into a few taxa. Others, however, e.g. most of the denitrifiers, are polyphyletic (or paraphyletic) and occur in many different taxa. Nevertheless, they share enzymes with a high similarity both in structure (based on their amino acid sequences) and in function, indicating that the responsible genes emerged early in evolution or were later on transferred laterally among different "species". Various physiotypes and taxa

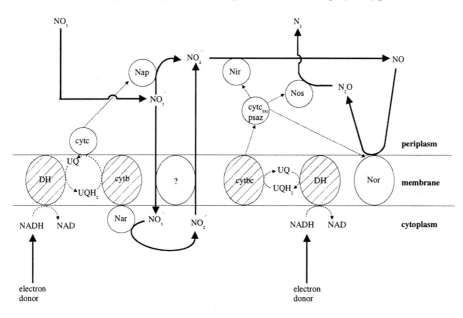

Figure 2. Topology and organisation of the enzymes and proteins involved in nitrogen flow (thick arrows) and electron flow (dashed arrows) during denitrification in *Paracoccus denitrificans* and *Pseudomonas stutzeri*. The hatched membrane-bound proteins are the sites of vectorial electron transport. DH = NADH dehydrogenase complex; Nar, Nap = nitrate reductases; Nir = nitrite reductase, Nor = NO reductase; Nos = N₂O reductase; UQ/UQH₂ = quinone cycle; cyt = cytochrome; psaz = pseudoazurin. Adapted from Berks et al. (1995) and Zumft (1997).

may be involved in the reduction of nitrate and the oxidation of ammonium. The relatively well studied pathways of the physiotypes of denitrification and autotrophic ammonium oxidation are shown in Fig. 2 and 3, respectively, together with the topology of the enzymes involved (see below).

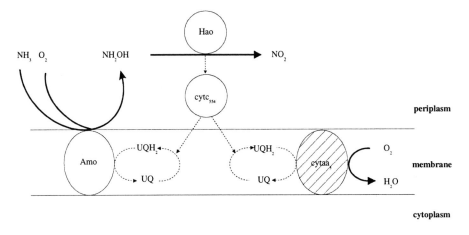

Figure 3. Topology and organisation of the enzymes and proteins involved in nitrogen flow (thick arrows) and electron flow (dashed arrows) during ammonium oxidation in *Nitrosomonas europaea*. Amo = ammonium monooxygenase; Hao = hydroxylamine oxidoreductase; for the other abbreviations see Fig. 1.1-2. Adapted from Berks et al. (1995).

3.1.1   Denitrification (sensu stricto)

Denitrifiers are aerobic microorganisms which in the absence of O_2 can switch to anaerobic denitrification (Tiedje 1988). Some denitrifiers can denitrify even in the presence of O_2 (Kuenen and Robertson 1988), including fungi (Zhou et al. 2001). Denitrification is the sequential reduction of

$$\text{nitrate} \rightarrow \text{nitrite} \rightarrow NO \rightarrow N_2O \rightarrow N_2,$$

which is driven by the oxidation of organic or inorganic (H_2, H_2S, Fe^{++}) substrates (electron donors). The reduction of nitrate, nitrite, or N_2O is coupled to energy conservation by electron transport phosphorylation, and allows the microorganisms to grow (Tiedje 1988; Berks et al. 1995; Fig. 2). The various denitrifier physiotypes can be distinguished by the type of electron donors used and by the extent to which nitrate, nitrite, and N_2O all serve as electron acceptors. Many denitrifiers exist which have truncated nitrate reduction pathways, e.g. reducing nitrate only to N_2O or using only nitrite or N_2O as electron acceptors. Denitrifiers are found in many prokaryotic taxa, including the domain *Archaea* (*Halobacteriaceae*), the phyla *Proteobacteria* (α, β, γ, ε subclass), *Cytophaga-Flavobacterium-Bacteroides*, and *Gram-positive Bacteria* (*Actinomycetales, Bacillus* sp.) (Zumft 1992; 1997). Denitrification is even found in eukaryotic fungi (Shoun et al. 1992). Production and consumption of NO and N$_2$O by denitrifiers is obvious, since both compounds are intermediates in the denitrifying pathway. Truncated denitrification pathways, with reduction of N$_2$O to N$_2$ lacking, are frequently observed in fungi and actinomycetes but also in many other bacteria (Zumft 1997; Albrecht et al. 1997). Truncated denitrification may substantially contribute to N$_2$O production in soil. In non-truncated denitrification, the extent of NO and N$_2$O turnover is basically a matter of enzyme regulation which, however, is highly diverse and complicated (see below).

3.1.2   Other nitrate reduction physiotypes

In contrast to denitrification, NO and N$_2$O are not intermediates in the other nitrate-reducing physiotypes, i.e. dissimilatory nitrate respiration, nitrate ammonification, nitrate fermentation, and assimilatory nitrate reduction. Nevertheless, production of traces of NO and N$_2$O have been observed in all of these physiotypes (Knowles 1985; Tiedje 1988). The biochemical mechanism of the production is not understood, but the production rates are sufficiently high to be of potential relevance for soil-atmosphere exchange.

Nitrate respiration is a physiotype in which nitrate is just reduced to nitrite which is excreted. Nitrate ammonification sequentially reduces

nitrate → nitrite → ammonia.

Similar to denitrifiers, nitrate respirers and nitrate ammonifiers are facultatively anaerobic microorganisms which are able to gain energy from the reduction process by electron transport phosphorylation. Nitrate respiration and nitrate ammonification is a common physiotype among the enterobacteria (γ-*Proteobacteria*), the sulfate/sulfur reducers (δ, ε-*Proteobacteria*) and the homoacetogenic *Clostridium thermoacticum* (Seifritz et al. 1993). It has been suggested that nitrate ammonification would take place in environments that are limited in nitrate but have a sufficient supply of organic carbon substrates as electron donors, while denitrification would take place in environments that are rather limited by electron donors (Tiedje 1988). However, nitrate respirers and nitrate ammonifiers are often numerous in forest soils and do produce NO and N$_2$O (Tiedje 1988; Kalkowski and Conrad 1991; Blösl and Conrad 1992). Nitrate-limited sediments, on the other hand, may be dominated by denitrification as well as nitrate ammonification (Rysgaard et al. 1993; 1996), suggesting that the reasons for the occurrence of the one or the other physiotype in nature are not understood.

Nitrate fermentation similarly as nitrate ammonification sequentially reduces nitrate via nitrite to ammonia. However, these reactions are not coupled to energy conservation, but just serve as electron balance during fermentation of organic substrates (Tiedje 1988; Allison and MacFarlane 1989). This physiotype is thus found among fermenting bacteria, such as clostridia (*Gram-positive Bacteria low G+C*) and propionibacteria (*Actinomycetales*).

Nitrate assimilation serves the incorporation of nitrogen into cell biomass and is widespread among microbial and plant life (Payne 1973; Cole 1988). Basically it is the same sequential reduction process as found in nitrate ammonification and nitrate fermentation, but here it serves assimilatory purposes and is typically expressed under ammonium-limiting conditions. In presence of ammonium, assimilatory nitrate reduction is switched off.

3.1.3   Nitrification

Autotrophic nitrification is an aerobic process which requires two physiotypes, one oxidizing ammonia and the other oxidizing nitrite (Wood 1986; Bock et al. 1991; Laanbroek and Woldendorp 1995). Both use CO_2 as the principal carbon source for biomass formation and gain the necessary energy from the oxidation of inorganic nitrogen with O_2. The first physiotype, the ammonium oxidizers, oxidizes

ammonia $\rightarrow$ hydroxylamine $\rightarrow$ nitrite

the latter of which is excreted (Fig. 3). This physiotype is found among two terrestrial genera (*Nitrosomonas*, *Nitrosospira*) within the subclass of β-*Proteobacteria*. The second physiotype, the nitrite oxidizers, oxidizes nitrite to nitrate, and is also found among only a few genera, e.g. *Nitrobacter* within the α-*Proteobacteria* and *Nitrospira* belonging to a new phylum (Ehrich et al. 1995). During the oxidation of ammonium via nitrite to nitrate, NO and N_2O should normally not be formed. However, ammonium oxidizers obviously denitrify some of the produced nitrite, thus forming NO and N_2O and sometimes even N_2 (Poth and Focht 1985; Remde and Conrad 1990; Zart and Bock 1998). They exhibit a denitrifying physiotype that is active under aerobic conditions. They possess all the necessary enzymes to do so (see below). However, it is largely unknown why they are denitrifying the produced nitrite. The following reasons have been suggested: saving O_2 required for the activation of ammonium (Poth and Focht 1985); regulating the pH by removing nitrous acid (Groeneweg et al. 1994); maintenance of an optimal redox poise (Wood 1986). Nitrite oxidizers are able to reduce nitrite to NO, but the environmental relevance of this reaction is unclear (Bock et al. 1991). We have to learn more about the regulation of enzyme expression and activity in autotrophic nitrifiers.

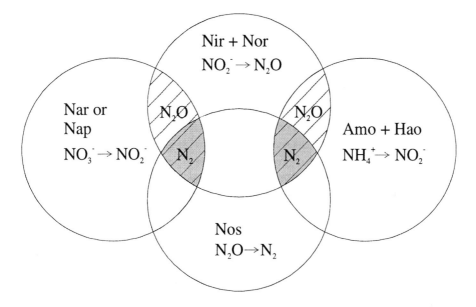

Figure 4. Modular organization of the enzymatic steps involved in denitrification and nitrification. The hatched and shaded areas indicate the intersections of different enzymes producing either N_2O or N_2, respectively, from reduction of nitrate (left side) or oxidation of ammonium (right side). Modified after Zumft (1997).

Heterotrophic nitrification is an ambiguous physiotype (see above). In general, heterotrophic nitrifiers oxidize organic compounds for energy conservation and use organic carbon for biomass synthesis. Those heterotrophic nitrifiers which oxidize ammonium seem to do this by using similar (homologous?) enzymes as the autotrophic nitrifiers (Ferguson 1998). They exhibit lower rates of ammonium oxidation than autotrophic nitrifiers and seem to oxidize ammonium up to nitrate. Similarly as the autotrophic ammonium oxidizers they are able to denitrify the intermediate nitrite, and thus sequentially produce NO and N_2O (Anderson et al. 1993). Heterotrophic nitrification is found among many bacterial taxa, similar to denitrification (Kuenen and Robertson 1988).

Instead of ammonium, organic nitrogen compounds may be oxidized to nitrate. This process is also called heterotrophic nitrification, but would probably depend on different enzymes, e.g. pyruvic oxime dioxygenase forming nitrite from pyruvic oxime (Ono et al. 1999). Little is known about the mechanisms involved in nitrification of organic nitrogen compounds and to which extent this process is involved in NO and N_2O turnover in soil.

Anaerobic nitrification has recently been discovered in anaerobic sludge digestors, the process being termed Anammox (Jetten et al. 1999; Jetten 2001). The overall reaction of this process is the synproportionation of ammonia with nitrite to N_2. The reaction mechanism seems to be the following:

ammonia + hydroxylamine $\rightarrow$ hydrazine   [1]

hydrazine + nitrite $\rightarrow$ hydroxalamine + N_2 [2]

The enzymology of the process is still largely unknown (Jetten 2001). NO or N_2O metabolism seem not to be involved. The microorganisms involved belong to the phylum *Planctomycetales* (Strous et al. 1999; Schmid et al. 2000; Egli et al. 2001). A similar reaction occurs in autotrophic nitrifiers with nitrite or NO_2 as oxidant. In this case, NO and N_2O seem to be products (Bock et al. 1995; Schmidt and Bock 1997; 1998).

3.2   Enzymes and their regulation

Our knowledge of the microbial enzymes and their genes responsible for production and consumption of NO and N_2O is still fragmentary. Especially the proteins and genes involved in the regulation of enzyme synthesis are still largely unknown. Nevertheless, knowledge has increased during the last 20 years and has recently been summarized and discussed in excellent and extensive reviews (Berks et al. 1995; Hooper et al. 1997; Zumft 1997; Ferguson 1998; Baker et al. 1998; Moreno-Vivian et al. 1999; Philippot and Højberg 1999).

3.2.1   Nitrate reductase

Different types of nitrate reductases catalyze the reduction of nitrate to nitrite. Dissimilatory nitrate reduction is achieved by two different types of nitrate reductases, a membrane-bound enzyme (Nar) and a periplasmatic enzyme (Nap). Nar is only induced in the absence of O_2, while Nap is also expressed under oxic conditions. Nap seems to be the nitrate reductase which is involved in the so-called "aerobic denitrification" and in redox balancing of the cell, while Nar is the enzyme that is involved in anaerobic nitrate reduction. Only Nar allows the generation of energy by electron transport phosphorylation. Many bacteria only possess Nar and reduce nitrate just to nitrite (so called nitrate respiration). Nitrate can also be reduced by a cytoplasmic nitrate reductase (Nas), which is synthesized for assimilatory purposes. This enzyme is activated under both oxic and anoxic conditions when the supply of ammonium is limited. All these nitrate reductases contain molybdopterin cofactors, which have to be synthesized by the cells to allow active nitrate reduction. Dissimilatory or assimilatory nitrate reduction is widely distributed in all microbial and plant life. None of the enzymes is expected to be directly involved in production or consumption of NO and N$_2$O, since nitrite is the only product. However, reports exist which claim that traces of NO and N$_2$O may be produced when nitrate reductase reacts with nitrite as non-physiological substrate (Cole 1988). In fact, NO production has been observed in many of the nitrate respirers, but the enzymatic basis of this reaction is still unclear.

3.2.2   Nitrite reductase

Different types of nitrite reductases catalyze the reduction of nitrite. There are two different types of dissimilatory nitrite reductase (NirS and NirK) that typically occur in the periplasm of denitrifiers and reduce nitrite to NO. The electrons for nitrite reduction are supplied by a membrane-bound electron transport chain thus allowing energy generation by electron transport phosphorylation (Fig. 2). NirS is a periplasmic cytochrome cd$_1$ enzyme which requires the biosynthesis of heme D for activity which is tightly regulated by environmental conditions. NirK is a periplasmic Cu-containing enzyme. Microorganisms with the denitrifying physiotype possess either NirS or NirK. NirS is believed to occur in the more abundant soil denitrifiers (e.g., *Pseudomonas fluorescens, P. stutzeri, P. aeruginosa*), but this conclusion requires confirmation. Cu-type nitrite reductases also occur in bacteria of the autotrophic and heterotrophic nitrifier physiotype, namely in *Nitrosomonas europaea* and *Alcaligenes faecalis*, in *Archaea*, and in eukaryotic fungi.

Another dissimilatory nitrite reductase (Nrf) reduces nitrite to ammonia, coupled to electron transport phosphorylation. It is a cytochrome c or hexaheme enzyme that typically occurs in the periplasm of enterobacteria and other Gram-negative nitrate ammonifiers. NO and N$_2$O are not intermediates in the reaction mechanism. However, the enzyme also catalyzes the reduction of hydroxylamine to ammonia and the reduction of NO to N$_2$O. In addition, production of NO and N$_2$O has been observed in cultures of various nitrate ammonifiers during utilization of nitrate, but the exact enzymology of these production processes is still unclear. A nine-electron reduction step from nitrite directly to ammonia is also catalyzed by the assimilatory nitrite reductase (Nra), a siroheme enzyme, that occurs in the cytoplasm of nitrate-assimilating microorganisms and does not support the generation of energy. The enzymatic mechanism of N$_2$O production that has often been observed in these microorganisms is also unclear, but may be ascribed to non-specific reactions of nitrate reductase (see above). The same applies for microorganisms which reduce nitrite to ammonia to serve as electron sinks during fermentation, e.g. clostridia and propionibacteria.

3.2.3   NO reductase

NO reductase (Nor) catalyses the reduction of 2 NO to N_2O. The bacterial Nor is a membrane-bound enzyme with either quinol (only in non-denitrifying pathogens) or cytochrome bc as electron donor (Hendriks et al. 2000). Nor structurally resembles cytochrome c oxidases, but does not translocate protons across the membrane (Fig. 2). In denitrifying eukaryotic fungi, the NO reductase is a cytochrome P-450 enzyme (P450-Nor). Nor is an enzyme that is characteristic for the denitrifying physiotype. Since NO is a compound that is potentially toxic to the cell, Nor is co-regulated with the NO-producing nitrite reductase and scavenges the NO when it is diffusing from the periplasm across the membrane into the cytoplasm. The affinity for NO is extremely high, so that the steady state concentration of NO is kept in the nanomolar range (Conrad 1996b). Knockout mutants of Nor are lethal. The high affinity of Nor to NO and its co-regulation with Nir may be the reason why NO production in soil is usually not due to the nitrate reduction processes but to ammonium oxidation (see above). It is unknown whether ammonium-oxidizing nitrifiers contain Nor. An absence of Nor would explain why most of the NO in soil is produced by nitrifiers. However, the absence of Nor would then require an alternative mechanism of the N_2O production which has been observed in all ammonium-oxidizing nitrifiers studied.

3.2.4 N_2O reductase

N_2O reductase (Nos) catalyses the reduction of N_2O to N_2. Nos is a periplasmic copper enzyme and typically occurs in microorganisms of the non-truncated denitrifying physiotype. However, it has also been found in bacteria that reduce nitrate to ammonium, e.g. *Wolinella succinogenes* (ε-*Proteobacteria*), *Escherichia coli* (γ-*Proteobacteria*), in which its physiological function is unknown. Reduction of N_2O to N_2 has also been observed in strains of the autotrophic nitrifier *Nitrosomonas* (β-*Proteobacteria*). The biosynthesis of the active enzyme requires the assembly of the copper centres involving many regulatory steps. The reduction of N_2O by Nos is coupled to electron transport phosphorylation (Fig. 2).

3.2.5   Ammonium monooxygenase

Ammonium monooxygenase (Amo) catalyses the oxidation of ammonia to hydroxylamine: $NH_3 + O_2 + 2\ e^- \rightarrow NH_2OH + H_2O$. Amo is the key enzyme of ammonium oxidizers and occurs in autotrophic and heterotrophic nitrifiers. It is a membrane-bound enzyme and probably contains copper. It has not yet been isolated from an autotrophic nitrifier, but recently from the heterotrophic *Paracoccus denitrificans* (α-*Proteobacteria*) which also belongs to the denitrifier physiotype (Moir et al. 1996). The gene (*amoA*) has homologies to the gene of the methane monooxygenase (*pmoA*) of methanotrophic bacteria and can be used as a phylogenetic marker for classification of ammonium-oxidizing nitrifiers (Holmes et al. 1995; Rotthauwe et al. 1997) which belong to the β-*Proteobacteria*. Amo does not allow electron transport phosphorylation and is not involved in production or consumption of NO and N₂O, but is required for the initiation of ammonia oxidation (Fig. 3). In *Nitrosomonas eutropha*, O_2 apparently can be replaced by NO_2 in activation of ammonium (Schmidt and Bock 1997; 1998), thus constituting a special case of anaerobic ammonium oxidation (Jetten 2001).

3.2.6   Hydroxylamine oxidoreductase

Hydroxylamine oxidoreductase (Hao) catalyses the oxidation of NH_2OH to nitrite. Hao is a periplasmic enzyme with three heme centres. The oxidation is believed to occur via two successive electron pair abstraction steps giving as intermediate a bound ferrous nitrosyl species. The electrons are transferred via cytochrome c554 and ubiquinone to Amo (2 electrons) and via electron transport chain to O_2 (2 electrons). Only the latter allows electron transport phosphorylation (Fig. 3). Hao is the second key enzyme of ammonium-oxidizing nitrifiers, both autotrophic and heterotrophic ones. The bound HNO intermediate may result in the release of NO and N₂O (Hooper 1984). However, since the ammonium-oxidizing nitrifiers also contain a Cu-type nitrite reductase (NirK), NO production is believed to be due to this enzyme. N₂O production in *N. europaea* has also been shown to be due to reduction of nitrite rather than chemical decomposition of the enzyme-bound nitrosyl (Poth and Focht 1985). Anaerobic ammonium oxidizers contain a novel type of Hao (Schalk et al. 2000).

3.2.7   Nitrite oxidoreductase

Nitrite oxidoreductase catalyzes both the oxidation of nitrite to nitrate and the reduction of nitrate to nitrite, and occurs in nitrite-oxidizing nitrifiers (Bock et al. 1991). The membrane-bound enzyme contains molybdopterin and couples the oxidation of nitrite to electron transport phosphorylation. The electron transport pathway is largely unknown. NO has been postulated to be an important intermediate in energy generation (Bock et al. 1991), but an unequivocal theory and systematic measurements of NO production and consumption in nitrite-oxidizing nitrifiers are so far lacking.

3.2.8   NO synthase

NO synthase catalyses the oxidation of L-arginine via N-hydroxy-L-arginine to L-citrulline plus NO. The NO synthase belongs to the cytochrome p-450 enzyme family and requires molecular O_2 as oxidant and NADPH as reductant (Stamler and Feelisch 1996). The enzyme has been found throughout the domain of *Eukarya* where it synthesizes NO as an important regulatory messenger molecule. Recently, it has also been detected in *Nocardia*, a Gram-positive soil bacterium (*Actinomycetales*). The role of this enzyme in the possible production of NO in soil is completely unknown.

3.2.9   Other enzymes

An uncharacterized enzymatic oxidation of NO to nitrate occurs in a *Pseudomonas* species (γ-*Proteobacteria*) and probably other heterotrophic bacteria under oxic conditions (Koschorreck et al. 1996). The enzymatic basis for this reaction is unclear but may involve non-specific oxidation with enzymatically produced O_2 radicals or by proteins containing transition metals. Although this process has a much lower affinity to NO than the nitric oxide reductase (Nor) of denitrifiers, it seems to be widespread in soil (Koschorreck and Conrad 1997; Dunfield and Knowles 1998). Similar reactions are important in higher life for degradation of the NO that has been produced by the NO synthase.

3.2.10   Regulation

Concerning regulation of enzyme activity the following aspects are noteworthy, and nicely reviewed by Zumft (1997). Enzymes are coded by their genes. The organization of genes can differ from one organism to the other and so does the organization of gene expression. For instance, the organization of denitrification genes is different even among the cytochrome cd_1-containing bacteria, i.e. *Pseudomonas stutzeri*, *P. aeruginosa* and *Paracoccus denitrificans* (all γ-*Proteobacteria*).

The expression of the genes requires a number of regulatory proteins that are expressed by other genes. Enzymes usually consist of several subunits and contain metals (e.g. Cu in Nos) or cofactors (e.g. Heme D in NirS).

Active enzyme is only obtained when the primary proteins are processed in the correct way. This processing requires again many gene activities which are regulated. Cofactors such as hemes have to be synthesized, again involving many different genes. It is very likely that the regulation patterns are different in some of the microbial taxa. This difference also applies to the kind of signals that are required to switch a particular regulatory cascade. In denitrification, for example, expression of denitrifying enzymes requires as signals at least O_2 and nitrogen oxides (e.g. nitrate) in a particular concentration. However, other environmental factors (temperature, pH, organic substrates, osmotic stress) may additionally trigger regulatory cascades that involve the expression of denitrification enzymes.

The various microbial taxa contain the enzymes involved in nitrate reduction and ammonium oxidation in all kinds of combinations (Fig. 4). Truncated denitrification pathways exist as well as combinations of nitrification/denitrification (Kuenen and Robertson 1988; Zumft 1997). The extent of NO, N_2O or N_2 being important products depends on the combination of the different enzyme activities.

The enzymes are arranged to each other within the microbial cell by a certain topology. The enzymes are located either in the cytoplasm, the membrane or the periplasm (in Gram negatives). Little is known about how variable the topology of the enzymes is among the different microbial taxa (Berks et al. 1995). Relatively well studied examples are the denitrifiers (Zumft 1997; Philippot and Højberg 1999; Moreno-Vivian et al. 1999). The topologies of the denitrification pathway, as found in *Paracoccus denitrificans*, and the nitrification pathway, as found in *Nitrosomonas europaea*, are shown in Fig. 2 and 3, respectively. Different topologies mean different regulation patterns. This applies in particular for the production and consumption of NO and N_2O which are usually only intermediates. A change in the activity of only one of the enzymes involved may substantially affect the production and/or consumption of these trace gases.

4.   SUMMARY AND OUTLOOK

Although most of the NO and N_2O in soil seems to be produced during ammonium oxidation and nitrate reduction, the exact production mechanism is usually not easily understood for a specific field situation. Release of NO and N_2O from the soil into the atmosphere is even more complicated, since consumption processes also take place and take control of the flux. It is obvious that production and consumption rates are regulated by environmental variables that affect the metabolism of the soil microorganisms involved. However, it is still impossible to describe the regulation patterns by basic mechanistic theory. Even relatively simple regulation patterns, such as the regulation of NO and N_2O production by soil temperature or soil moisture, are only described on the basis of empirical observations in soils and microbial cultures, but not by a profound understanding of the molecular and enzymatic level of microbial metabolism.

The last 20 years of research have created an enormous wealth of data which nicely document more or less reproducible patterns of soil behavior upon changing environmental conditions. These observations helped to create empirical and process-oriented models which allow us to make to some extent predictions of NO and N_2O emission. In parallel, a reasonable inventory of microbial enzymes and regulatory proteins involved in nitrogen metabolism has been created. However, the observed patterns of soil behavior are not understood on the basis of microbial taxa, microbial enzymes and regulatory networks. Although there seem to be some basic features in nitrogen enzymology and regulation, there is also an exciting diversity which is not yet sufficiently comprehended. Without a better knowledge of this biochemical diversity it will not be possible to interpret empirical observations in soils by molecular processes operating in the microbial cells. For example, the question has been asked whether or not microbial diversity matters with respect to the release of NO and N_2O into the atmosphere (Schimel and Gulledge 1998). Recent investigations have shown now that denitrifier diversity in soil indeed is important for production and emission of N_2O (Cavigelli and Robertson 2000; 2001).

Therefore, future research will have to focus on understanding processes on a microscopic scale, i.e. the scale of the individual microorganisms in soil. This will require taxonomic identification of individual microorganisms and assessing their potential and actual enzyme activities. Steps in this direction are the assessment of the microbial and genetic inventory involved in enzymatic turnover of NO and N_2O in samples of soil and soil microorganisms (Smith and Tiedje 1992; Linne von Berg and Bothe 1992; Kloos et al. 1998; Braker et al. 1998; 2000; Bruns et al. 1998; 1999; Scala and Kerkhof 1999; Bothe et al. 2000; Horz et al. 2000; Grüntzig et al. 2001) and the in-situ identification of individual members of diverse nitrifying communities (Wagner et al. 1998; Lee et al. 1999). As soon as the actual diversity of microorganisms that are possibly involved in NO and N_2O metabolism in soil is known, the dominant species may be isolated, enzymatically and genetically characterized, supplemented with reporter systems, and brought back into the soil environment to study the regulation of enzyme expression under in-situ conditions.

REFERENCES

Albrecht A, Ottow JCG, Benckiser G, Sich I & Russow R (1997) Incomplete denitrification (NO and N₂O) from nitrate by *Streptomyces violaceoruber* and *S. nitrosporeus* revealed by acetylene inhibition and ¹⁵N gas chromatography quadrupole mass spectrometry analyses. Naturwiss. 84: 145-147

Allison C & MacFarlane GT (1989) Dissimilatory nitrate reduction by *Propionibacterium acnes*. Appl Environ Microbiol. 55: 2899-2903

Amann RI, Ludwig W & Schleifer KH (1995) Phylogenetic identification and in situ detection of individual microbial cells without cultivation. Microbiol Rev 59: 143-169

Ambus P & Christensen S (1994) Measurement of N₂O emission from a fertilized grassland: An analysis of spatial variability. J Geophys Res 99: 16549-16555

Anderson IC & Levine JS (1987) Simultaneous field measurements of biogenic emissions of nitric oxide and nitrous oxide. J Geophys Res 92: 965-976

Anderson IC, Poth M, Homstead J & Burdige D (1993) A comparison of NO and N₂O production by the autotrophic nitrifier *Nitrosomonas europaea* and the heterotrophic nitrifier *Alcaligenes faecalis*. Appl Environ Microbiol 59: 3525-3533

Arah JRM, Smith KA, Crichton IJ & Li HS (1991) Nitrous oxide production and denitrification in Scottish arable soils. J Soil Sci 42: 351-367

Baker SC, Ferguson SJ, Ludwig B, Page MD, Richter OMH & VanSpanning RJM (1998) Molecular genetics of the genus *Paracoccus*: metabolically versatile bacteria with bioenergetic flexibility. Microbiol Molec Biol Rev 62: 1046-1078

Bauhus J, Meyer AC & Brumme R (1996) Effect of the inhibitors nitrapyrin and sodium chlorate on nitrification and N₂O formation in an acid forest soil. Biol Fertil Soils 22: 318-325

Beauchamp EG (1997) Nitrous oxide emission from agricultural soils. Can J Soil Sci 77: 113-123

Berg P, Klemedtsson L & Rosswall T (1982) Inhibitory effect of low partial pressures of acetylene on nitrification. Soil Biol Biochem 14: 301-303

Berks BC, Ferguson SJ, Moir JWB & Richardson DJ (1995) Enzymes and associated electron transport systems that catalyse the respiratory reduction of nitrogen oxides and oxyanions. Biochim Biophys Acta 1232: 97-173

Blackmer AM & Cerrato ME (1986) Soil properties affecting formation of nitric oxide by chemical reactions of nitrite. Soil Sci Soc Am J 50: 1215-1218

Blösl M & Conrad R (1992) Influence of an increased pH on the composition of the nitrate-reducing microbial populations in an anaerobically incubated acidic forest soil. Syst Appl Microbiol 15: 624-627

Bock E, Koops HP, Harms H & Ahlers B (1991) The biochemistry of nitrifying organisms. In: Shively JM & Barton LL (eds) Variations in Autotrophic Life, pp 171-200. Academic Press, London, U.K.

Bock E, Schmidt I, Stüven R & Zart D (1995) Nitrogen loss caused by denitrifying *Nitrosomonas* cells using ammonium or hydrogen as electron donors and nitrite as electron acceptor. Arch Microbiol 163: 16-20

Bollmann A & Conrad R (1997a) Acetylene blockage technique leads to underestimation of denitrification rates in oxic soils due to scavenging of intermediate nitric oxide. Soil Biol Biochem 29: 1067-1077

Bollmann A & Conrad R (1997b) Enhancement by acetylene of the decomposition of nitric oxide in soil. Soil Biol Biochem 29: 1057-1066

Bollmann A & Conrad R (1997c) Recovery of nitrification and production of NO and N$_2$O after exposure of soil to acetylene. Biol Fertil Soils 25: 41-46

Bollmann A & Conrad R (1998) Influence of O$_2$ availability on NO and N$_2$O release by nitrification and denitrification in soils. Global Change Biology 4: 387-396

Bollmann A, Koschorreck M, Meuser K & Conrad R (1999) Comparison of two different methods to measure the nitric oxide turnover in soils. Biol Fertil Soils 29: 104-110

Bothe H, Jost G, Schloter M, Ward BB & Witzel KP (2000) Molecular analysis of ammonia oxidation and denitrification in natural environments. FEMS Microbiol Rev 24: 673-690

Bouwman AF, van der Hoek KW & Olivier JGJ (1995) Uncertainties in the global source distribution of nitrous oxide. J Geophys Res 100: 2785-2800

Braker G, Fesefeldt A & Witzel KP (1998) Development of PCR primer systems for amplification of nitrite reductase genes (*nirK* and *nirS*) to detect denitrifying bacteria in environmental samples. Appl Environ Microbiol 64: 3769-3775

Braker G, Zhou JZ, Wu LY, Devol AH & Tiedje JM (2000) Nitrite reductase genes (*nirK* and *nirS*) as functional markers to investigate diversity of denitrifying bacteria in Pacific Northwest marine sediment communities. Appl Environ Microbiol 66: 2096-2104

Breitenbeck GA, Blackmer AM & Bremner JM (1980) Effects of different nitrogen fertilizers on emission of nitrous oxide from soil. Geophys Res Lett 7: 85-88

Bremner JM, Blackmer AM & Waring SA (1980) Formation of nitrous oxide and dinitrogen by chemical decomposition of hydroxylamine in soils. Soil Biol Biochem 12: 263-269

Bruns MA, Fries MR, Tiedje JM & Paul EA (1998) Functional gene hybridization patterns of terrestrial ammonia-oxidizing bacteria. Microb Ecol 36: 293-302

Bruns MA, Stephen JR, Kowalchuk GA, Prosser JI & Paul EA (1999) Comparative diversity of ammonia oxidizer 16S rRNA gene sequences in native, tilled, and successional soils. Appl Environ Microbiol 65: 2994-3000

Butterbach-Bahl K, Gasche R, Breuer L & Papen H (1997) Fluxes of NO and N$_2$O from temperate forest soils: impact of forest type, N deposition and of liming on the NO and N$_2$O emissions. Nutr Cycl Agroecosys 48: 79-90

Cavigelli MA & Robertson GP (2000) The functional significance of denitrifier community composition in a terrestrial ecosystem. Ecology 81: 1402-1414

Cavigelli MA & Robertson GP (2001) Role of denitrifier diversity in rates of nitrous oxide consumption in a terrestrial ecosystem. Soil Biol Biochem 33: 297-310

Cohan FM (1994) Genetic exchange and evolutionary divergence in prokaryotes. Tree 9: 175-180

Cole CV, Duxbury J, Freney J, Heinemeyer O, Minami K, Mosier A, Paustian K, Rosenberg N, Sampson N, Sauerbeck D & Zhao Q (1997) Global estimates of potential mitigation of greenhouse gas emissions by agriculture. Nutr Cycl Agroecosys 49: 221-228

Cole JA (1988) Assimilatory and dissimilatory reduction of nitrate to ammonia. In: Cole JA & Ferguson S (eds) The Nitrogen and Sulphur Cycles, pp 281-329. Cambridge University Press, Cambridge, U.K.

Conrad R (1990) Flux of NO$_x$ between soil and atmosphere: Importance and soil microbial metabolism. In: Revsbech NP & Sørensen J (eds) Denitrification in Soil and Sediment, pp 105-128. Plenum, New York, U.S.A.

Conrad R (1994) Compensation concentration as critical variable for regulating the flux of trace gases between soil and atmosphere. Biogeochem 27: 155-170

Conrad R (1995) Soil microbial processes involved in production and consumption of atmospheric trace gases. Adv Microb Ecol 14: 207-250

C9onrad R (1996a) Metabolism of nitric oxide in soil and soil microorganisms and regulation of flux into the atmosphere. In: Murrell JC & Kelly DP (eds) Microbiology of Atmospheric Trace Gases: Sources, Sinks and Global Change Processes, pp 167-203. Springer Verlag, Berlin, Germany

Conrad R (1996b) Soil microorganisms as controllers of atmospheric trace gases (H_2, CO, CH_4, OCS, N_2O, and NO). Microbiol Rev 60: 609-640

Conrad R & Dentener FJ (1999) The application of compensation point concenpts ion scaling of fluxes. In: Bouwman AF (ed) Approaches to Scaling of Trace Gas Fluxes in Ecosystems, pp 205-216. Elsevier, Amsterdam, The Netherlands

Conrad R & Seiler W (1980) Field measurements of the loss of fertilizer nitrogen into the atmosphere as nitrous oxide. Atmos Environ 14: 555-558

Crutzen PJ (1979) The role of NO and NO_2 in the chemistry of the troposphere and stratosphere. Ann Rev Earth Planet Sci 7: 443-472

Daum M, Zimmer W, Papen H, Kloos K, Nawrath K & Bothe H (1998) Physiological and molecular biological characterization of ammonia oxidation of the heterotrophic nitrifier *Pseudomonas putida*. Curr Microbiol 37: 281-288

Davidson EA (1991) Fluxes of nitrous oxide and nitric oxide from terrestrial ecosystems. In: Rogers JE & Whitman WB (eds) Microbial Production and Consumption of Greenhouse Gases: Methane, Nitrogen Oxides, and Halomethanes, pp 219-235. American Society for Microbiology, Washington, DC, U.S.A.

Davidson EA (1992) Sources of nitric oxide and nitrous oxide following wetting of dry soil. Soil Sci Soc Am J 56: 95-102

Davidson EA & Hackler JL (1994) Soil heterogeneity can mask the effects of ammonium availability on nitrification. Soil Biol Biochem 26: 1449-1453

Davidson EA & Kingerlee W (1997) A global inventory of nitric oxide emissions from soils. Nutr Cycl Agroecosys 48: 37-50

Davidson EA & Schimel JP (1995) Microbial processes of production and consumption of nitric oxide, nitrous oxide and methane. In: Matson PA & Harriss RC (eds) Biogenic Trace Gases: Measuring Emissions from Soil and Water, pp 327-357. Blackwell, Oxford, U.K.

Davidson EA & Verchot LV (2000) Testing the hole-in-the-pipe model of nitric and nitrous oxide emissions from soils using the TRAGNET database. Glob Biogeochem Cycle 14: 1035-1043

De Boer W, Klein Gunnewiek PA & Laanbroek HJ (1995) Ammonium-oxidation at low pH by a chemolithotrophic bacterium belonging to the genus *Nitrosospira*. Soil Biol Biochem 27: 127-132

Dunfield PF & Knowles R (1997) Biological oxidation of nitric oxide in a humisol. Biol Fertil Soils 24: 294-300

Dunfield PF & Knowles R (1998) Organic matter, heterotrophic activity, and NO consumption in soils. Global Change Biology 4: 199-207

Egli K, Fanger U, Alvarez PJJ, Siegrist H, VanderMeer JR & Zehnder AJB (2001) Enrichment and characterization of an anammox bacterium from a rotating biological contactor treating ammonium-rich leachate. Arch Microbiol 175: 198-207

Ehrich S, Behrens D, Lebedeva E, Ludwig W & Bock E (1995) A new obligately chemolithoautotrophic, nitrite-oxidizing bacterium, *Nitrospira moscoviensis* sp nov and its phylogenetic relationship. Arch Microbiol 164: 16-23

Ferguson SJ (1998) Nitrogen cycle enzymology. Curr Opinion Chem Biol 2: 182-193

Firestone MK & Davidson EA (1989) Microbiological basis of NO and N_2O production and consumption in soil. In: Andreae MO & Schimel DS (eds) Exchange of Trace Gases between Terrestrial Ecosystems and the Atmosphere. Dahlem Konferenzen, pp 7-21. John Wiley & Sons Ltd., Chichester, New York, U.S.A.

Frolking SE, Mosier AR, Ojima DS, Li C, Parton WJ, Potter CS, Priesack E, Stenger R, Haberbosch C, Dörsch P, Flessa H & Smith KA (1998) Comparison of N_2O emissions from soils at three temperate agricultural sites - simulations of year-round measurements by four models. Nutr Cycl Agroecosys 52: 77-105

Gasche R & Papen H (1999) A 3-year continuous record of nitrogen trace gas fluxes from untreated and limed soil of a N-saturated spruce and beech forest ecosystem in Germany: 2. NO and NO_2 fluxes. J Geophys Res 104: 18505-18520

Gödde M & Conrad R (1998) Simultaneous measurement of nitric oxide production and consumption in soil using a simple static incubation system, and the effect of soil water content on the contribution of nitrification. Soil Biol Biochem 30: 433-442

Gödde M & Conrad R (1999) Immediate and adaptational temperature effects on nitric oxide production and nitrous oxide release from nitrification and denitrification in two soils. Biol Fertil Soils 30: 33-40

Gödde M & Conrad R (2000) Influence of soil properties on the turnover of nitric oxide and nitrous oxide by nitrification and denitrification at constant temperature and moisture. Biol Fertil Soils 32: 120-128

Granli T & Bøckman OC (1994) Nitrous oxide from agriculture. Norwegian J Agric Sci, Supplement 12: 7-128

Groeneweg J, Sellner B & Tappe W (1994) Ammonia oxidation in Nitrosomonas at NH_3 concentrations near K_m: effects of pH and temperature. Water Res 28: 2561-2566

Grüntzig V, Nold SC, Zhou JZ & Tiedje JM (2001) *Pseudomonas stutzeri* nitrite reductase gene abundance in environmental samples measured by real-time PCR. Appl Environ Microbiol 67: 760-768

Henault C & Germon JC (1995) Quantification de la dénitrification et des émissions de protoxyde d'azote (N_2O) par les sols. Agronomie 15: 321-355

Hendriks J, Oubrie A, Castresana J, Urbani A, Gemeinhardt S & Saraste M (2000) Nitric oxide reductases in bacteria. Biochim Biophys Acta 1459: 266-273

Holmes AJ, Costello A, Lidstrom ME & Murrell JC (1995) Evidence that particulate methane monooxygenase and ammonia monooxygenase may be evolutionarily related. FEMS Microbiol Lett 132: 203-208

Hooper AB (1984) Ammonia oxidation and energy transduction in the nitrifying bacteria. In: Strohl WR & Tuovinen OH (eds) Microbial Chemoautotrophy, pp 133-167. Ohio State University Press, Columbus, U.S.A.

Hooper AB, Vannelli T, Bergmann DJ & Arciero DM (1997) Enzymology of the oxidation of ammonia to nitrite by bacteria. Ant Leeuwenhoek 71: 59-67

Horz HP, Rotthauwe JH, Lukow T & Liesack W (2000) Identification of major subgroups of ammonia-oxidizing bacteria in environmental samples by T-RFLP analysis of *amoA* PCR products. J Microbiol Methods 39: 197-204

Hyman MR & Arp DJ (1992) $^{14}C_2H_2$- and $^{14}CO_2$-labeling studies of the *de novo* synthesis of polypeptides by *Nitrosomonas europaea* during recovery from acetylene and light inactivation of ammonia monooxygenase. J Biol Chem 267: 1534-1545

Hyman MR & Wood PM (1985) Suicidal inactivation and labelling of ammonia mono-oxygenase by acetylene. Biochem J 227: 719-725

Jetten MSM (2001) New pathways for ammonia conversion in soil and aquatic systems. Plant Soil 230: 9-19

Jetten MSM, Strous M, van de Pas-Schoonen KT, Schalk J, van Dongen UGJM, van de Graaf AA, Logemann S, Muyzer G, van Loosdrecht MCM & Kuenen JG (1999) The anaerobic oxidation of ammonium. FEMS Microbiol Rev 22: 421-437

Johansson C (1989) Fluxes of NO$_X$ above soil and vegetation. In: Andreae MO & Schimel DS (eds) Exchange of Trace Gases between Terrestrial Ecosystems and the Atmosphere. Dahlem Konferenzen, pp 229-246. John Wiley & Sons Ltd., Chichester, New York, U.S.A.

Kalkowski I & Conrad R (1991) Metabolism of nitric oxide in denitrifying *Pseudomonas aeruginosa* and nitrate-respiring *Bacillus cereus*. FEMS Microbiol Lett 82: 107-112

Keller M & Reiners WA (1994) Soil atmosphere exchange of nitrous oxide, nitric oxide, and methane under secondary succession of pasture to forest in the atlantic lowlands of Costa Rica. Glob Biogeochem Cycle 8: 399-409

Killham K (1986) Heterotrophic nitrification. In: Prosser JI (ed) Nitrification, pp 117-126. IRL-Press, Oxford, U.K.

Klemedtsson L, Hanson G & Mosier A (1990) The use of acetylene for the quantification of N$_2$ and N$_2$O production from biological processes in soil. In: Revsbech NP & Sørensen J (eds) Denitrification in Soil and Sediment, pp 167-180. Plenum, New York, U.S.A.

Klemedtsson L, Svensson BH & Rosswall T (1988) A method of selective inhibition to distinguish between nitrification and denitrification as sources of nitrous oxide in soil. Biol Fertil Soils 6: 112-119

Kloos K, Hüsgen UM & Bothe H (1998) DNA-probing for genes coding for denitrification, N$_2$-fixation and nitrification in bacteria isolated from different soils. Z Naturforsch C 53: 69-81

Knowles R (1985) Microbial transformations as sources and sinks for nitrogen oxides. In: Caldwell DE, Brierley JA & Brierley CL (eds) Planetary Ecology, pp 411-426. Van Nostrand Reinhold, New York, U.S.A.

Koschorreck M & Conrad R (1997) Kinetics of nitric oxide consumption in tropical soils under oxic and anoxic conditions. Biol Fertil Soils 25: 82-88

Koschorreck M, Moore E & Conrad R (1996) Oxidation of nitric oxide by a new heterotrophic *Pseudomonas* sp.. Arch Microbiol 166: 23-31

Kuenen JG & Robertson LA (1988) Ecology of nitrification and denitrification. In: Cole JA & Ferguson S (eds) The Nitrogen and Sulphur Cycles, pp 161-218. Cambridge University Press, Cambridge, U.K.

Laanbroek HJ & Woldendorp JW (1995) Activity of chemolithotrophic nitrifying bacteria under stress in natural soils. Adv Microb Ecol 14: 275-304

Lee N, Nielsen PH, Andreasen KH, Juretschko S, Nielsen JL, Schleifer KH & Wagner M (1999) Combination of fluorescent in situ hybridization and microautoradiography - a new tool for structure-function analyses in microbial ecology. Appl Environ Microbiol 65: 1289-1297

Li CS, Aber J, Stange F, Butterbach-Bahl K & Papen H (2000) A process-oriented model of N$_2$O and NO emissions from forest soils: 1. Model development. J Geophys Res 105: 4369-4384

Li CS, Narayanan V & Harriss RC (1996) Model estimates of nitrous oxide emissions from agricultural lands in the United States. Glob Biogeochem Cycle 10: 297-306

Linne von Berg KHL & Bothe H (1992) The distribution of denitrifying bacteria in soils monitored by DNA-probing. FEMS Microbiol Ecol 86: 331-340

Logan JA (1983) Nitrogen oxides in the troposphere: global and regional budgets. J Geophys Res 88: 10785-10807

Ludwig J, Meixner FX, Vogel B & Forstner J (2001) Soil-air exchange of nitric oxide: An overview of processes, environmental factors, and modeling studies. Biogeochem 52: 225-257

Martikainen PJ (1985) Nitrous oxide emission associated with autotrophic ammonium oxidation in acid coniferous forest soil. Appl Environ Microbiol 50:1519-1525

Martikainen PJ & de Boer W (1993) Nitrous oxide production and nitrification in acidic soil from a Dutch coniferous forest. Soil Biol Biochem 25: 343-347

McKenney DJ, Drury CF & Wang SW (1997) Reaction of nitric oxide with acetylene and oxygen - implications for denitrification assays. Soil Sci Soc Am J 61: 1370-1375

McKenney DJ, Johnson GP & Findlay WI (1984) Effect of temperature on consecutive denitrification reactions in Brookston clay and Fox sandy loam. Appl Environ Microbiol 47: 919-926

Meixner FX (1994) Surface exchange of odd nitrogen oxides. Nova Acta Leopoldina 70: 299-348

Moir JWB, Crossman LC, Spiro S & Richardson DJ (1996) The purification of ammonia monooxygenase from *Paracoccus denitrificans*. FEBS Lett 387: 71-74

Moreno-Vivian C, Cabello P, Martinez-Luque M, Blasco R & Castillo F (1999) Prokaryotic nitrate reduction: molecular properties and functional distinction among bacterial nitrate reductases. J Bacteriol 181: 6573-6584

Nägele W & Conrad R (1990) Influence of pH on the release of NO and N_2O from fertilized and unfertilized soil. Biol Fertil Soils 10: 139-144

Nevison CD, Esser G & Holland EA (1996) A global model of changing N_2O emissions from natural and perturbed soils. Climatic Change 32: 327-378

Nielsen TH, Nielsen LP & Revsbech NP (1996) Nitrification and coupled nitrification-denitrification associated with a soil-manure interface. Soil Sci Soc Am J 60: 1829-1840

Ono Y, Enokiya A, Masuko D, Shoji K & Yamanaka T (1999) Pyruvic oxime dioxygenase from the heterotrophic nitrifier *Alcaligenes faecalis*: purification, and molecular and enzymatic properties. Plant Cell Physiol 40: 47-52

Papen H & Butterbach-Bahl K (1999) A 3-year continuous record of nitrogen trace gas fluxes from untreated and limed soil of a N-saturated spruce and beech forest ecosystem in Germany: 1. N_2O emissions. J Geophys Res 104: 18487-18503

Papen H, Hellmann B, Papke H & Rennenberg H (1993) Emission of N-oxides from acid irrigated and limed soils of a coniferous forest in Bavaria. In: Oremland RS (ed) Biogeochemistry of Global Change, pp 245-260. Chapman & Hall, New York, U.S.A.

Papen H, von Berg R, Hinkel I, Thoene B & Rennenberg H (1989) Heterotrophic nitrification by *Alcaligenes faecalis*: NO_2^-, NO_3^-, N_2O, and NO production in exponentially growing cultures. Appl Environ Microbiol 55: 2068-2072

Parkin TB (1987) Soil microsites as a source of denitrification variability. Soil Sci Soc Am J 51: 1194-1199

Parkin TB, Sexstone AJ & Tiedje JM (1985) Adaption of denitrifying populations to low soil pH. Appl Environ Microbiol 49: 1053-1056

Parton WJ, Mosier AR, Ojima DS, Valentine DW, Schimel DS, Weier K & Kulmala AE (1996) Generalized model for N_2 and N_2O production from nitrification and denitrification. Global Biogeochem Cycle 10: 401-412

Payne WJ (1973) Reduction of nitrogenous oxides by microorganisms. Bacteriol Rev 37: 409-452

Pedersen H, Dunkin KA & Firestone MK (1999) The relative importance of autotrophic and heterotrophic nitrification in a conifer forest soil as measured by ^{15}N tracer and pool dilution techniques. Biogeochem 44: 135-150

Philippot L & Højberg O (1999) Dissimilatory nitrate reductases in bacteria. Biochim Biophys Acta 1446: 1-23

Poth M & Focht DD (1985) ^{15}N kinetic analysis of N₂O production by *Nitrosomonas europaea*: an examination of nitrifier denitrification. Appl Environ Microbiol 49: 1134-1141

Potter CS, Matson PA, Vitousek PM & Davidson EA (1996) Process modeling of controls on nitrogen trace gas emissions from soils worldwide. J Geophys Res 101: 1361-1377

Remde A & Conrad R (1990) Production of nitric oxide in *Nitrosomonas europaea* by reduction of nitrite. Arch Microbiol 154: 187-191

Remde A & Conrad R (1991a) Metabolism of nitric oxide in soil and denitrifying bacteria. FEMS Microbiol Ecol 85: 81-93

Remde A & Conrad R (1991b) Role of nitrification and denitrification for NO metabolism in soil. Biogeochem 12: 189-205

Robertson GP (1989) Nitrification and denitrification in humid tropical ecosystems: Potential controls on nitrogen retention. In: Proctor J (ed) Mineral Nutrients in Tropical Forest and Savanna Ecosystems, pp 55-69. Blackwell Scientific, Boston, U.S.A.

Robertson GP & Tiedje JM (1987) Nitrous oxide sources in aerobic soils: nitrification, denitrification and other biological processes. Soil Biol Biochem 19: 187-193

Robertson K (1994) Nitrous oxide emission in relation to soil factors at low to intermediate moisture levels. J Environ Qual 23: 805-809

Rotthauwe JH, Witzel KP & Liesack W (1997) The ammonia monooxygenase structural gene *amo*A as a functional marker: molecular fine-scale analysis of natural ammonia-oxidizing populations. Appl Environ Microbiol 63: 4704-4712

Rudolph J, Koschorreck M & Conrad R (1996a) Oxidative and reductive microbial consumption of nitric oxide in a heathland soil. Soil Biol Biochem 28: 1389-1396

Rudolph J, Rothfuss F & Conrad R (1996b) Flux between soil and atmosphere, vertical concentration profiles in soil, and turnover of nitric oxide.

1. Measurements on a model soil core. J Atmos Chem 23: 253-273

Ryden JC (1981) N₂O exchange between a grassland soil and the atmosphere. Nature 292: 235-237

Rysgaard S, Risgaard-Petersen N & Sloth NP (1996) Nitrification, denitrification, and nitrate ammonification in sediments of two coastal lagoons in Southern France. Hydrobiologia 329: 133-141

Rysgaard S, Risgaard-Petersen N, Nielsen LP & Revsbech NP (1993) Nitrification and denitrification in lake and estuarine sediments measured by the ^{15}N dilution technique and isotope pairing. Appl Environ Microbiol 59: 2093-2098

Saad OALO & Conrad R (1993a) Adaptation to temperature of nitric oxide-producing nitrate-reducing bacterial populations in soil. Syst Appl Microbiol 16: 120-125

Saad OALO & Conrad R (1993b) Temperature dependence of nitrification, denitrification, and turnover of nitric oxide in different soils. Biol Fertil Soils 15: 21-27

Sahrawat KL & Keeney DR (1986) Nitrous oxide emission from soils. Adv Soil Sci 4: 103-148

Sanhueza E, Hao WM, Scharffe D, Donoso L & Crutzen PJ (1990) N₂O and NO emissions from soils of the northern part of the Guayana Shield, Venezuela. J Geophys Res 95: 22481-22488

Scala DJ & Kerkhof LJ (1999) Diversity of nitrous oxide reductase (*nosZ*) genes in continental shelf sediments. Appl Environ Microbiol 65: 1681-1687

Schalk J, de Vries S, Kuenen JG & Jetten MSM (2000) Involvement of a novel hydroxylamine oxidoreductase in anaerobic ammonium oxidation. Biochem 39: 5405-5412

Schimel JP & Gulledge J (1998) Microbial community structure and global trace gases. Global Change Biology 4: 745-758

Schimel JP, Firestone MK & Killham KS (1984) Identification of heterotrophic nitrification in a Sierran forest soil. Appl Environ Microbiol 48: 802-806

Schmid M, Twachtmann U, Klein M, Strous M, Juretschko S, Jetten M, Metzger JW, Schleifer KH & Wagner M (2000) Molecular evidence for genus level diversity of bacteria capable of catalyzing anaerobic ammonium oxidation. Syst Appl Microbiol 23: 93-106

Schmidt I & Bock E (1997) Anaerobic ammonia oxidation with nitrogen dioxide by *Nitrosomonas eutropha*. Arch Microbiol 167: 106-111

Schmidt I & Bock E (1998) Anaerobic ammonia oxidation by cell-free extracts of *Nitrosomonas eutropha*. Ant Leeuwenhoek 73: 271-278

Seifritz C, Daniel SL, Gossner A & Drake HL (1993) Nitrate as a preferred electron sink for the acetogen *Clostridium thermoaceticum*. J Bacteriol 175: 8008-8013

Seiler W & Conrad R (1981) Field measurements of natural and fertilizer induced N_2O release rates from soils. J Air Poll Contr Ass 31: 767-772

Shoun H, Kim DH, Uchiyama H & Sugiyama J (1992) Denitrification by fungi. FEMS Microbiol Lett 94: 277-281

Sich I & Russow R (1998) Cryotrap enrichment of nitric oxide and nitrous oxide in their natural air concentration for ^{15}N analysis by GC-QMS. Isotopes Environ Health Stud 34: 279-283

Skiba U, Fowler D & Smith KA (1997) Nitric oxide emissions from agricultural soils in temperate and tropical climates: sources, controls and mitigation options. Nutr Cycl Agroecosys 48:139-153

Slemr F & Seiler W (1984) Field measurements of NO and NO_2 emissions from fertilized and unfertilized soils. J Atmos Chem 2: 1-24

Smith GB & Tiedje JM (1992) Isolation and characterization of a nitrite reductase gene and its use as a probe for denitrifying bacteria. Appl Environ Microbiol 58: 376-384

Stackebrandt E & Goebel BM (1994) Taxonomic note: A place for DNA-DNA reassociation and 16S rRNA sequence analysis in the present species definition in bacteriology. Int J Syst Bact 44: 846-849

Stackebrandt E, Tindall B, Ludwig W & Goodfellow M (1999) Prokaryotic diversity and systematics. In: Lengeler JW, Drews G & Schlegel HG (eds) Biology of the Prokaryotes, pp 674-720. Thieme, Stuttgart, Germany

Stamler JS & Feelisch M (1966) Biochemistry of nitric oxide and redox-related species. In: Feelisch M & Stamler JS (eds) Methods in Nitric Oxide Research, pp 19-27. John Wiley & Sons Ltd., Chichester, New York, U.S.A.

Stange F, Butterbach-Bahl K, Papen H, Zechmeister-Boltenstern S, Li CS & Aber J (2000) A process-oriented model of N_2O and NO emissions from forest soils: 2. Sensitivity analysis and validation. J Geophys Res 105: 4385-4398

Stark JM & Firestone MK (1995) Isotopic labeling of soil nitrate pools using nitrogen-15-nitric oxide gas. Soil Sci Soc Am J 59: 844-847

Stevens RJ, Laughlin RJ & Malone JP (1998) Soil pH affects the processes reducing nitrate to nitrous oxide and di-nitrogen. Soil Biol Biochem 30: 1119-1126

Stevens RJ, Laughlin RJ, Burns LC, Arah JRM & Hood RC (1997) Measuring the contributions of nitrification and denitrification to the flux of nitrous oxide from soil. Soil Biol Biochem 29: 139-151

Strous M, Fuerst JA, Kramer EHM, Logemann S, Muyzer G, van de Pas-Schoonen KT, Webb R, Kuenen JG & Jetten MSM (1999) Missing lithotroph identified as new planctomycete. Nature 400: 446-449

Tiedje JM (1988) Ecology of denitrification and dissimilatory nitrate reduction to ammonia. In: Zehnder AJB (ed) Biology of Anaerobic Microorganisms, pp 179-244. John Wiley & Sons Ltd., Chichester, New York, U.S.A.

Torsvik V, Goksøyr J & Daae FL (1990) High diversity in DNA of soil bacteria. Appl Environ Microbiol 56: 782-787

Venterea RT & Rolston DE (2000) Mechanisms and kinetics of nitric and nitrous oxide production during nitrification in agricultural soil. Glob Change Biol 6: 303-316

Wagner M, Noguera DR, Juretschko S, Rath G, Koops HP & Schleifer KH (1998) Combining fluorescent in situ hybridization (FISH) with cultivation and mathematical modeling to study population structure and function of ammonia-oxidizing bacteria in activated sludge. Water Sci Tech 37: 441-449

Wang YP, Meyer CP, Galbally IE & Smith CJ (1997) Comparisons of field measurements of carbon dioxide and nitrous oxide fluxes with model simulations for a legume pasture in Southeast Australia. J Geophys Res 102: 28013-28024

Williams EJ, Hutchinson GL & Fehsenfeld FC (1992) NO_X and N_2O emissions from soil. Global Biogeochem Cycle 6: 351-388

Woese CR (1987) Bacterial evolution. Microbiol Rev 51: 221-271

Wood PM (1986) Nitrification as a bacterial energy source. In: Prosser JI (ed) Nitrification, pp 39-62. IRL Press, Oxford, U.K.

Zart D & Bock E (1998) High rate of aerobic nitrification and denitrification by *Nitrosomonas eutropha* grown in a fermentor with complete biomass retention in the presence of gaseous NO_2 or NO. Arch Microbiol 169: 282-286

Zhou ZM, Takaya N, Sakairi MAC & Shoun H (2001) Oxygen requirement for denitrification by the fungus *Fusarium oxysporum*. Arch Microbiol 175: 19-25

Zumft WG (1992) The denitrifying prokaryotes. In: Balows A, Trüper HG, Dworkin M, Harder W & Schleifer KH (eds) The Prokaryotes, Vol 1, pp 554-582. Springer, New York, U.S.A.

Zumft WG (1993) The biological role of nitric oxide in bacteria. Arch Microbiol 160: 253-264

Zumft WG (1997) Cell biology and molecular basis of denitrification. Microb Molec Biol Rev 61: 533-616

Chapter 1.2

NO₂, NO and HNO₃ Uptake by Trees

Alan R. Wellburn[†]
Department of Biological Sciences, Lancaster University, Lancaster LA1 4YQ,, U.K.

Concentrations of NO_2 in the troposphere rarely cause visible injury to trees or any other vegetation. Accidental or uncontrolled emissions, however, do occur from time to time which can affect some species more than others. Table 1 lists those tree species that are known to be sensitive to NO_2 injury and those that have been found to be more tolerant at similar acute concentrations. Sometimes, the injury which develops later in sensitive tree species is that of necrosis especially at the the tips of leaves and coniferous needles. More common is a distinct greening of leaves and needles, which cannot be described as an injury, across a wide range of vegetation including trees exposed over long periods to lower atmospheric NO_2 concentrations especially in situations where levels of soil N are low (Wellburn 1994).

Some of the additional N which brings about this greening enters trees by the indirect route of first entering the soil from the troposphere in a variety of different N forms (NO_2, NH_3, etc.) and is then taken up by roots as nitrate or ammonium. This indirect soil/root pathway is considered elsewhere in this volume. However, there is increasing evidence that a direct route of uptake, mainly of NO_2, from troposphere to leaf or needle also accounts for part of this extra N. The proportions of N entering indirectly or directly from the troposphere into the plant are currently of some moment because tropospheric O_3 concentrations are rising well above the intentions set by numerous international authorities. The only way of restricting the formation of O_3 is to remove the sources of O_3; namely nitric oxide (NO) and its surrogate, NO_2. Consequently, if direct uptake mechanisms for NO and NO_2 from the troposphere can be encouraged then rising O_3 levels may be

---

[†] The author passed away shortly after submission of the manuscript

R. Gasche et al. (eds.), Trace Gas Exchange in Forest Ecosystems, 35–52.
© *2002 Kluwer Academic Publishers. Printed in the Netherlands.*

circumvented. A case for manipulating and improving NO and NO_2 uptake (or NO_X-fixing) capabilities of vegetation has already been made (Wellburn 1998) on the basis that (a) considerable natural variation for direct uptake exists even within as well as between species, (b) many species are able to acclimate to the presence of tropospheric NO and NO_2 in a positive manner and (c) it is possible to select out from mutated populations, individuals that have a high capacity to take up and use tropospheric NO_2 even when well supplied with N in the soil.

Table 1. Relative sensitivity of tree species to acute concentrations of NO_2 (adapted from Taylor et al. 1989 and EPA 1991)

Sensitive	Intermediate	Tolerant	
Eucalyptus camaldulensis Dehn.	*Acer palmatum* Thunb. *A. platanoides* L.	*Carpinus betulus* L. *Fagus sylvatica* L.	*Pinus nigra* Arnold
Fraxinus pennsylvanica Marsh.	*Betula pendula* Roth. *Citrus aurantium* L.	*Fraxinus americana* L. *Ginkgo biloba* L.	*P. rigida* Mill.
Malus pumila Mill.	*Tilia cordata* Mill.	*Liquidambar styraciflua* L.	*P. taeda* L.
Pyrus communis L.		*Populus nigra* L. *Quercus alba* L.	*P. thunbergiana* Franco
	Abies concolor (Gord.) Lindl. ex. Hildbr.	*Q. robur* L. *Sambucus nigra* L.	*P. virginiana* Mill.
Larix decidua Mill.	*A. homolepis* Siebold and Zucc.	*Robinia pseudoacacia* L *Ulmus glabra* Huds.	*Taxus baccata* L.
Pinus strobus L	*Picea abies* (L.) Karst. *P. pungens* Engelm.	*Zelkova serrata* (Thunb.) Mak.	
Pseudotsuga menziesii (Mirb.) Franco	*P. glauca* (Moench) Voss		

Although these NO_X-fixing characteristics have been identified in cereals, grasses and horticultural crops, there is no reason to believe why they might not equally apply to tree species and there is some evidence to support this. Firstly, there is the possibility of acclimation. The EPA (1991) criteria document cites the main observations from a large number of experimental NO_2 fumigations of trees. Leaving aside the large proportion of the earlier studies which used concentrations of NO_2 well above those found in the urban environment (> 300 nL L^{-1}), the remaining 22 separate tree fumigations fall into 2 groups in terms of overall dosage but similar mean NO$_2$ concentrations. One group of 13 was of fairly short duration, often just of a few weeks having a mean dose of 33 µL L^{-1} h^{-1} NO$_2$, while 9 others were for a season or more (221 µL L^{-1} h^{-1} mean dose). The difference in terms of above-ground change in biomass between the 2 groups of fumigation studies is striking (Table 2). Those fumigated for a relatively short time had 11% less growth than trees growing in clean air while those exposed to NO$_2$ for longer were 13% bigger than their ambient compatriots. This is fairly convincing evidence for NO$_2$-fixing acclimation by trees which could be a significant feature of the natural environment where exposure goes on year after year especially near roads and large conurbations.

Table 2. Summary of effects on above-ground biomass of trees with different doses of NO$_2$

Number of fumigation studies	Mean dose [µL L^{-1} h^{-1}]	% change from clean air controls
13	33 [a]	-1
9	221 [b]	+13

[a]: 16.8-114.4 µL L^{-1} h^{-1}; [b]: 216-228.8 µL L^{-1} h^{-1}

Evidence for intra-species differences in the effects of NO$_2$ on tree growth also occurs. For example, 8 clones of *Pinus strobus* L. showed differences in length and mass of needles (Yang et al. 1983a, b) after 4h exposure to 100–300 nL L^{-1} NO$_2$ for 35 d while 3 hybrid crosses of poplar had different increases in foliar mass and area in response to similar fumigations (Okano et al. 1989). Likewise, 2 clones of *Betula pendula* L. showed differences in biomass when exposed to 8h of 62 nL L^{-1} NO$_2$ for 150 d (Wright 1987).

Differences in uptake rates of certain atmospheric N compounds into trees have also been detected. The atmosphere contains many N-containing compounds with different physical and chemical properties which are summarized in Table 3. All of them have different fluxes towards tree leaves and needles both through the free turbulence of the troposphere above and the relatively still band of air around leaves and needles known as the boundary layer. Once at the leaf/needle surface, they can remain there until washed off by rain or enter the leaf mainly through the stomatal complexes. The fluxes of ammonia and ammonium compounds by such routes are covered in another chapter of this volume and those for N$_2$O and the minor oxidized compounds are not of major concern. However, effort has been extended to assess the characteristics of such fluxes for NO$_2$, NO and HNO$_3$ but they are not easy to measure accurately and have been considered in detail elsewhere in this volume. Briefly, fluxes downwards towards canopies can be determined using either eddy correlation or flux gradient micrometerological techniques. The first measures vertical turbulent flux using the mean covariance between pollutant concentration and wind velocity while the latter estimates from eddy exchange coefficients across a concentration profile. On a smaller scale, mass balance techniques are also applied in environmental chambers to determine fluxes to leaves but, increasingly, the use of labelled $^{15}NO_2$, ^{15}NO and $H^{15}NO_3$ is becoming more popular. In both cases, leaf-washing to determine the proportion of gas not entering the plant is also determined.

Table 3. N-containing compounds in the atmosphere arranged in general order of measured concentrations [highest first]

Compound	Name	Molecular weight	Oxidation state of N	Solubility [g or mL 100 mL^{-1}]
N_2	Dinitrogen	28.01	0	2.33 ml
N_2O	Nitrous oxide	44.01	+1	130 ml
NO_2	Nitrogen dioxide	46.01	+4	s, d
NH_3	Ammonia	17.03	-3	89.9 g
NH_4HCO_3	Ammonium hydrogen carbonate	76.06	-3	11.9 g
$(NH_4)_2CO_3.H_2O$	Ammonium carbonate	114.1	-3	100 g
NH_4NO_3	Ammonium nitrate	80.04	-3	118.3 g
NO	Nitric oxide	30.01	+2	7.34 mL
$(NH_4)_2SO_4$	Ammonium sulphate	132.13	-3	70.6 g
HNO_3	Nitric acid	63.01	+5	infinite
N_2O_5	Dinitrogen pentoxide	108.01	+5	s
HNO_2	Nitrous acid	47.01	+3	d
N_2O_4	Dinitrogen tetroxide	92.01	+4	s, d
NO_3^-	Nitrate radical	62	+5	s, d
N_2O_3	Dinitrogen trioxide	76.01	+3	s
NO^+	Nitrosonium ion	30.01	+3	s

s, soluble; d, decomposes

The flux to a canopy (F_c) measured in nmol m^{-2} s^{-1} is defined as:

$$F_c = V_d * (C_z - C_0) \qquad [1]$$

where V_d is the deposition velocity in m s^{-1} and C_z and C_0 are the concentrations (in nmol m^{-3}) at the height of measurement and at the receptor site in the canopy, respectively. A very similar equation describes the flux to leaves (F_l) for the smaller scale chamber measurements:

$$F_l = K_l * (C_a - C_i) \qquad [2]$$

where K_l is the conductance of the pollutant into or onto the leaf; C_a, the concentration of pollutant in the atmosphere around the leaf and C_i, the concentration in or on the leaf, which is often taken to be zero. The units are the same in both equations. If the leaf area index (LAI) is also known then the 2 fluxes can be equated:

$$V_d = K_l * LAI \qquad\qquad\qquad [3]$$

although V_d is nearly always > than K_l.

Table 4 lists typical deposition velocities and conductances for conifers but it must be emphasized that these values are representative and vary by over an order of magnitude according to species and many other factors (Table 5). For example, they change with concentration of NO_2, NO or HNO_3. Figures 1 and 2 illustrate this using conductance data for NO_2 and HNO_3 collected from all those tree studies (EPA 1991) which used concentrations experienced in the field omitting many that have been done at higher levels. In the case of NO_2, conductances rise with increasing concentrations (Fig. 1) while those for HNO_3 have a minimum around 45 nL L^{-1} (Fig. 2).

Table 4. Conductances of and deposition velocities to tree surfaces for NO_2, NO and HNO_3

[mm s^{-1}]	NO_2 [a]	NO [b]	HNO_3 [b]
Deposition velocity [V_d] to tree leaf surface	27	9	3.8
Conductance [K_l] of tree leaf surfaces [mm s^{-1}]	1.2	0.3	0.03

[a]: at 20 nL L^{-1}; [b]: 10 nL L^{-1}

Table 5. Factors known to influence dry deposition to trees of N-containing compounds (after Sehmel 1980)

Micrometerology	Particles	Gases	Receptor
Aerodynamic resistance	Size	Solubility	Stomatal conductance
Friction velocity	Diffusion	Concentration	Mesophyll metabolism
Wind velocity & turbulence	Impaction	Chemical activity	Surface adsorption
Temperature	Gravity	Diffusion	Surface absorption
Relative humidity	Electrostatic charge		Leaf & needle area
Precipitation			Previous loading
Solar radiation			Dew, exudates,
			wax & pubescence
			Snow & frost

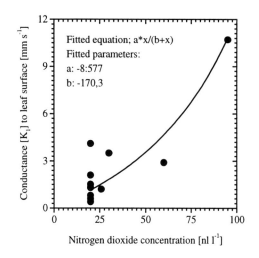

Figure 1. Variation of conductance to leaf/needle surfaces in trees with increasing atmospheric concentrations of NO_2 (data summarized from EPA 1991).

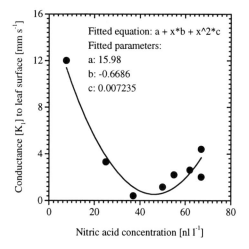

Figure 2. Variation of conductance to leaf/needle surfaces in trees with increasing atmospheric concentrations of HNO_3 (data summarized from EPA 1991).

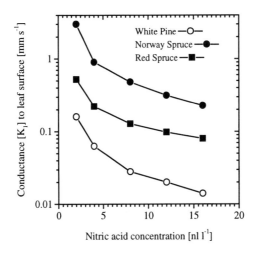

Figure 3. Changes in needle conductance of HNO₃ to different coniferous tree species with time (re-plotted data from Cadle et al. 1991)

The nature of the uptake surface also affects conductance (see Table 5). The plotted data of Cadle et al. (1991) schow higher rates for uptake of HNO₃ by clean fresh surfaces than when they are partially saturated after time (Fig. 3). The same figure also shows how individual species may differ by more than an order of magnitude in rate at each stage. As a general rule, conductances for spruces are higher than those for pines but lower than broad-leaved trees. Temperature effects on uptake by trees are also interesting because, even at low temperatures over winter (+7 to –12°C), red spruce take up NO2 (and SO2) and accumulate nitrite (and sulphite) in their xylem sap (Wolfenden et al. 1991).

The chemical reactivity of NO₂, NO or HNO₃ are very different (Table 5). NO₂ reacts with water once inside needles or leaves after diffusion through the stomatal complexes while HNO₃ mainly adsorbs onto the cuticular surfaces (Vose and Swank 1990). This means that stomatal opening and closing mainly affects NO₂ uptake (Hanson et al. 1989) but NO exchanges slowly with both cuticles and inner leaf spaces. Cuticular deposition of NO₂ is not negligible but can be as much as 2 orders of magnitude less than stomatal uptake although the difference between cuticular and stomatal uptake of NO is not so great. Actual emissions of NO from vegetation are also known to occur (Dean and Harper 1986) but have not been studied in trees.

The main reactions associated with NO_2, NO and water when the gaseous and aqueous phases are in contact are:

$$2\,NO_2 + H_2O \xrightarrow{K_{eq}=2.44\,x10^2} 2\,H^+ + NO_2^- + NO_3^- \qquad [4]$$

$$NO_2 + NO + 2\,H_2O \xrightarrow{K_{eq}=3.28\,x10^{-5}} 4\,H^+ + NO_2^- + NO_3^- \qquad [5]$$

$$3\,NO_2 + H_2O \xrightarrow{K_{eq}=1.81\,x10^{-9}} 2\,H^+ + 2\,NO_3^- + NO \qquad [6]$$

which means that Reaction 4 predominates to produce both nitrite as well as nitrite from NO_2. Furthermore, small amounts of NO can be taken up into the aqueous phase (Reaction 5) and very small amounts of NO can be formed from NO_2 by dissolution (Reaction 6). All these reactions generate acidity which may be significant in unbuffered locations but this is unlikely to be of significance in the apoplastidic fluid.

Both NO_2 and NO are free radicals but NO_2 presents no real problem because it readily dissolves to form anions (Reaction 4) which are not free radicals and the concentration of NO_2 beyond the apoplast is close to zero. However, NO will persist for much longer, travel further into the cell and may chelate to transition metals. For example, NO reacts with the copper-containing laccases of the oriental lacquer tree, *Rhus vercifera* (Martin et al. 1981). A case has been made elsewhere (Wellburn 1990) which states that much of the toxicity associated with NO_X is caused by NO although the potential toxicity of nitrite formed by Reactions 4 and 5 must also be considered.

Nitrate is transported across membranes by a nitrate transporter but there is little evidence for a nitrite transporter in plants. Nitrite is in equilibrium with nitrous acid (HNO_2, Reaction 7), a weak acid:

$$HNO_2 \xleftrightarrow{pK=3.3} H^+ + NO_2^- \qquad [7]$$

$$[NO_2^-]_a + [H^+]_a \Leftrightarrow [HNO_2]_{cm} \Leftrightarrow [NO_2^-]_c + [H^+]_c$$
$$\Leftrightarrow [HNO_2]_{pe} \Leftrightarrow [NO_2^-]_p + [H]_p \qquad [8]$$

which is the un-ionized form able to diffuse across both the cell membrane (*cm*) and the plastid envelopes (*pe*, Reaction 8). Consequently, there is likely to be little difference in HNO_2 concentrations either side of membranes so the ionization of the weak HNO_2 will be totally dominated by the $[H^+]$ in each compartment which the nitrite concentration must follow. Some estimates have put nitrite concentrations of the apoplast/cell walls (*a*), which have a pH of about 4.3, at around 50 μM and those of nitrate at 2.25 mM when plants are exposed to 50 nL L^{-1} NO_2 (EPA 1991). Consequently, if the cytosol (*c*) has a pH of about 7, the nitrite concentrations theoretically would be approximately 500 times higher (i.e. 25 mM) in the absence of metabolism because the $[H^+]_c$ is correspondingly lower. Following this argument, similar concentrations of nitrite would exist in the plastids (*p*) in the dark (pH about 7) but theoretically could go even higher in illuminated chloroplasts (pH about 9) if they were not removed promptly (Reaction 8). In practice, direct measurements of nitrate and nitrite in plastids of clean-air grown plants gave concentrations of 2.46 and 0.103 mM, respectively. However, when they were fumigated (3d with 280 nL L^{-1} NO_2), the levels of nitrate actually declined by 48% by the second day but the nitrite concentration rose by 45% (Wellburn, 1985). By the end of the third day, both concentrations were similar to clean-air values.

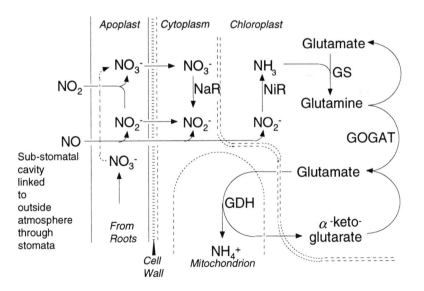

Figure 4. Uptake and metabolic pathways of nitrogen oxides into plant tissues from the atmosphere, through the stomata, sub-stomatal cavity, apoplastidic fluid, cell walls, cell membranes and cytoplasm into chloroplasts. Enzymes involved include nitrate reductase (NaR), nitrite reductase (NiR), glutamine synthetase (GS), glutamate synthase (GOGAT) and glutamate dehydrogenase (GDH). Reproduced with permission from Wellburn (1994).

In the above-ground parts of trees, nitrate from NO_2 is indistinguishable from nitrate arriving from the roots and is metabolized by the normal pathway involving nitrate reductase (NaR), nitrite reductase (NiR), glutamine synthetase (GS) and glutamate synthase (GOGAT, Fig. 4). Most of the nitrite from NO_2 and NO is also reduced by means of the last 3 enzymes but it is possible that as HNO_2 it could react with the primary (and secondary) amine groups of proteins to form N-nitrosamine groups (Reaction 9):

$$\text{Protein-NH}_3^+ + HNO_2 \Rightarrow \text{Protein-N-N=O} + 2\,H_2O \qquad\qquad [9]$$

which could disturb both enzymic and translocator functions. This may partly account for the toxicity associated with NO and nitrite. A fuller discussion of the metabolic effects of NO_2 and NO is available elsewhere (Wellburn 1990) but concentrations of NO_2 are unlikely ever to be high enough to affect lipids as has been suggested in the early literature (Malhotra and Kahn 1984).

There are now a number of studies which report a 3- to 10-fold induction of NaR activity upon experimental exposure of trees to NO_2 and HNO_2 (Table 6) but the variability and diurnal changes associated with this enzyme prevent its use as a bioindicator of the presence and uptake of NO_2 or HNO_2 in the forest (Tjoelker et al. 1992). Maximum levels of NaR activity are achieved 1–3 days after exposure commences but, subsequently, the rates decline even though the fumigation continues (Table 6). No doubt this later stage is part of the acclimation process but it is important to understand the nature of the induction of NaR and subsequent enzymes involved in the reduction of nitrite, etc. NaR is an enzyme that exhibits a strong diurnal rhythm, being formed afresh each morning with activity rising to a maximum around noon and declining thereafter. Both it and the following enzyme, NiR, are induced by nitrate but the turnover of NiR is much slower although, generally, endogenous rates of activity exceed those of NaR. As a result, NiR activities are much slower in responding to nitrite arising from NO_2 (Thoene et al. 1991). What is of significance is that NiR activity is directly light-driven and not induced by nitrite which means nitrite can only be removed in the light and that early inbalances of nitrite due to NO_2 and NO will not be immediately acted upon. This could be a partial explanation of why short exposures to NO_X appear to be more detrimental than longer ones. In the latter, fluxes of nitrate to ammonia and onwards are amply covered by elevated or fully-integrated N-assimilation activities throughout although inbalances may exist in the early stages of exposure.

Table 6. Effects of NO_2 and HNO_3 fumigations on needle/leaf nitrate reductase (NaR) activities

Pollutant	Tree species	Exposure	% increase in NaR activity	Maximum activity (days)	50% decline from maximum (d)	Reference
NO₂	*Pinus sylvestris* L.	75 nL L^{-1} + 10 nL L^{-1} NO for 10 d	1000	3	7	Wingsle et al. (1987)
	Picea rubens Sarg.	75 nL L^{-1} for 7 d	300	1	7[a]	Norby et al. (1989)
	Picea abies L. Karst	60 nL L^{-1} for 7 d	550	2	1	Thoene et al. (1991)
	Picea abies L. Karst	50 nL L^{-1} for 2 d	330	2		von Ballmoos et al. (1998)
	Populus x euroamericana (Dode) Guiner. cv. Dorskamp	80-135 nL L^{-1} for 84 d; low, medium & high soil N	200 low N 292 med N 165 high N	[b]	[b]	Schmutz et al. (1995)
HNO₃	*Picea rubens* Sarg.	75 nL L^{-1} for 3 d	300	2	1	Norby et al. (1989)
	Quercus kelloggii Newb. *Quercus chrysolepsis* Liebm. *Pinus ponderosa* Doug.	65-80 nL L^{-1} for 1d	elevated significantly in all species	[c]	[c]	Krywult and Bytnerowicz (1997)

[a]: young tissues only; [b]: only measured after 84 d; [c]: only measured after 24 h

The cellular distribution of enzymes is also important. NaR is located in the cytoplasm and ideally positioned for the nitrate arriving from the apoplast via the nitrate translocator in the cell membrane (Fig. 4). The immediate source of reductant is NADH, although NaR isoforms are known which can use NADPH, and these are maintained in reduced state by shuttles operating from both the mitochondria and plastids. NiR and subsequent N-assimilation enzymes are chloroplastidic which means that nitrite molecules from both NaR processing and from NO and NO_2 directly have to cross the cytoplasm to the plastid envelopes, form HNO_2, enter plastids and cross their stroma to the NiR on the thylakoid surfaces; ample opportunity for inadvertant nitrosamine formation on the way.

Six electrons are required to reduce one molecule of nitrite and GS imposes a further energy demand of 1 ATP molecule. If the NAD(P)H of NaR is included, this means the cost of dealing with NO_2 and NO pollution by a plant is relatively expensive; perhaps as much as 6 ATPs per NO_2 not counting the bioenergetic cost of incorporating extra amino acids into protein. This is no problem in strong sunlight but it may be one of the reasons why NO_2 and especially NO can limit growth at low light intensities.

At relatively high concentrations of NO_2, Norway spruce seedlings have elevated amounts of glutamate and increased GS activities (Tischner et al. 1988) but this is less obvious at lower levels (Thoene et al. 1991). Indeed, glutamate, glutamine and total free amino acid contents in Scots pine are lower after NO_2 fumigation than clean-air controls although total needle N contents are higher (Wingsle et al. 1987). Arginine is well established as an N-storage compound in conifers (Durzan and Steward 1983) and has also been found to be 6-fold lower after NO_2 fumigation with 75 nL L^{-1} NO_2 plus 10 nL L^{-1} NO for 10 d (Wingsle et al. 1987). However, at lower concentrations (25 nL L^{-1} NO_X for 40 d), higher concentrations of glutamine and arginine were found if the seedlings had mycorrhizal associations (Näsholm et al. 1991). Presumably, the extra N from NO_2 is mobilized more readily into protein, along with existing stored N, which accounts for the extra growth during long term studies of NO_2 exposure (Table 2).

One of the great merits of using labelled $^{15}NO_2$, ^{15}NO and $H^{15}NO_3$ to measure uptake rates is that the fates of the label once inside the leaves/needles can also be determined. The application of such techniques to trees came well after the first use of $^{15}NO_2$ on herbaceous plants (Durmishidze and Nutsubidze 1976) and demonstrated that the total $^{15}NO_2$-N absorbed by *Populus nigra x P. maximoiczii* hybrids was more than 4-fold those of *Quercus myrsinaefolia* Blume when simultaneously exposed to 300 nL L^{-1} NO_2 for 30 d (Okano et al. 1989). Five other tree species fumigated at the same time were intermediate between these two. This variation may be explained almost entirely by a close correlation between total $^{15}NO_2$-N absorbed and stomatal conductance.

More recently, a wider survey of 217 taxa of plants exposed to very high concentrations of NO_2 (4 μL L^{-1} for 8 h) revealed more than a 600-fold difference in uptake of labelled $^{15}NO_2$ (Morikawa et al. 1998) but it is not clear if stomatal conductance accounts for all this variation. Figure 5 shows the relationship between total $^{15}NO_2$-N content and reduced N from $^{15}NO_2$ for just the trees in this survey. As may be observed there is a fairly close correlation between the 2 parameters. The lowest uptake rate was recorded for *Citrus tachibana* L. while the highest was for *Populus nigra* L.; a 28-fold difference (Fig. 5). When 6-year old *Picea abies* (L.) Karst. trees were fumigated with 60 nL L^{-1} of ^{15}NO$_2$ for 4 d, ^{15}N label accumulated mainly in the free glutamate of needles as well as in the glutamate of bark and roots (Nussbaum et al. 1993).

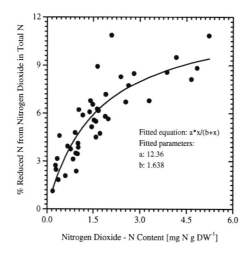

Figure 5. Relationship between total N content acquired from ^{15}NO$_2$ and the proportion converted into reduced forms by a range of tree species. Each data point represents a different species (re-plotted from Morikawa et al. 1998).

Canopy uptake studies using labelled ^{15}N-nitrate and ^{15}N-ammonium have also provided evidence for direct foliar uptake amounting to 1% of the total uptake of nitrate and 5% of total ammonium which represents direct canopy uptake amounting to 2–8% of the total N requirement of *Picea rubens* Sarg. in high elevation forests (Boyce et al. 1996). It is presumed that this 1% nitrate is taken up directly by the same trans-cuticular route identified using H^{15}NO$_3$ in *Pinus strobus* L. and *Picea abies* (L.) Karst. trees (Cadle et al. 1991) although much of the surface-deposited nitrate could be washed off the cuticles in both cases.

The net effect of all these observations would lead one to conclude that tropospheric NO$_2$ at concentrations up to 60 nL L^{-1} with 10–20 nL L^{-1} NO over the long term do not adversely affect the majority of trees. Futhermore, the extra N that is taken up by the direct pathways may be usefully incorporated into amino-acids, proteins and, hence, growth. The problem is that NO$_2$ with NO rarely exist in the troposphere in isolation from other pollutants like SO_2 and O_3 and, in these more common mixed situations, this positive situation disappears.

There have been many mixed fumigations, along with many comparisons between the effects of unfiltered and charcoal-filtered semi-urban air, on trees. The trouble with many of these is that all the relevant controls and single pollutant exposures were not carried out alongside the mixed fumigations. Many of the properly blocked out and replicated fumigations, showing more-than-additive (synergistic) adverse effects on tree growth, etc., are listed in Table 7. Other studies have identified detrimental changes due to mixtures of $SO_2 + NO_2$ in individual physiological parameters such as photosynthesis (Strand 1993) and transpiration (Neighbour et al. 1988). While studies with *Populus x euroamericana* (Dode) Guiner. have demonstrated that the availability of soil N modulates plant response to NO_2

Table 7. Significant more-than-additive effects on visible injury and depression of growth (*) in trees caused by mixtures of air pollutant gases including NO_2

Species	Pollutant mixture	Exposure	Reference
Alnus incana L.	$NO_2 + SO_2$ *	8 h of 62 nL L^{-1} of each gas for 150 d	Whitmore and Freer-Smith (1982)
"	"	8 h of 100-110 nL L^{-1} of each gas for 150 d	Freer-Smith (1984)
Betula pendula Roth.	"	8 h of 62 nL L^{-1} of each gas for 150 d	Whitmore and Freer-Smith (1982); Wright (1987)
"	"	8 h of 100-110 nL L^{-1} of each gas for 150 d	Freer-Smith (1984)
B. pubescens Ehrh.	"	8 h of 62 nL L^{-1} of each gas for 150 d	Whitmore and Freer-Smith (1982); Wright (1987)
"	"	8 h of 100-110 nL L^{-1} of each gas for 150 d	Freer-Smith (1984)
Fraxinus americana L.	$NO_2 + O_3$ *	6 h of 100 nL L^{-1} NO_2 + 50 nL L^{-1} O_3 for 28 d	Kress and Skelly (1982)
F. pennsylvanica Marsh.	"	"	"
Liquidambar styraciflua L.	"	"	"
Malus domestica L.	$NO_2 + SO_2$ *	8 h of 100-110 nL L^{-1} of each gas for 150 d	Freer-Smith (1984)
Pinus rigida Mill.	$NO_2 + O_3$	6 h of 100 nL L^{-1} NO_2 + 50 nL L^{-1} O_3 for 28 d	Kress and Skelly (1982)

Species	Pollutant mixture	Exposure	Reference
P. sitchensis (Bong.) Carr.	$NO_2 + SO_2$ *	8 h of 62 nL L^{-1} of each gas for 56 d	Freer-Smith and Mansfield (1987)
P. strobus L	$NO_2 + SO_2$ *	4h of 50 nL L^{-1} of each gas for 35 d	Yang et al. (1982)
"	$NO_2 + O_3$ *	"	"
"	$NO_2 + SO_2 + O_3$ *	"	"
P. taeda L.	$NO_2 + SO_2 + O_3$	6 h of 100 nL L^{-1} NO_2, 140 nL L^{-1} SO_2 + 50 nL L^{-1} O_3 for 28 d	Kress et al. (1982b)
"	$NO_2 + O_3$	6 h of 100 nL L^{-1} NO_2 + 50 nL L^{-1} O_3 for 28 d	Kress and Skelly (1982)
P. virginiana Mill.	"	"	"
Populus occidentalis L.	$NO_2 + SO_2 + O_3$ *	6 h of 100 nL L^{-1} NO_2, 140 nL L^{-1} SO_2 + 50 nL L^{-1} O_3 for 28 d	Kress et al. (1982a)
P. nigra L.	$NO_2 + SO_2$ *	8 h of 62 nL L^{-1} of each gas for 150 d	Whitmore and Freer-Smith (1982)
"	$NO_2 + SO_2$ *	8 h of 100-110 nL L^{-1} of each gas for 150 d	Freer-Smith (1984)
Pseudotsuga menziesii (Mirb.) Franco	$NO_2 + SO_2$ *	48 nL L^{-1} of each gas for 140 d	van Hove et al. (1992)
Tilia cordata L.	"	8 h of 100-110 nL L^{-1} of each gas for 150 d	Freer-Smith (1984)

plus O_3 (Schmutz et al. 1995). Exposure of *Populus nigra* to SO_2 during dormancy delayed leaf growth in the following spring, but the extra presence of NO_2 did not influence this effect (Freer-Smith, 1984). However with SO_2 exposure during the growing season, these depressions of growth did not occur to *Tilia cordata, Malus domestica, Betula pendula* and *P. nigra* until the second year of exposure, but in *Betula pubescens* and *Alnus incana* decreases of dry weight occurred in the first year. NO_2 alone initially stimulated the shoot growth of *T. cordata, B. pendula* and *A. incana*, but these effects were lost for *T. cordata* and *B. pendula* in the second season of exposure. Inhibitory effects on a number of parameters developed more rapidly in $SO_2 + NO_2$ mixtures than in SO_2 alone. The responses to SO_2 and NO_2 were variable, depending on species, on the duration of exposure and the time of year. Sequential observations of the shoots suggested that the physiology of growth influenced pollution sensitivity. Foliar blemish, senescence and abscission were the main symptoms of injury with SO_2 + NO_2, and the severity of these effects was well correlated with the magnitude of the decreases in dry weight (Freer-Smith 1984).

Combinations of NO_2 with either SO_2 or O_3 are metabolically harmful and this may be traced to the free radical-based injury known to take place in these mixed situations (Wellburn 1987). Trees, like other plants, are known to produce a wide range of additional free radical scavengers to protect themselves before visible injury occurs (Wellburn and Wellburn 1997). For example, field exposures of *Pinus sylvestris* L. to both NO_2 (10–60 nL L^{-1}) and SO_2 (15–50 nL L^{-1}) over 50 d produced a 2.3-fold increase in chloroplastidic and cytoplasmic CuZn-superoxide dismutase (SOD) mRNA (Karpinski et al. 1992). However, if the concentrations of NO_2 and SO_2 are much lower (10-15 nL L^{-1} each) over longer periods (2 years), additional free radical scavenging enzyme activities such as the SODs or glutathione reductase are not induced and are less than those in needles from Scots pine growing in clean air (Wingsle and Hallgren 1993). In other words, the protection mechanisms are not switched on and the long term consequences are reflected in synergistic depressions of growth (Table 7). The overall implications of pollutant mixtures for all types of vegetation have recently been considered in some detail and a generalized model of the mode of action of pollutant combinations has been developed (Barnes and Wellburn 1998).

REFERENCES

Barnes JD & Wellburn AR (1998) Air pollutant combinations. In: de Kok LJ & Stulen I (eds) Responses of Plant Metabolism to Air Pollution and Climate Change, pp. 147-164. Bachhuys Publishers, Leiden, The Netherlands

Boyce RL, Friedland AJ, Chamberlain CP & Poulson SR (1996) Direct canopy nitrogen uptake from ^{15}N-labeled wet deposition by mature red spruce. Can J For Res 26: 1539-1547

Cadle SH, Marshall JD & Mulawa PA (1991) A laboratory investigation of the routes of HNO_3 dry deposition to coniferous seedlings. Environ Pollut 72: 287-305

Dean JV & Harper JE (1986) Nitric oxide and nitrous oxide production by soybean and winged bean during the in vivo nitrate reductase assay. Plant Physiol 82: 718-723

Durmishidze SV & Nutsubidze NN (1976) Absorption and conversion of nitrogen dioxide by higher plants. Dokl Biochem 227: 104-107

Durzan DJ & Steward FC (1983) Nitrogen metabolism. In: Steward FC (ed) Plant Physiology: A Treatise, Vol 8, pp. 55-265. Academic Press, New York, U.S.A.

EPA, US Environmental Protection Agency (1991) Air Quality Criteria for Oxides of Nitrogen. EPA/600/8-91/049a-cA. Washington, U.S.A.

Freer-Smith PH (1984) The responses of six broadleaved trees during long-term exposure to SO_2 and NO_2. New Phytol 97: 49-61

Freer-Smith PH & Mansfield TA (1987) The combined effects of low temperature and SO$_2$ + NO$_2$ pollution on the new season's growth and water relations of *Picea sitchensis*. New Phytol 106: 237-250

Karpinski S, Wingsle G, Karpinska B & Hallgren J-E (1992) Differential expression of CuZn-superoxide dismutases in *Pinus sylvestris* needles exposed to SO$_2$ and NO$_2$. Physiol Plant 85: 689-696

Kress LW & Skelly JM (1982) Response of several eastern forest tree species to chronic doses of ozone and nitrogen dioxide. Plant Disease 66: 1149-1152

Kress LW, Skelly JM & Hinkelmann KH (1982a) Growth impact of O$_3$, NO$_2$ and/or SO$_2$ on *Platanus occidentalis*. Agric Envir 7: 265-274

Kress LW, Skelly JM & Hinkelmann KH (1982b) Growth impact of O$_3$, NO$_2$ and SO$_2$ on *Pinus taeda*. Env Mon Assess 1: 229-239

Krywult M & Bytnerowicz A (1997) Induction of nitrate reductase activity by nitric acid vapor in California black oak (*Quercus kelloggii*), canyon live oak (*Quercus chrysolepis*), and ponderosa pine (*Pinus ponderosa*) seedlings. Can J For Res 27: 2101-2104

Mathotra SS & Khan AA (1984) Biochemical and physiological impact of major pollutants in: Treshow M (ed) Air pollution and plant life, pp. 113-157. John Wiley & Sons Ltd., Chichester, New York, U.S.A.

Martin CT, Morse RH, Kanne RM, Gray HB, Mälmstrom BG & Chan SI (1981) Reactions of nitric oxide with tree and fungal laccase. Biochem 20: 5147-5155

Morikawa H, Higaki A, Nohno M, Takahashi M, Kamada M, Nakata M, Toyohara G, Okamura Y, Matsui K, Kitani K, Fujita K, Irifune K & Goshima N (1998) More than a 600-fold variation in nitrogen dioxide assimilation among 217 plant taxa. Plant Cell Environ 21: 180-190

Näsholm T, Högberg P & Edfast A-B (1991) Uptake of NO$_X$ by mycorrhizal and non-mycorrhizal Scots pine seedlings. Quantities and effects on amino acid and protein concentrations. New Phytol 119: 83-92

Neighbour EA, Pearson M & Mehlhorn H (1990) Purafil-filtration prevents the development of ozone-induced frost injury: A potential role for nitric oxide. Atmos Environ 24: 711-715

Norby RJ, Weerasurija Y & Hanson PJ (1989) Induction of nitrate reductase activity in red spruce needles by NO$_2$ and HNO$_3$ vapour. Can J For Res 19: 889-896

Nussbaum S, von Ballmoos P, Gfeller H, Schlunegger UP, Fuhrer J, Rhodes D & Brunold C (1993) Incorporation of atmospheric ^{15}NO$_2$-nitrogen into free amino acids by Norway spruce (*Picea abies* L. Karst). Oecologia 94: 408-414

Okano K, Machida T & Totsuka T (1989) Differences in ability of NO$_2$ absorption in various broad-leaved tree species. Environ Pollut 58: 1-17

Schmutz P, Tarjan D, Gunthardt-Goerg MS, Matyssek R & Bucher JB (1995) Nitrogen dioxide - A gaseous fertilizer of poplar trees. Phyton 35: 219-232

Sehmel GA (1980) Particle and gas dry deposition: A review. Atmos Environ 14: 983-1011

Strand M (1993) Photosynthetic activity of Scots pine (*Pinus sylvestris* L.) needles during winter is affected by exposure to SO$_2$ and NO$_2$ during summer. New Phytol 123: 133-141

Taylor HJ, Ashmore MR & Bell JNB (1989) Air Pollution Injury to Vegetation. HM Health and Safety Executive, IEHO, London

Theone B, Schröder P, Papen H, Egger A & Rennenberg H (1991) Absorption of atmospheric NO$_2$ by spruce (*Picea abies* L. Karst) trees I. NO$_2$ influx and its correlation with nitrate reduction. New Phytol 117: 575-585

Tischner R, Peuke A, Godbold DL, Feig R, Merg G & Huttermann A (1988) The effect of NO$_2$ fumigation on aseptically grown spruce seedlings. J Plant Physiol 133: 243-246

Tjoelker MG, McLaughlin SB, Dicosty RJ, Lindberg SE & Norby RJ (1992) Seasonal variation in nitrate reductase activity in needles of high-elevation red spruce trees. Can J For Res 22: 375-380

van Hove LWA, Bossen ME, Mensink MGJ & van Kooten 0 (1992) Physiological effects of a long term exposure to low concentrations of NH_3, NO_2 and SO_2 on Douglas fir (*Pseudotsuga menziesii*). Physiol Plant 86: 559-567

von Ballmoos P, Ammann M, Egger A, Suter M & Brunold C (1998) NO_2-induced nitrate reductase in needles of Norway spruce (*Picea abies*) under laboratory and field conditions. Physiol Plant 102: 596-604

Vose JM & Swank WT (1990) Preliminary estimates of foliar absorption of ^{15}N labeled nitric acid vapor (HNO_3) by mature eastern white pine (*Pinus strobus* L.). Can J For Res 20: 34-45.

Wellburn AR (1985) Ion chromatographic determination of levels of anions in plastids from fumigated and non-fumigated barley seedlings. New Phytol 100: 329-339

Wellburn AR (1987) Biochemical mechanisms of combined action of atmospheric pollutants upon plants. In: Vouk VB, Butler GC, Upton AC, Parke DV, Asher SC (eds) Methods for Assessing the Effects of Mixtures of Chemicals. SCOPE Series No. 30, SGOMSEC Series No.3, pp. 813-831. John Wiley & Sons Ltd., Chichester, New York, U.S.A.

Wellburn AR (1990) Why are atmospheric oxides of nitrogen usually phytotoxic and not alternative fertilizers? Tansley Review No. 24, New Phytol 115: 395-429

Wellburn AR (1994) Air Pollution and Climate Change: The Biological Impact. [Second Edition], Longmans Scientific, Harlow, Essex, U.K.

Wellburn AR (1998) Atmospheric nitrogenous compounds and ozone. Is NO_X fixation by plants a possible solution? New Phytol 139, 5-9

Wellburn AR, Wellburn FAM (1997) Air pollution and free radical protection responses of plants. In: Scandalios JG (ed) Oxidative Stress and the Molecular Biology of Antioxidant Defenses, pp. 861-876. Cold Spring Harbor Lab. Press, Cold Spring Harbor, U.S.A.

Whitmore ME, Freer-Smith PH (1982) Growth effects of SO_2 and/or NO_2 on woody plants and grasses during spring and summer. Nature (London) 300: 55-57

Wingsle G, Hällgren J-E (1993) Influence of SO_2 and NO_2 exposure on glutathione, superoxide dismutase and glutathione reductase activities in Scots pine needles. J Exp Bot 44: 463-470

Wingsle G, Näsholm T, Lundmark T & Ericsson A (1987) Introduction of nitrate reductase in needles of Scots pine seedlings by NO_X and NO_3^-. Physiol Plant 70: 399-403

Wolfenden J, Pearson M & Francis BJ (1991) Effects of overwinter fumigation with sulphur and nitrogen dioxides on biochemistryical parameters and spring growth in red spruce (*Picea rubens* Sarg.). Plant Cell Environ 14: 35-45

Wright EA (1987) Effects of sulphur dioxide and nitrogen dioxide, singly and in mixtures, on the macroscopic growth of three birch clones. Environ Poll 46: 209-221

Yang Y-S, Skelly JM & Chevone BI (1982) Clonal response of eastern white pine to low doses of O_3, SO_2, and NO_2, singly and in combination. Can J For Res 12: 803-808

Yang Y-S, Skelly JM & Chevone BI (1983a) Sensitivity of eastern white pine clones to acute doses of ozone, sulfur dioxide, or nitrogen dioxide. Phytopathology 73: 1234-1237

Yang Y-S, Skelly JM & Chevone BI (1983b) Effects of pollutant combinations at low doses on growth of forest trees. Aquilo Ser Bot 19: 406-418

Chapter 1.3

Production and consumption of NH_4^+ and NH_3 in trees

John Pearson[1]; Janet Woodall[1]; Clough, Elisabeth C.M.[1]; Kent H. Nielsen[2]; and Jan K. Schjørring[2]
[1] *Department of Biology, University College London, Gower Street, London, WCIE 6BT, U.K.*
[2] *Plant Nutrition Laboratory, Department of Agricultural Sciences, Royal Veterinary and Agricultural University, Thorvaldsensvej 40, 8, 5., DK-1871 Frederiksberg C, Denmark*

1.   INTRODUCTION

Of all the nutrients acquired from the soil for plant growth, nitrogen (N) is generally required in the greatest amount. The availability of N is dependent on the global N cycle, which relies on the formation of combined inorganic N from atmospheric N_2, as very little N is made available to soils from weathering of substratum (Sprent 1987). Climate and soil organisms interact with soil building processes in such a way that the N cycle is very dynamic and has an important impact on habitat development and plant succession. Plants form a critical part of the dynamics of the N cycle in that they act as a large store for N, but also release much of their N back to the global cycle as tissues senesce and decay. However, as plant succession progresses to climax, more N is locked up in relatively larger and longer-lived species and N becomes a growth-limiting nutrient in most habitats. In such a situation, the N economy of a plant is likely to be under strong selection pressure. In the case of trees, they are faced with balancing acquisition of new or primary N, against retention and recycling of secondary N from old/storage tissue to new growth. Thus in senescing deciduous trees the maximum re-absorption of N for recycling is about 70%, with a slightly smaller value for evergreen species (Aerts 1996).

R. Gasche et al. (eds.), Trace Gas Exchange in Forest Ecosystems, 53–77.

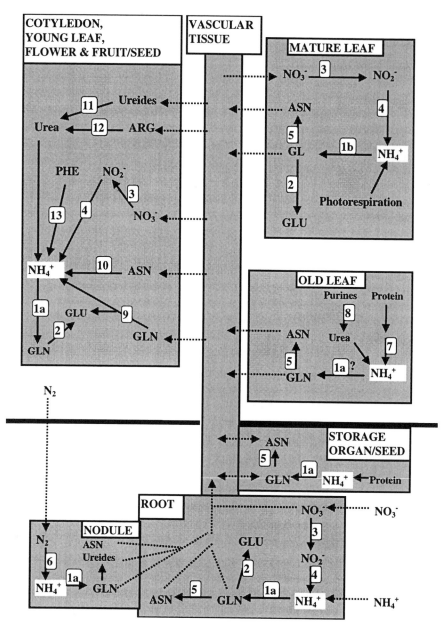

Figure 1. Schematic of metabolic pathways leading to ammonium production showing the major tissues of localization in higher plants. The enzymes are: (1a) cytosolic GS; (1b) chloroplastic GS; (2) glutamate synthase; (3) nitrate reductase; (4) nitrite reductase; (5) asparagine synthetase; (6) nitrogenase – bacterial enzyme complex; (7) glutamate dehydrogenase; (8) uricase; (9) glutaminase; (10) asparaginase; (11) allantoicase; (12) arginase; (13) phenylalanine lyase. Solid lines represent biochemical steps, broken lines indicate passage of transport compounds into or out of the organ types.

As important as N recovery is during the senescent phase for the nitrogen economy of plants, the turnover of N for remobilization and background cell maintenance is vital at all stages of plant growth (Fig. 1) (Raven et al. 1992a). This reflects the relative scarcity of N and the wide range of requirements for N in many types of organic molecules, a large number of which are modulated by, or required for, short-term responses to environmental fluctuation. This includes mobilization of N from seeds/storage tissue; background turnover for *de-novo* synthesis, or repair of proteins and other N-containing compounds; the production of stress metabolites, for example proline when faced with drought, or of secondary metabolites such as CN^-, which acts against insect predation; re-translocation of N between cell compartments as in photorespiration; or between cells and tissues for new growth.

Although nitrogen retention and turnover is a vital process in plants they still possess the potential to lose N to the atmosphere. A full description of how this occurs as gaseous NH_3 volatilization, via pH-dependent equilibration of $NH_4^+ \Leftrightarrow NH_3$ in the apoplastic fluid of leaves, is given in Chapter 3.1. Here we focus mainly on the role of NH_4^+ production and consumption. The role of NH_3 uptake is covered mainly with regard to uptake and assimilation from atmospheric sources. Its interaction with uptake of N from the soil and maintenance of cell pH is also considered.

The nitrogen molecule has the potential to appear in many organic forms during its time in the plant. In this Chapter, we will emphasize that in order for these transformations of N to occur the processes of de-amination/de-amidation, with the production of NH_4^+, are vital. Subsequent re-assimilation of this NH_4^+ is a key feature of the N economy, recovery and turnover of N, and all of this takes place through the enzyme glutamine synthetase (GS; EC 6.3.1.2). Although transamination is also of great importance in the further transformation of organic N from the product of NH_4^+ assimilation by GS, glutamine, all roads in N metabolism inevitably lead to NH_4^+ (Woodall et al. 1996a).

2.   PRIMARY PROCESSES GENERATING NH_4^+/NH_3 IN TREES

For trees there is a constant need to assimilate N for continued growth as well as to replenish N, which is not only lost through senescent tissue, but through shoots as NH_3/NO_X volatilization, or through leaching in throughfall and stemflow (see Chapters 3.1 and 3.2 in this volume) (Pearson and Stewart 1993; Pearson et al. 1998). This can be termed primary N assimilation, that is inorganic N absorbed from outside the plant and then assimilated to organic form. Often this inorganic N is soil-derived, but uptake and assimilation of atmospheric NH_X and NO_X can also be included in this category. Nevertheless, all primary N assimilation takes place through NH_4^+ incorporation into organic form via GS.

2.1   Inorganic nitrogen use

Of the three forms of inorganic N only NO_3^- and NH_4^+ are potentially available to all species, whereas N_2 is only available to those species which can develop a root symbiosis with N_2-fixing soil bacteria (Fig. 1). In the latter instance this includes some members of the legume family which have symbioses with bacteria of the genus *Rhizobia*, as well as actinorhizal trees with associations with *Frankia* bacteria. Not all leguminous tree species are necessarily nodulated or possess the ability to fix N_2, for example nodulation is low in members of the legume sub-family Caesalpinoideae (Sprent, 1987; Frioni et al. 1998). Assimilation of N_2 by these species is via the bacterial nitrogenase enzyme complex, which produces NH_4^+. The NH_4^+ is subsequently assimilated by the plant enzyme GS (see below). Di-nitrogen assimilation is the most energy expensive means of primary assimilation (Sprent 1987), and particularly when NO_3^- is in reasonable supply then nodulation of tree roots may be inhibited (Sprent 1987; Thomas et al. 2000). In *Gliricidia sepium*, a tropical leguminous tree species, when fed 10 mM NO_3^- there was a significant reduction in nodulation and nitrogenase activity compared to trees fed only 0.1 mM NO_3. When trees were grown on high NO_3^- and high CO_2 then this inhibition was not as marked, possibly indicating that C allocation/energy has a role to play in the main form of N assimilated (Thomas et al. 2000). In many N_2-fixing legumes the ureides (allantoin, allantoic acid and citrulline) are the main transport compounds produced following NH_4^+ incorporation. The metabolism of ureides at their sink tissue does proceed through NH_4^+ production, but the exact pathway for this NH_4^+ production is open to some debate and probably does not involve urea, hence the question mark in Fig. 1 (Lea and Ireland 1999).

All soils will have some NH_4^+ and NO_3^- present but these will be available in different proportions depending on soil type (for the purposes of this section organic N is not covered in detail). As a generalization, acidic soils tend to have more NH_4^+ than NO_3^- and *vice versa* for basic soils. Trees and woody species have different preferences for the form that they acquire for the majority of their inorganic N. Many species take up NH_4^+ as the preferred source and some take up NH_4^+ almost exclusively. In the latter case plants often form a large group within families such as the Proteaceae, Ericaceae and many gymnosperms (Stewart et al. 1988). For example, Gessler et al. (1998) showed that spruce and beech favoured NH_4^+ uptake to that of NO$_3^-$, the difference being almost an order of magnitude in favour of NH$_4^+$ across a range of temperatures. Spruce was almost exclusively an NH$_4^+$ assimilator, whereas beech would use some NO$_3^-$ when soil conditions allowed. Fire-prone species of wallum heath in Australia showed an order of preference for N as NH$_4^+$ > glycine > NO$_3^-$. Even though following a fire, soil NO$_3^-$ increased 60-fold, there was little evidence for much NO$_3^-$ utilization by many of the species of this habitat (Stewart et al.1993; Schmidt and Stewart 1997).

Uptake of NH$_4^+$ by tree roots is via both a high affinity (HATS) and low affinity (LATS) transport system, both HATS and LATS for NH$_4^+$ have been found to be constitutive in *Picea glauca, Populus tremuloides* and *Pinus contorta* (Kronzucker et al. 1996; Min et al. 2000). Thus a a biphasic uptake system for NH$_4^+$ exists at low and relatively high NH$_4^+$ concentrations respectively. For HATS in *Picea glauca* the V_{max} for NH$_4^+$ was between 1.9 to 2.4 μmol g^{-1} FWt h^{-1} and the K_m 20–40 μM, and overall the capacity for uptake of NH$_4^+$ was much higher than for NO$_3^-$ in this species (Kronzucker et al. 1996). A linear uptake rate for NH$_4^+$ across a 0.5 to 50 mM concentration range was found for LATS in *Picea glauca*. However, soil NH$_4^+$ is very unlikely to attain the higher concentrations used in this work. Estimates will obviously vary with soil type, but a general range for soil NH$_4^+$ of 0.01 to 0.1 mM is often quoted (Sprent 1987; Fredeen and Field 1992), and most soils, in an ecological rather agricultural context, will probably lie to the lower end of this scale, i.e. 10–50 μM (Freeden and Field, 1992). There is some debate as to the form (NH$_4^+$ vs. NH$_3$) and the effect of pH on NH$_4^+$ uptake. It is often stated that the unprotonated form, NH$_3$, crosses the plasmalemma more readily than NH$_4^+$. However, Forde and Clarkson (1999) stress that at most soil pHs there will be very little uptake as the unprotonated form, and that equilibrium with cytoplasmic concentrations means that efflux from roots across the plasmalemma as NH$_3$ is more likely. Some studies suggest that NH$_4^+$ uptake is favoured at high pH, but Forde and Clarkson (1999) also point to the conflicting evidence for this. It is possible that pH plays an indirect, though important, role through its effect on rhizosphere nitrification potential. Species assimilating NH$_4^+$ have been shown to inhibit nitrification in their rhizosphere, when compared to bulk soil nitrification rate, and *vice versa* for species assimilating mainly NO$_3^-$ (Ollsson and Falkengren-Gerup 2000). Nitrifying bacteria are known to be inhibited at low pH and NH$_4^+$ assimilation would lead to a lower pH in the rhizosphere (see below).

2.2   Tissue localization

Once taken up by the root, NH_4^+ is rapidly assimilated to organic form in the root via the enzyme GS. The assumption is that NH_4^+ is toxic in relatively small amounts and is rapidly incorporated in the roots to avoid problems of phyto-toxicity (Givan 1979). The NH_4^+ ion is thought to uncouple electron transport in the chloroplast, though there is little evidence for uncoupling of electron transport in the mitochondria. Pearson and Stewart (1993) discuss some of the evidence for and against this explanation of NH_4^+ toxicity.

Although it used to be thought that trees did not assimilate much NO_3^-, and if they did it was confined mostly to roots, there is now good evidence that certain species have a strong NO_3^- preference (Smirnoff et al. 1984). Nitrate reduction proceeds via nitrate and nitrite reductase to produce NH_4^+. The NH_4^+ is then assimilated by the enzyme GS (Fig. 1). Some examples of highly nitrophilous temperate tree species are: *Sambucus nigra; Betula spp and Populus spp* (Smirnoff et al. 1984; Soares et al. 1995) (Table 1), see also Stewart et al. (1988, 1993) for examples of Australian tropical and sub-tropical species. Trees can be broadly divided into two groups, as climax or pioneer species. Pioneer species are opportunists and tend to invade new and disturbed ground, while climax species are dominant in stable habitats and represent the end of the stage of succession for a given climate/environment. Much work to elucidate their N use has been carried out by George Stewart and colleagues. In studies on many species from very varied habitats one broad generalization seems to apply: climax species tend to favour NH_4^+ (and organic N) over NO_3^-, while pioneer species prefer NO_3^-. Nitrate reductase (NR) activity is a substrate inducible enzyme and can be used to infer NO_3^- utilization. Thus, in Australian rain forest (Stewart et al. 1988, 1990), in Brazilian cerrado (Stewart et al. 1992) and in temperate deciduous woodland (Clough et al. 1989; Pearson et al. 1989), pioneer plants of open and disturbed habitats have consistently higher leaf nitrate reductase (NR) activity compared to climax species. Even when trees or cut stems are fed NO_3^- the NR activity is only significantly induced in pioneer plants with little or no evidence for induction or high rates of incorporation in climax species (Stewart et al. 1993). Clough (1993) showed that in pot-grown saplings fertilized with NO_3^-, greatest induction of leaf NR occurred in the pioneer *Betula pendula*. Maximum induction occurred 48 h after application and was still higher than in controls 8 days following application. By comparison, in the climax species *Quercus petraea,* only a small increase in activity was detected (Fig. 2). Xylem sap analysis also confirms the low rates of NO_3^- uptake in these climax species and that this correlates with the low induction of NR (Stewart et al. 1993; Schmidt and Stewart 1997). In the majority of nitrophilous trees, root uptake leads to translocation of most of the NO_3^- in the xylem to their leaves, where it undergoes reduction to NH_4^+, which is in turn rapidly assimilated by leaf GS (Fig. 2) (Stewart et al. 1988, 1989, 1993). These findings are corroborated by Min et al. (1998) who showed that the pioneer, trembling aspen, was much more efficient at taking up NO_3^- through roots which is assimilated largely via NR in the leaves. By comparison the gymnosperm, lodgepole pine, was very inefficient at NO_3^- assimilation. Leaf NR activity was not detected in lodgepole pine and root NR, though measurable, was low compared to aspen.

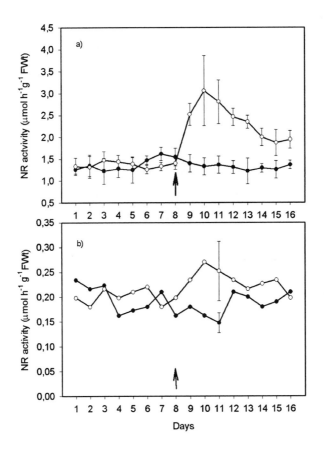

Figure 2. Leaf nitrate reductase activity (*in vivo*) in two tree species of contrasting ecology, the pioneer *Betula pendula* (a) and climax species *Quercus petraea* (b). Two-year-old saplings were grown in pots placed outside during summer. At day 8 (arrow) one group was treated with a one-off 5mM KNO_3 application to the soil (-o-), and compared to controls with a water only addition (-•-). Values are the mean for three replicates taken at midday on each day and error bars are for 95% confidence limits, except in b), where most error bars are omitted for clarity.

Further confirmation for some of these points comes from labelling studies (e.g Stewart et al.1992; Wallenda et al. 2000). Figure 3 shows one such comparative experiment, again contrasting pioneer against climax species. The pioneer, *Betula pendula,* shows a marked preference for $^{15}NO_3^-$ over $^{15}NH_4^+$. When fed NO_3^- there is much more label evident in the leaves than when fed NH_4^+, and with plants fed NH_4^+, either singly, or in mixture, there is much more label in root tissue compared to shoots (Fig. 3a). For the climax species, *Quercus petraea,* when fed either $^{15}NO_3^-$ or $^{15}NH_4^+$ singly or in mixture, the rate of uptake of NH_4^+ is twice that of NO_3^- uptake (Fig. 3b). As the roots retain similar amounts of label much of the increase is accounted for by increased label translocation through stem to leaves.

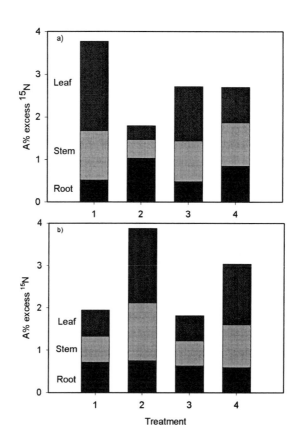

Figure 3. Incorporation of ^{15}N to root, stem, and leaf of two tree saplings chosen to represent pioneer, *Betula pendula* (a), and climax species, *Quercus petraea* (b). Treatments 1–4 were applied as $K^{15}NO_3$, $^{15}NH_4SO_4$ or $^{15}NO_3NH_4$, $NO_3^{15}NH_4$. Treatments were applied to the soil as a 300 ml solution in dd H_2O (5mM) and 99% ^{15}N-enriched. Plants were harvested 48 h following application and values are the mean of three replicates.

For the primary assimilation of inorganic N the energy costs in biochemical terms are $N_2 > NO_3^- > NH_4^+$, and for N_2-fixation will be higher still if investment in nodules and loss of C to the bacteroids is accounted for (Sprent 1987). There are also energy costs or rate limiting factors involved to provide a C skeleton for primary N assimilation and to transport this C, if needed, to the tissue or site of assimilation. For root assimilators of NH_4^+ this clearly involves transport through the phloem, whereas leaf NO_3^- assimilators have their C and N acquisition in close proximity in the chloroplast (Pearson and Stewart 1993). Further implications of this are discussed in later sections.

3.   SECONDARY PROCESSES GENERATING NH_4^+/NH_3

Tissue and organelle localization also has important implications in secondary production of NH_4^+ (Fig. 1).

3.1   Photorespiration

Photorespiration is an aspect of C_3 photosynthesis whereby the oxygenase activity of Rubisco leads to the production of NH_4^+ and stoichiometric amounts of CO_2. Photorespiration is often called the C_2 pathway, because the reaction of Rubisco with O_2 produces the 2-carbon phosphoglycollate. Photorespiration is very much dependent on environmental conditions, and is primarily controlled by light, temperature and the internal CO_2/O_2 ratio. With high light and high temperatures and with a low internal CO_2 to O_2 ratio, conditions will favour the oxygenase activity of Rubisco. Although the photorespiratory pathway is initiated in the chloroplast, the subsequent pathway and metabolism of phosphoglycolate is fairly complex and not only involves enzymes in the chloroplast, but in two other sub-cellular compartments, the peroxisome and mitochondrion. The NH_4^+ and CO_2 is released in the mitochondria from two molecules of glycine that also produce serine, a reaction catalysed by an enzyme complex that involves glycine decarboxylase and serine hydroxymethyltransferase. Detailed reviews of the biochemistry involved can be found in Leegood et al. (1995) and Lea and Ireland (1999).

In fact, very little work on photorespiration in tree leaves has been carried out. Trees are often difficult to work on, but in any case the measurement of the rate of photorespiration is not an easy undertaking in any species, so that very few studies have been able to quantify the contribution that the photorespiratory pathway plays in plants. Nevertheless, in a qualitative sense we know it makes a significant contribution in many C_3 photosynthetic species and it is estimated that the production of NH_4^+ is 10-fold greater than that produced from primary NO_3^- assimilation (Wallsgrove et al. 1983; Raven et al. 1992a). That environmental conditions play a significant role in regulating photorespiration can lead us to speculate that in a tree canopy, leaf aspect and self-shading will lead to heterogeneous conditions with respect to light and temperature and thus to wide variation in photorespiration and the rate of NH_4^+ production from leaf to leaf. For simplicity we have shown photorespiration only occurring in mature leaves (Fig. 1). However, it is likely that all photosynthetic tissue at all stages of development will experience some degree of photorespiration.

The role of photorespiration was open to some debate, but it is now thought to act as a safety valve for e^- transport and reductant use under conditions when PSII is fully quenched (Osmond et al. 1997). Of importance with regard to the NH_4^+ produced is that nearly all of this is quickly recovered by the chloroplast enzyme GS (see below for further discussion). The recovery of NH_4^+ is generally thought to be very efficient. In barley mutants, which lacked a small proportion of the chloroplast GS activity compared to wild type, NH_4^+ was detected under photorespiratory conditions, but was otherwise barely detected in wild type, except under high N concentrations in the growth media (Hausler et al. 1996; Lea and Ireland 1999). This quick re-assimilation of the secondarily-produced NH_4^+ via the GS/GOGAT cycle provides glutamate, which through transamination can lead to production of glycine. In this way much of the recovered NH_4^+ can be recycled through the C_2 pathway many times. Enzymes of transamination are ubiquitous in plant tissues and often non-specific with regard to substrates. Glutamate in particular is also a substrate for several different transamination enzymes (Ireland and Lea 1999). Therefore, some NH_4^+ recovered from photorespiration has the potential to be diverted towards N compounds for transport, or growth maintenance. The extent to which this may occur is unclear, particularly as the photorespiratory NH_4^+ is re-assimilated in the background of other sources of NH_4^+ (Fig. 1). Nevertheless, in many tree species that are characterized by root assimilation of inorganic or organic N, the background NH_4^+ produced in leaves may be low and photorespiratory NH_4^+ production assume a comparatively greater part of N turnover in leaves.

Species which are C_4 photosynthetic largely overcome photorespiration, but so far, no C_4 photosynthetic trees or shrubs are known. It is interesting to speculate on the selection pressures that might have led to this. Here we list a number of possibilities:

1. Woody species are likely to experience shading with rapid changes in light intensity, especially at the seedling stage. To tolerate this fluctuating environment, leaf structure and development must be flexible. Production of bundle sheath tissue and use of PEPC for initial CO$_2$ fixation is part of the C$_4$ syndrome and may constrain the ability to respond to changing light intensity.

2. As a corollary to the above, many woody species are root assimilators of N and may be restricted with regard to N supply and turnover for leaves; the N supply being very much dependent on the rate of transpiration. Under conditions of high temperature and high light, water availability may be even more constrained and supply of N in xylem fluid restricted. Under such conditions, photorespiratory NH$_4^+$ may provide a relatively rapid turnover for potential new N compounds, some of which can be directed to requirements for a fluctuating environment e.g. for repair of photodamage or for compounds for water shortage such as proline and glycine betaine. Schmidt and Stewart (1998) have shown that in a wide range of woody Australian plants the majority are root N assimilators, but that during the dry season or periods of drought the few leaf NO$_3^-$ assimilators lose the ability to take up much NO$_3^-$ and reduce it in their leaves.

3. "Woodiness" may exert a selection pressure by diverting N to phenylalanine and lignin production at the expense of investment in secondary thickening in bundle sheath cells.

Nitrogen use efficiency as we have indicated is very much linked to NH$_4^+$ production at primary and secondary levels. In leaves with a restriction in primary assimilation, photorespiration may represent the main source of NH$_4^+$ for what can be termed the "N response" to a changing environment. The connection may lie in the trade off between secondary growth or sclerophylly and tissue water. Stewart et al. (1990) and Roderick et al. (1999) have shown that the mass of N per unit mass of water was relatively constant in leaves with differing structure and that the N:C ratio is positively correlated with liquid content. In this way there may be a link, either by direct or indirect selection pressure, between photosynthesis, and both water and nitrogen use efficiency.

3.2   Phenylpropanoid pathway

The metabolism of phenylalanine by the enzyme phenylalanine ammonia-lyase (PAL) produces NH$_4^+$ and trans-cinnamic acid. This enzyme links the primary metabolic pathway of shikimic acid metabolism to that of the phenylpropanoid pathway, the latter of which is responsible for the production of a number of secondary phenolic metabolites such as lignin and flavonoids. As such, PAL is generally thought to play a minor role in NH_4^+ production (Joy 1988). However, a lot of the work on this enzyme is confined to herbaceous or crop plants and it is possible it may play a more significant contribution to NH_4^+ production in woody (lignified) plants.

In most plants PAL is encoded by a multi-gene family, for example in *Pinus banksiana* at least 8–10 loci have been found (Butland et al. 1998). The enzyme is present in all tissues, but its activity or regulation is subject to a number of abiotic and biotic factors, among which are light, temperature, ethylene, pollution or pathogen attack (Biagioni et al. 1997; Reglinski et al. 1998). It is localized in the cytoplasm, chloroplast, plastids and mitochondria. In trees the production of lignin is not only important in wood formation and secondary thickening but also in the ripening processes in fruit/seed development, and in storage tissue. The activity of PAL in *Robinia pseudoaccacia* is confined to living and developing xylem tissue, but is absent from the dead heartwood (Magel 1997). The NH_4^+ produced from PAL activity has been shown by a combination of ^{15}N NMR studies of metabolites, and inhibitors, to produce first glutamine and then glutamate, which is consistent with the NH_4^+ released being re-assimilated by GS (Singh et al. 1998).

3.3   Nitrogen mobilization, seed-set, and senescence

The movement of N between tissues is a constant process at all stages of plant development. Nevertheless, we can identify four types of mobilization that differ in importance with the stage of development:
1. the remobilization of storage reserves from seeds during germination or from bark or storage organs at the start of a new season
2. movement to new growing parts from mature tissues or as a result of primary assimilation and transport
3. translocation and/or storage due to times of plenty or excess
4. removal of N from mature and senescing tissues to seeds or storage organs

There is very little evidence for inorganic N being found in the vasculature for the purpose of recycling N, although we note that NO_3^- in xylem may be high in leaf assimilators and may be stored in leaf vacuoles prior to primary assimilation. Nitrate has also been found in small amounts in phloem of Norway spruce subject to high NO_X pollution (Rennenberg et al. 1998). Enzymes to oxidize amino-N or NH_4^+ to NO_3^- have not been found in plants. In contrast to NO_3^- there are, as this chapter asserts, many processes that produce NH_4^+. Despite this, there is no evidence of NH_4^+

translocation between tissues playing any significant role in N transport. By far the greater amount of N translocation takes place in the form of amino acids and amides. Although nearly all of the amino acids can be detected in xylem or phloem sap, certain amino acids are usually present in much greater concentrations and are known as transport N compounds. These are amino acids which are characterized by their high N to C ratio, solubility and ability to move across membranes; two important amides in this respect are glutamine and asparagine.

The catabolism and turnover of functional and structural N compounds involves proteolytic and hydrolytic enzymes which eventually yield constituent amino acids or secondary N compounds. These processes are particularly important during end of season seed-set and senescence. Deamination plays an important role in converting N to a suitable form for transport, it is in this context a type of "repackaging". The NH_4^+ produced is re-assimilated by GS and the indications are that the cytosolic isoform may be important in this function (Kawakami and Watanabe 1988; Pearson and Ji 1994). In particular an increase in cytosolic GS is associated with phloem companion/transfer cells (Edwards et al. 1990; Carvalho et al. 1992). Circumstantial evidence for the importance of NH_4^+ production comes from an increase in tissue NH_4^+ in senescent tissue (Schoerring et al. 1998). Beech has been shown to significantly increase tissue NH_4^+ in senescing leaves when compared to mid-phase leaves (Nielsen and Schjoerring unpublished). In addition, the activity of glutamate dehydrogenase (GDH) increases in tissues that are senescent or rely on rapid N turnover (Lea and Ireland 1999). Glutamate dehydrogenase is ubiquitous in plant tissues and although there has been some debate about its role (Oaks 1995), it has been shown to be important in deamination (Robinson et al. 1992; Fox et al. 1995). In leaves of *Sambucus nigra* and *Quercus* sp. there was a four-fold increase in GDH in senescent leaves compared to middle-aged leaves (Pearson not published). This places the substrate for GDH, glutamate, in a paradoxical position in that it may act as NH_4^+ source for GS via deamination, but it is also required as a substrate by GS for the NH_4^+ to be re-assimilated. However, glutamate can be produced by many processes, including several transamination reactions (e.g. aspartate and alanine aminotransferase) and by metabolism of all amino acids (Lea and Ireland 1999). In addition the production of asparagine is catalysed by the enzyme asparagine synthetase from glutamine and aspartate to yield asparagine and glutamate. This step involves GS, and still generates glutamate for continued production of the two main transport amides, glutamine and asparagine. It is likely to be of importance in NH_4^+ re-assimilation during N mobilization, and is associated with low GOGAT activities (Ireland and Lea 1999). Compartmentalization will also be important throughout, for example if the GDH is in the mitochondria and the

GS in the cytosol then there will be two separate pools of glutamate to act as source or substrate.

We know very little about the mechanisms that underlie loading of the vasculature with this transport amino-N, or whether NH_4^+ or amino acids themselves move short distances to the vasculature via symplasm or apoplasm and undergo interconversions in transfer cells. For example, the characteristic amino acids of N transport are also found in both seed and vegetative storage proteins (Staswick 1994). Beardmore et al. (1996) showed that in poplar both seed and vegetative storage proteins had a strong homology and therefore did not seem to be tissue specific. Similarly, during times of excess N the soluble amino pool can change and is often seen as an increase in characteristic amides or arginine, all with high N to C ratios. Does a build up of soluble reserves, or catabolism of storage reserves, yield amino-N in a transportable form, with little need for inter-conversions? There is little evidence to answer this question. However, further evidence that may have some bearing on this, is that in sink tissues, such as seeds and fruits, there is a similar high level of GS and GDH activity along with other enzymes involved in N transformation. Is this because whether source or sink, the N molecule has to undergo substantial "re-packaging" via NH_4^+ even if it arrives in a form already compatible with transport or storage? It is possible that these conversion processes are a necessary function of loading and unloading N at source or sink respectively.

3.4   Uptake of organic N

Organic N can be taken up by plant roots via mycorrhizal associations or directly through uptake of amino acids, peptides or proteins, termed dissolved organic N (DON). For trees the contribution that mycorrhizae play in N uptake is likely to be relatively small in vesicular arbuscular associations, but may play a more significant contribution in ectomycorrhizal and ericoid-types of association (Lambers et al. 1998). Recent work has shown that uptake of DON could play a potentially significant role in the uptake of N by trees (Nasholm et al. 1998; Schmidt and Stewart 1999). These authors also affirm that the uptake and utilization of DON by woody species is not necessarily confined to a select group of species nor to specific habitats, but is of widespread occurrence.

This is an attracting area of research for the coming years. From the early evidence several points can be made, but may need later modification:

1. glycine, as the smallest and simplest amino acid, seems to be more readily assimilated by roots than other amino acids.
2. Uptake of glycine/glutamine does not interfere with uptake of NH_4^+ or NO_3^-, or *vice versa* (Schmidt and Stewart 1999; Wallenda et al. 2000).

3. For glycine, transamination is the first major step once taken up by roots.

Questions remain about the relative proportions and types of amino acids in soils and about the exact amounts taken up by roots relative to inorganic N. There is also the question of whether deamination plays a big role in re-packaging the amino-N for translocation as discussed in Chapter 3.3.

4.   GLUTAMINE SYNTHETASE

All of the NH_4^+ produced in plants, whether from primary or secondary sources, is primarily (re-)assimilated through the enzyme GS (Lea et al. 1990; Fox et al. 1995). Glutamine synthetase is encoded in the nucleus by a small multi-gene family, which gives rise to two main groups, the chloroplastic and cytosolic isoforms. Both are octameric enzymes with the sub-units ranging in size from 36–45 kDa; the cytosolic polypeptide is usually between 36–40 kDa and the chloroplastic 41–45 kDa. However, purified GS from roots of *Pseudotsuga menziesii* has been reported to have two sub-units of molecular mass of 54 and 64 kDa (Bedell et al. 1995). Several cytosolic isoforms have been reported across a range of species including trees, and while the vast majority of species studied to date have only one chloroplast isoform, there are one or two exceptions, for example, the non-woody perennial herb, *Trientalis europaea,* where both a small and large sub-unit have been isolated from chloroplasts (Woodall et al.1996b). A standard technique for separation of the two (or more) isoforms has been through ion-exchange chromotography (IEC). In the past it was commonly assumed that cytosolic GS always eluted before (at a lower salt concentration) the chloroplastic GS isoform. This has been shown not to be the case for a number of species, where the profile is reversed and chloroplast GS has eluted before the cytosolic GS, including *Lotus japonicus, L. corniculatus, Trientalis europaea,* and *Pinus sylvestris* (Elmlinger and Mohr 1992; Woodall et al. 1996b). Elmlinger and Mohr (1992) carried out IEC separation of GS from *P. sylvestris* needles and speculated that the first peak to elute was chloroplastic as this isoform was present in greater amounts and was modulated or regulated by light, whereas the smaller second peak was not. We have subsequently carried out western blots of IEC peaks from *P. sylvestris* needles and can confirm that the first peak to elute is composed entirely of the large sub-unit polypeptide and the second peak to elute had the small sub-unit (unpublished).

The GS enzyme is present in all tissues throughout the plant (Lea et al. 1990). In the 20 or so tree species we have looked at GS activity is always higher in the leaves than the roots, at least during the young to mid-phase of leaf development. For example, in the young leaves of *Sambucus nigra,* GS activity was 147 µmol hr^{-1} g^{-1} FWt, whereas in roots this was 8 µmol hr^{-1} g^{-1} FWt (Woodall et al. 1996b). Generally, we have found the range of total GS activity for leaves falls between 50–250 µmol hr^{-1} g^{-1} FWt and in roots between 2–10 µmol hr^{-1} g^{-1} FWt, when using the semi-biosynthetic assay (Pearson and Ji 1994) (Table 1). Environmental factors have been shown to alter GS activity, such as light, N supply and temperature (Margolis et al. 1988; Lea et al. 1990; Seith et al. 1994; Woodall et al. 1996b). Seith et al. (1994) found that GS activity was higher in the cotyledonary whorl than in the roots of Scots pine. Seith et al. (1994) also found no clear relation between N supply and needle GS activity, which is in agreement with the data in Table 1, which compares leaf GS and leaf NR activity.

Table 1. Glutamine synthetase and nitrate reductase (*in vivo*) activities in trees collected from the field (F), or grown in plant pots fertilized with 3mM KNO$_3$ (P). All values are average of three replicates.

Species	GS leaf (µmol h^{-1} g^{-1} FWt)	GS root (µmol h^{-1} g^{-1} FWt)	NR leaf (µmol h^{-1} g^{-1} FWt)
Picea sitchensis (P)	67	3	0.25
Picea abies (F)	50	ND	0.21
Abies alba (F)	82	ND	0.09
Pinus sylvestris (F)	61	ND	0.47
Quercus petraea (P)	220	3	0.34
Fagus sylvatica (F)	95	ND	0.19
Populus deltoides (P)	54	5	3.60
Populus tremula (F)	73	ND	2.70
Betula pendula (P)	58	7	4.10
Sambucus nigra (F)	165	ND	15.80

The range of activity in the leaf is usually the total activity of several isoforms, and interpretation of changes in total activity often does not distinguish between whether it is the cytosolic or chloroplast enzymes that are affected. Pearson and Ji (1994) found that in the leaves of several temperate and deciduous trees the activity of the two main GS isoforms, chloroplast and cytosolic, was dependant on season. In the early season, chloroplastic activity predominated and was high, whereas cytosolic activity was not noticeable till mid- or late-season, and increased in activity in autumn while that of the chloroplastic isoform declined, though the latter was never completely lost. The importance of this change to the different functions of the isozymes is mentioned above, with chloroplastic GS thought to be responsible for assimilation of NH$_4^+$ produced from NO$_3^-$ reduction and from photorespiration, while the cytosolic isoform(s) are involved in recovery of NH_4^+ produced in N turnover/translocation and senescence (Pearson and Ji 1994).

The biochemical properties of GS have been reviewed by Lea et al. (1990). Briefly, GS has a pH optimum between pH 7.0–8.0, has a high affinity for NH_4^+ (K_m 10–20 µM) and a lower affinity for glutamate (K_m 1–15 mM) and ATP (K_m 0.1–2 mM). At the time no specific information on tree GS was available, however since 1990 a limited number of studies have confirmed that GS from woody species broadly agrees with these K_m ranges (Table 1). The leaf GS of *P. banksiana* is perhaps worth mention as having a slightly higher K_m, lower affinity, than its root isoform (Vézina and Margolis 1990). In *P. banksiana* the pH optimum for both leaf and root isoforms was lower than commonly reported by about 1 unit, at pH 6.5 (Vézina and Margolis 1990). Partially purified GS from needles of *Pinus sylvestris* had a pH optimum between 7.5–7.7 (Schlee et al. 1994), and of purified GS from roots of *Pseudotsuga menziesii* of pH 7.6 (Bedell et al. 1995).

Table 2. Apparent values of substrate K_m for glutamine synthetase in trees

Species and tissue	Ammonium (µM)	Glutamate (mM)	ATP (mM)	Reference
Pinus banksiana – Leaf	33	1.1	2.9	Vézina and Margolis (1990)
Pinus banksiana – Root	19	1.8	2.3	*ibid.*
Pseudotsuga menziesii – Root	11	2.6	0.5	Bedell et al. (1995)

At the primary level of assimilation in nodules of N_2-fixers, where the bacteroid nitrogenase assimilates atmospheric N_2 to form NH_4^+, GS is known to assimilate this NH_4^+ to organic form (Baker and Parsons, 1997; Woodall et al. 1996a). Labelling studies carried out over very short time courses (seconds) show that glutamine is the most heavily labelled product in nodules of the shrub *Myrica gale* when fed [15]N_2 (Baker and Parsons, 1997). Unfortunately, much work on nodules and GS is carried out on crop plants and there is very little in the literature on nodulated trees, with the possible exception of the actinorrhizal association in *Alnus* spp. Nodules contain a relatively large proportion of GS activity when compared to roots, and in *Phaseolus vulgaris* GS can account for 1–2% of total nodule protein (Lea et al. 1990). Even though many bacterial symbionts possess GS activity, the host root tissue of the nodule is responsible for NH_4^+ assimilation; and GS mRNA transcripts and activity are high in infected cells and in the cells of the vascular pericycle (Espin 1994; Guan et al. 1996).

In the case of primary NH_4^+ uptake from the soil, the cytosolic root GS is known to be important for most of the assimilation. Reports vary as to whether root or soil NH_4^+ treatment induces significant change in GS activity

(Margolis et al. 1988; Lea et al. 1990; Truax et al. 1994). Truax et al. (1994) showed that both red oak and red ash significantly increased root GS activity when fed either high NO_3^- or high NH_4^+ (25 mg week^{-1}). Root NO_3^- assimilators may also rely on cytosolic GS for NH_4^+ produced via NO_3^- reduction. Woodall and Forde (1996) found that in roots of some tropical woody legume species, when they are not fixing N_2 but are fed NO_3^-, a root plastidic isoform is produced. In this case there are perhaps some analogies with leaf NO_3^- reduction and assimilation of the NH_4^+ produced by chloroplastic GS. Although these roles for cytosolic and plastidic/ chloroplastic GS is presented as a firm division of labour, it is possible and even likely that some overlap of roles does occur. In herbaceous species the increase in cytosolic GS during senescence has been shown to be associated with an increase in GS in phloem companion cells. Woodall et al. (1996b) excised the major veins from a number of woody species and compared them to the lamina cells from the same leaf. Measurement of GS activity in these two parts of the leaf showed that the veins had about 80% of the total cytosolic GS activity compared to the lamina.

Glutamine synthetase is an important enzyme at all phases of plant growth and in the overall nitrogen economy of the plant. For perennial species GS helps retain N as part of a system of plant N recycling. It is the enzyme ultimately responsible for withholding much N from the greater global N recycling scheme.

5.   PHYSIOLOGICAL RESPONSES TO NH_4^+/NH_3 AND FOREST DECLINE

Physiological responses of plants to NH_4^+/NH_3 are reviewed by (Raven et al. 1992b; Pearson and Stewart 1993; Pearson et al. 1998), and there has been mention of the ecological consequences in Section 2.2. Here we intend to focus on the associated problem of acid-base regulation (which occurs with inorganic N assimilation) and in particular, the primary uptake of NH_4^+ or NH_3 and their subsequent assimilation which impose constraints on regulating H^+, or problems of acidity. Even though gaseous NH_3 is readily soluble and forms a base when in solution (NH_4^+ + OH^-), the overall net effect following assimilation by GS is the production of protons (H^+). Solubilization of NH_3 followed by assimilation generates $1/3$ H^+, whereas direct uptake and assimilation of NH_4^+ generates $1^1/_3$ H^+. The opposite occurs with NO_3^- assimilation, which overall generates OH^- as part of the pathway of reduction and assimilation to organic form (Raven et al. 1992b; Pearson and Soares 1995; Pearson et al. 1998). Generally, dealing with alkalinity in the form of OH^- is not as problematic for cells or tissues, as organic acids are readily generated as part of the pH homeostatic mechanism (Raven et al. 1992b; Pearson et al. 1998). Pearson et al. (1998) showed that in *Populus deltoides* and *Quercus robur*, both malic and citric acid increased in leaves when fed NO$_3^-$ through the soil or direct to the foliage. Conversely, treating the plants with NH$_4^+$ caused a reduction in the organic acid concentration. Dealing with acidity is more problematic for plants and one reason why root assimilation of NH$_4^+$ is thought to be preferred is that H$^+$ is readily lost to the rhizopshere (Raven et al. 1992b).

In this chapter we have covered aspects of uptake of NH$_4^+$ from the soil, but bi-directional exchange of atmospheric NH$_X$ through foliage can and does occur. For factors regulating this bi-directional exchange with the atmosphere see Chapter 3.2 in this volume. Nevertheless, the contribution that foliar/canopy uptake can make to total plant nitrogen should not be underestimated. Harrison et al. (2000) have shown that mixed atmospheric N sources taken up by a spruce canopy in Waldstein, Germany, accounted for between 16–42% of total plant N. Generally, NH$_3$ is more soluble than the oxides of nitrogen and more reactive, making its contribution to foliar uptake more likely where it is present as the main or in a mixed source of atmospheric N (Pearson and Stewart 1993; Harrison et al. 2000).

The regulation of pH homeostasis is particularly problematic with regard to atmospheric NH$_X$ pollution, notably with regard to foliar uptake. Foliar uptake can occur as the gas, NH$_3$, or as uptake from rainfall as NH$_4^+$ (Pearson and Stewart 1993). Evidence suggests that the stomata are the main route for uptake of NH$_3$ but it may also readily solubilize on wet leaf surfaces as well as in the apoplastic fluid surrounding cells of the stomatal air spaces. Uptake of NH$_4^+$ from rainfall or stem throughfall may also be through leaf stomata, or even through the bark of twigs and branches (Harrison et al. 2000). Solubilization of gaseous NH$_3$ will initially lead to alkalization, but any subsequent uptake across the plasma membrane followed by assimilation will generate net H$^+$. Gaseous NH$_3$ is more soluble the lower the pH. Thus, any tendency to reduce pH through assimilation may increase further uptake. In addition, Soares et al. (1995) have shown that tree species, which are root assimilators of N, tend to have leaves that have a low overall acidity compared to leaf NO$_3^-$ assimilators. Root assimilators include many conifers, and broad-leaved species such as oak and beech, species often reported in the literature as suffering damage from N pollution/acid rain.

Table 3. The pH of whole leaf extracts in water and buffering capacity index (BCI) of several tree species selected as primarily root N assimilators, or as leaf nitrate assimilators on the basis of leaf NR activity. See Soares et al. (1995) for details.

Species	initial pH	BCI	Leaf NR activity (μmol h^{-1} g^{-1} FWt)
Picea sitchensis	2.9	0.12	0.25
Pinus sylvestris	4.2	0.14	0.38
Quercus robur	4.2	0.13	0.41
Fagus sylvatica	4.5	0.27	0.48
Corylus avellana	5.1	0.54	1.10
Populus deltoides	5.4	0.64	8.00
Betula pendula	5.5	0.93	0.90
Sambucus nigra	6.2	6.35	7.50

Table 3 shows that measurement of the initial pH of whole leaf extracts and comparison to leaf NR activity showed good correlation between high NR activity and a high leaf pH. A number of other physiological measurements such as higher cation content was also positively correlated with leaf NR activity (Soares et al. 1995). The buffering capacity index was measured by titration against H$^+$ of the extract over 1 pH unit (BCI). The BCI was also low in root assimilators, thus low base cation status and a low potential to generate OH$^-$ through NR activity in leaves may account for susceptibility of such tree species to N pollution as well as acid rain (Soares et al. 1995; Pearson and Soares, 1995). When two-year-old saplings were misted with a fairly high concentration of NH$_4$Cl (6 mM) for 15 minutes and the leaves sampled 24 h later a decrease in whole leaf pH was found (Table 4). Generally, the ready mobility of NH$_4^+$ to cross membranes and the plants capacity to remove NH$_4^+$ by assimilation to avoid toxicity may account for these changes (vanHove et al. 1989). As all the applications were carried out at pH 5, the changes are also consistent with the assimilation of NH$_4^+$ and production of H$^+$. However, as the measurements were carried out on whole leaves it is impossible to say where the acidity may have been compartmentalized, if at all.

Table 4. Whole leaf pH in control plants misted with water at pH 5 only (HCl), and with 6 mM NH$_4$Cl at pH 5. Leaves were sampled 24 h after the application, which was applied to the foliage only and not the soil.

Species	Control	6mM NH$_4$Cl
Picea sitchensis	2.84	2.58
Pinus sylvestris	4.27	3.92
Fagus sylvatica	4.54	4.21

Increasing atmospheric N pollution means that trees live in an inverted world in which foliar uptake of N may **not** act as additional fertilizer, but rather it circumvents or short-circuits the more "normal" form of N acquisition by trees via the root (Rennenberg et al. 1998; Pearson et al. 1998). Rennenberg et al. (1998) estimate that in N polluted environments foliar uptake of N reduced root uptake by 20–30% in beech and spruce. Thus foliar uptake of NH_X may be more damaging, not only because certain species have a low capacity to maintain pH homeostasis, but also because it may reduce uptake of other counter ions which are also important nutrients (Soares et al. 1995; Pearson and Soares 1995).

6.　　INTERACTIONS BETWEEN C AND NH_4^+ METABOLISM

The interactions between C and NH_4^+ metabolism are manifold and a full discussion is beyond the scope of this particular chapter. Some of the aspects have been covered briefly above. Here we will confine ourselves to a brief mention of one aspect of C and N metabolism as it relates to NH_4^+ assimilation. Incorporation of inorganic N into organic form requires a C skeleton as oxaloacetate, which is ultimately derived from CO_2 fixation. For root assimilation of NH_4^+ this C has to be translocated from shoots to roots for incorporation of the N. This process will rely on phloem flow as well as the soil availability of N. For NH_4^+ supply this is likely to be slow, because it is more tightly held by soil cation exchange sites. The distinction can be carried further with respect to growth of climax trees, which tend to have a steady, slow growth, and are often long-lived, whereas pioneer trees are often faster growing, in keeping with their need to rapidly colonize open areas. They also tend to be relatively short-lived species. Although many climax trees are efficient assimilators of NH_4^+, there seems to be rate-limiting processes involved that has led to selection for a slow but steady growth rate which relates to the steady supply of NH_4^+ in most soils. This hard gained NH_4^+ is then withheld in these longer-lived species from the global N cycle. Thus we return to the points raised in the introduction, where the uptake and retention of N by trees can be related to plant ecology and succession (Stewart et al. 1988; Min et al. 1998). Undoubtedly, NH_4^+ plays a key role in both global and plant N cycles.

REFERENCES

Aerts R (1996) Nutrient resorption from senescing leaves of perennials: Are there general patterns? J Ecol 84: 597-608

Baker A & Parsons R (1997) Rapid assimilation of recently fixed N_2 in root nodules of *Myrica gale*. Physiol Plantarum 99: 640-647

74 *J. Pearson* et al.

Beardmore T, Wetzel S, Burgess D & Charest PJ (1996) Characterization of seed storage proteins in *Populus* and their homology with *Populus* vegetative storage proteins. Tree Physiol 16: 833-840

Bedell J-P, Chalot M, Brun A & Botton B (1995) Purification and properties of glutamine synthetase from Douglas fir roots. Physiol Plantarum 94: 597-604

Biagioni M, Nali C, Heimler D & Lorenzini G (1997) PAL activity and differential ozone sensitivity in tobacco, bean and poplar. J Phytopathol 145: 533-539

Butland SL, Chow ML & Ellis BE (1998) A diverse family of phenylalanine ammonia-lyase genes expressed in pine trees and cell cultures. Plant Mol Biol 37: 15-24

Carvalho H, Pereira S, Sunkel C & Salema R (1992) Detection of a cytosolic glutamine synthetase in leaves of *Nicotiana tabacum* L. by immunocytochemical methods. Plant Physiol 100: 1591-1594

Clough ECM (1993) Ecological aspects of nitrate utilization in woody plants. PhD thesis, University College London

Clough ECM, Pearson J & Stewart GR (1989) Nitrate utilisation and nitrogen status in English woodland communities. Ann Sci Forest 46: 669-672

Edwards J, Walker E & Coruzzi G (1990) Cell-specific expression in transgenic plants reveals nonoverlapping roles for chloroplast and cytosolic glutamine synthetase. Proc Nat Acad of Sci USA 87: 3459-3463

Elmlinger MW & Mohr H (1992) Glutamine-synthetase in scots pine-seedlings and its control by blue-light and light absorbed by phytochrome. Planta 188: 396-402

Espin G, Moreno S & Guzman J (1994) Molecular-genetics of the glutamine synthetases in rhizobium species. Crit Rev Microbiol 20: 117-123

Forde BG & Clarkson DT (1999) Nitrate and ammonium nutrition of plants: Physiological and molecular perspectives. Adv Bot Res 30: 1-90

Fox G, Ratcliffe R, Robinson S & Stewart GR (1995) Evidence for deamination by glutamate dehydrogenase in higher plants: commentary. Can J Bot 73: 1112-1115

Fredeen AL & Field CB (1992) Ammonium and nitrate uptake in gap, generalist and understory species of the genus *Piper*. Oecologia 92: 207-214

Frioni L, Dodera R, Malates D & Irigoyen I (1998) An assessment of nitrogen fixation capability of leguminous trees in Uruguay. App Soil Ecol 7: 271-279

Gessler A, Schneider S, von Sengbusch D, Weber P, Hanemann U, Huber C, Rothe A, Kreutzer K & Rennenberg H (1998) Field and laboratory experiments on net uptake of nitrate and ammonium by the roots of spruce (*Picea abies*) and beech (*Fagus sylvatica*) trees. New Phytol 138: 275-285

Givan CV (1979) Metabolic detoxification of ammonia in tissues of higher plants. Phytochem 18: 375-382

Guan CG, Ribeiro A, Akkermans ADL, Jing YX, vanKammen A, Bisseling T & Pawloski K (1996) Nitrogen metabolism in actinorhizal nodules of *Alnus glutinosa*: Expression of glutamine synthetase and acetylornithine transaminase. Plant Mol Biol 32: 1177-1184

Harrison AF, Schulze E-D, Gebauer G & Bruckner G (2000) Canopy uptake and utilization of atmospheric pollutant nitrogen In: Schulze E-D (ed) Carbon and Nitrogen Cycling in European Forest Ecosystems, pp 171-188. Springer Verlag, Berlin, Germany

Hausler RE, Bailey KJ, Lea PJ & Leegood RC (1996) Control of photosynthesis in barley mutants with reduced activities of glutamine synthetase and glutamate synthase. 3. Aspects of glyoxylate metabolism and effects of glyoxylate on the activation state of ribulose-1,5-bisphosphate carboxylase-oxygenase. Planta 200: 388-396

Ireland RJ & Lea PJ (1999) The enzymes of glutamine, glutamate, asparagine and aspartate metabolism. In: Singh BK (ed) Plant Amino Acids, pp 49-109. Marcel Dekker Inc., New York, U.S.A.

Joy K (1988) Ammonia, glutamine and asparagine: A carbon-nitrogen interface. Can J Bot 66: 2103-2109

Kawakami N & Watanabe A (1988) Senescence-specific increase in cytosolic glutamine synthetase and its mRNA in radish cotyledons. Plant Physiol 88: 1430-1434

Kronzucker HJ, Siddiqi MY & Glass ADM (1996) Kinetics of NH₄⁺ influx in spruce. Plant Physiol 110: 773-779

Lambers H, Chapin F & Pons T (1998) Plant Physiological Ecology. Springer, New York, U.S.A.

Lea PJ, Blackwell R, Chen F-L & Hecht U (1990) Enzymes of ammonia assimilation. In: Lea PJ (ed) Enzymes of Primary Metabolism, pp 257-276. Academic Press, London, U.K.

Lea PJ & Ireland RJ (1999) Nitrogen metabolism in higher plants. In: Singh BK (ed) Plant Amino Acids, pp 1-47. Marcel Dekker Inc., New York, U.S.A.

Leegood RC, Lea PJ, Adcock MD & Hausler RE (1995) The regulation and control of photorespiration. J Exp Bot 46: 1397-1414

Magel E & Hubner B (1997) Distribution of phenylalanine ammonia lyase and chalcone synthase within trunks of *Robinia pseudoacacia* L.. Bot Acta 110: 314-322

Margolis H, Vézina L & Ouimet R (1988) Relation of light and nitrogen source to growth, nitrate reductase and glutamine synthetase activity of jack pine seedlings. Physiol Plant 72: 790-795

Min X, Siddiqi MY, Guy RD, Glass ADM & Kronzucker HJ (1998) Induction of nitrate uptake and nitrate reductase activity in trembling aspen and lodgepole pine. Plant Cell Environ 21: 1039-1046

Min X, Siddiqi MY, Guy RD, Glass ADM & Kronzucker HJ (2000) A comparative kinetic analysis of nitrate and ammonium influx in two early-successional tree species of temperate and boreal forest ecosystems. Plant Cell Environ 23: 321-328

Nasholm T, Ekblad A, Nordin A, Giesler R, Hogberg M & Hogberg P (1998) Boreal forest plants take up organic nitrogen. Nature 392: 914-916

Oaks A (1995) Evidence for deamination by glutamate dehydrogenase in higher plants: reply. Can J Bot 73: 1116-1117

Olsson MO & Falkengren-Grerup U (2000) Potential nitrification as an indicator of preferential uptake of ammonium or nitrate by plants in an oak woodland understorey. Ann Bot 85: 299-305

Osmond B, Badger M, Maxwell K, Bjorkman O & Leegood R (1997) Too many photons: Photorespiration, photoinhibition and photooxidation. Trends Plant Sci 2: 119-121

Pearson J, Clough ECM & Kershaw JL (1989) Comparative-study of nitrogen assimilation in woodland species. Ann Sci Forest 46: 663-665

Pearson J, Clough ECM, Woodall J, Havill DC & Zhang XH (1998) Ammonia emissions to the atmosphere from leaves of wild plants and *Hordeum vulgare* treated with methionine sulphoximine. New Phytol 138: 37-48

Pearson J & Ji YM (1994) Seasonal-variation of leaf glutamine-synthetase isoforms in temperate deciduous trees strongly suggests different functions for the enzymes. Plant Cell Environ 17: 1331-1337

Pearson J & Soares A (1995) A hypothesis of plant susceptibility to atmospheric pollution based on intrinsic nitrogen metabolism: Why acidity really is the problem. Water Air Soil Poll 85: 1227-1232

Pearson J & Soares A (1998) Physiological responses of plant leaves to atmospheric ammonia and ammonium. Atmos Environ 32: 533-538

Pearson J & Stewart GR (1993) The deposition of atmospheric ammonia and its effects on plants. New Phytol 125: 283-305

Raven J, Wollenweber B & Handley L (1992a) A comparison of ammonium and nitrate as nitrogen sources for photolithotrophs. New Phytol 121: 19-32

Raven J, Wollenweber B & Handley L (1992b) Ammonia and ammonium fluxes between photolithotrophs and the environment in relation to the global nitrogen cycle. New Phytol 121: 5-18

Reglinski T, Stavely FJL & Taylor JT (1998) Induction of phenylalanine ammonia lyase activity and control of *Sphaeropsis sapinea* infection in *Pinus radiata* by 5-chlorosalicylic acid. Eur J Forest Pathol 28: 153-158

Rennenberg H, Kreutzer K, Papen H & Weber P (1998) Consequences of high loads of nitrogen for spruce (*Picea abies*) and beech (*Fagus sylvatica*) forests. New Phytol 139: 71-86

Robinson SA, Stewart GR & Phillips R (1992) Regulation of glutamate-dehydrogenase activity in relation to carbon limitation and protein catabolism in carrot cell-suspension cultures. Plant Physiol 98: 1190-1195

Roderick ML, Berry SL, Saunders AR & Noble IR (1999) On the relationship between the composition, morphology and function of leaves. Funct Ecol 13: 696-710

Schjoerring JK, Husted S & Mattsson M (1998) Physiological parameters controlling plant-atmosphere ammonia exchange. Atmos Environ 32: 491-498

Schlee D, Tintemann H, Thöringer C, Jung K & Förstel H (1994) Aktivitäten und Eigenschaften von Glutaminsynthetase und Glutamatdehydrogenase aus Nadeln von *Pinus sylvestris* in Abhängigkeit vom Standort. Angewandte Botanik 68: 89-94

Schmidt S & Stewart GR (1997) Waterlogging and fire impacts on nitrogen availability and utilization in a subtropical wet heathland (wallum). Plant Cell Environ 20: 1231-1241

Schmidt S & Stewart GR (1999) Glycine metabolism by plant roots and its occurrence in Australian plant communities. Aust J Plant Physiol 26: 253-264

Schmidt S, Stewart GR, Turnbull MH, Erskine PD & Ashwath N (1998) Nitrogen relations of natural and disturbed plant communities in tropical Australia. Oecologia 117: 95-104

Seith B, Setzer B, Flaig H & Mohr H (1994) Appearance of nitrate reductase, nitrite reductase and glutamine synthetase in different organs of the scots pine (*Pinus sylvestris*) seedling as affected by light, nitrate and ammonium. Physiol Plant 91: 419-426

Singh S, Lewis NG & Towers GHN (1998) Nitrogen recycling during phenylpropanoid metabolism in sweet potato tubers. J Plant Physiol 153: 316-323

Smirnoff N, Todd P & Stewart GR (1984) The occurrence of nitrate reduction in the leaves of woody-plants. Ann Bot 54: 363-374

Soares A, Ming JY & Pearson J (1995) Physiological indicators and susceptibility of plants to acidifying atmospheric-pollution - a multivariate approach. Environ Poll 87: 159-166

Sprent JI (1987) The Ecology of the Nitrogen Cycle. Cambridge, Cambridge University Press, UK.

Staswick PE (1994) Storage proteins of vegetative plant-tissue. Annu Rev Plant Phys 45: 303-322

Stewart GR, Gracia CA, Hegarty EE & Specht RL (1990) Nitrate reductase-activity and chlorophyll content in sun leaves of subtropical australian closed-forest (Rain-forest) And open-forest communities. Oecologia 82: 544-551

Stewart GR, Hegarty EE & Specht RL (1988) Inorganic nitrogen assimilation in plants of Australian rainforest communities. Physiol Plant 74: 26-33

Stewart GR, Joly CA & Smirnoff N (1992) Partitioning of inorganic nitrogen assimilation between the roots and shoots of cerrado and forest trees of contrasting plant-communities of south east Brazil. Oecologia 91: 511-517

Stewart GR, Pate JS & Unkovich M (1993) Characteristics of inorganic nitrogen assimilation of plants in fire-prone mediterranean-type vegetation. Plant Cell Environ 16: 351-363

Stewart GR, Pearson J, Kershaw JL & Clough ECM (1989) Biochemical aspects of inorganic nitrogen assimilation by woody-plants. Ann Sci Forest 46: 648-653

Thomas RB, Bashkin MA & Richter DD (2000) Nitrogen inhibition of nodulation and N$_2$ fixation of a tropical N$_2$ fixing tree (*Gliricidia sepium*) grown in elevated atmospheric CO$_2$. New Phytol 145: 233-243

Truax B, Lambert F, Gagnon D & Chevrier N (1994) Nitrate reductase and glutamine synthetase activities in relation to growth and nitrogen assimilation in red oak and red ash seedlings; effects of N-forms, N concentration and light intensity. Trees 9: 12-18

Van0 Hove LWA, Vankooten O, Adema EH, Vredenberg WJ & Pieters GA (1989) Physiological-effects of long-term exposure to low and moderate concentrations of atmospheric NH$_3$ on poplar leaves. Plant Cell Environ 12: 899-908

Vézina L-P & Margolis H (1990) Purification and properties of glutamine synthetase in leaves and roots of *Pinus banksiana* Lamb. Plant Physiol 94: 657-664

Wallenda T, Stober C, Högbom L, Schinkel H, George E, Högberg P & Read DJ (2000) Nitrogen uptake processes in roots and mycorrhizas. In: Schulze E-D (ed) Carbon and Nitrogen Cycling in European Forest Ecosystems, pp 122-143. Springer Verlag, Berlin, Germany

Wallsgrove RM, Keys AJ, Lea PJ & Miflin BJ (1983) Photosynthesis, photo-respiration and nitrogen-metabolism. Plant Cell Environ 6: 301-309

Woodall J, Boxall JG, Forde BG & Pearson J (1996a) Changing perspectives in plant nitrogen metabolism: the central role of glutamine synthetase. Science Progress 79: 1-26

Woodall J &Forde BG (1996) Glutamine synthetase polypeptides in the roots of 55 legume species in relation to their climatic origin and the partitioning of nitrate assimilation. Plant Cell Environ 19: 848-858

Woodall J, Havill DC & Pearson J (1996b) Developmental changes in glutamine synthetase isoforms in *Sambucus nigra* and *Trientalis europaea*. Plant Physiol Biochem 34: 697-706

Chapter 1.4

Isoprene and terpene biosynthesis

Hartmut K. Lichtenthaler and Johannes G. Zeidler
Botany II, University of Karlsruhe, D-76128 Karlsruhe, Germany

1.   INTRODUCTION

A major part of the volatile organic compounds (VOC) in the atmosphere originates from isoprenoids emitted from plants and in particular from trees (Sharkey et al. 1991; Helas et al. 1997; Kesselmeier and Staudt 1999). The largest proportions of the terpenoids emitted from vegetation consist of the hemiterpene isoprene, various monoterpenes and, much less important, certain sesquiterpenes. These days great importance is placed on monoterpene and isoprene emission due to their impact on atmospheric chemistry and ozone formation (Trainer et al. 1987; Lerdau et al. 1997). In order to obtain better estimates of biogenic terpene emissions, it is essential to understand the biochemical and physiological background involved in terpene biosynthesis and emission by plants.

The biosynthesis of terpenoids follows the "biogenetic isoprene rule" seen by Wallach (1885) and established by Ruzicka et al. (1953) and Eschenmoser et al. (1955). At first, the active isoprenic C_5 unit has to be synthesized yielding isopentenyl diphosphate (IPP) and its isomer dimethylallyl diphosphate (DMAPP). Both C_5 units are in equilibrium through the action of IPP isomerase. DMAPP is then transformed into longer prenyl diphosphates such as geranyl diphosphate (GPP), farnesyl diphosphate (FPP) and geranylgeranyl diphosphate (GGPP) by a consecutive elongation with IPP in a head-to-tail condensation (Fig. 1). Then the basic C-skeleton of these prenyl diphosphates can undergo various isomerizations, cyclizations and redox reactions to finally yield the great variety of ten

R. Gasche et al. (eds.), Trace Gas Exchange in Forest Ecosystems, 79–99.
© 2002 *Kluwer Academic Publishers. Printed in the Netherlands.*

thousands of different isoprenoids and terpenoids of the secondary plant metabolism as found in particular plants and many plant families.

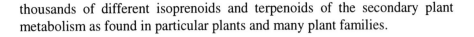

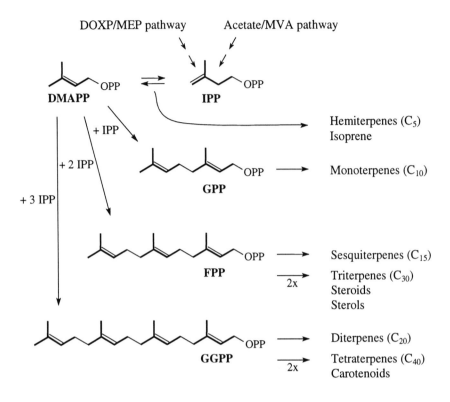

Figure 1. Biogenetic relationship of plant isoprenoids and terpenoids which are formed from the active C_5-units dimethylallyl diphosphate (DMAPP) and isopentenyl diphosphate (IPP) by head-to-tail condensation. Depending on the type of isoprenoid, the isoprenic C_5-units of IPP are derived either from the cytosolic acetate/mevalonate pathway or the plastidic 1-deoxy-D-xylulose-5-phosphate (DOXP)/methylerythritol phosphate (MEP) pathway. (GPP = geranyl diphosphate, FPP= farnesyl diphosphate, GGPP = geranylgeranyl diphosphate).

Various reviews have dealt with the biosynthesis of plant terpenoids from the active C_5-unit IPP in reflecting the increasing knowledge in this field (Goodwin 1965; Spurgeon and Porter 1981; Gershenzon and Croteau 1993; Chappell 1995; Cane 1998; McCaskill and Croteau 1998). A periodical survey of terpenes and their biosynthesis is given by Dewick (1999). The biosynthetic C_5-unit in plants is, however, made via two independent pathways as is outlined below.

After the groups of K. Bloch and F. Lynen in 1958 had proved that the biosynthesis of cholesterol in mammals and fungi proceeds via the acetate/mevalonate (MVA) pathway of IPP formation (Chaykin et al. 1958;

Lynen et al. 1958), it was generally accepted on the basis of the incorporation of ^{14}C-labeled precursors by Goodwin and other groups that photosynthetic algae and higher plants make their isoprenoids also via the acetate/MVA pathway of IPP biosynthesis (see Goodwin 1965, and the review of Lichtenthaler 1999). The fact that the situation in plants is more complex had only been established since 1995 by the cooperation of the groups of H. K. Lichtenthaler and M. Rohmer using ^{13}C-labeling studies coupled with high resolution NMR spectroscopy (see Lichtenthaler et al. 1995, 1997a, b; Schwender et al. 1995, 1996). In fact, higher plants possess two separate IPP producing pathways: (i) the cytosolic acetate/MVA pathway for sterol biosynthesis and (ii) the plastidic 1-deoxy-D-xylulose-5-phosphate (DOXP)/methylerythritol-phosphate (MEP) pathway for the biosynthesis of plastidic isoprenoids such as carotenoids, phytol or isoprene and apparently also monoterpenes and diterpenes (see Lichtenthaler et al. 1997a, and the reviews Lichtenthaler 1999, 2000). For this reason the question whether a plant isoprenoid is made (i) either via the acetate/MVA pathway or (ii) via the DOXP/MEP pathway of IPP biosynthesis has to be reinvestigated for each individual terpenoid present in a plant.

In this review on terpenoid biosynthesis an emphasis is made (i) on *isoprene biosynthesis* as a highly significant pathway regarding the carbon budget of plants and (ii) on the *alternative DOXP/MEP pathway* of IPP and isoprenoid biosynthesis as a newly discovered major pathway of primary plant metabolism and its significance for the emission of volatile terpenoids.

2. THE ACETATE/MVA PATHWAY OF IPP BIOSYNTHESIS

The classical acetate/MVA pathway of IPP and DMAPP formation is well established in plants, e.g. for the biosynthesis of sterols. A recent review on this pathway is given by Bach et al. (1997). It starts from 3 acetate, proceeds via hydroxymethylglutaryl-CoA (HMG-CoA) and mevalonate (MVA) to finally yield IPP. The different enzymatic steps and intermediates of this MVA pathway are summarized in Figure 2A. This cytosolic IPP biosynthesis can highly specifically be inhibited by mevinolin (see Fig. 3A) by blocking the cytosolic microsomal enzyme HMG-CoA reductase (Bach and Lichtenthaler 1982, 1983). This inhibitor efficiently blocks the cytosolic sterol accumulation in plants, whereas the formation of plastidic isoprenoid pigments such as carotenoids and chlorophylls (with the diterpene phytol as side-chain) are little or not affected.

A. Classical acetate/
mevalonate pathway

B. Deoxyxylulose-5-phosphate/
MEP pathway

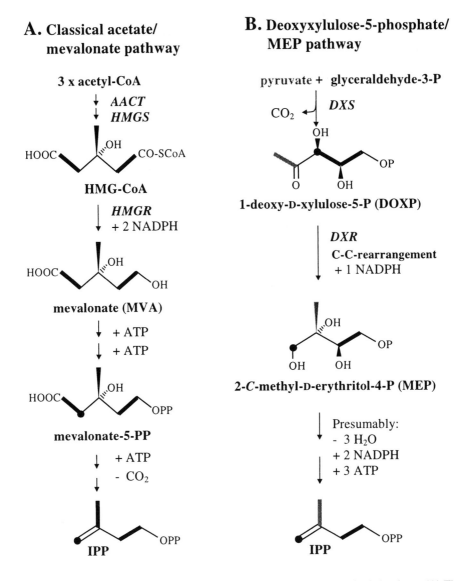

Figure 2. The two pathways for isopentenyl diphosphate (IPP) biosynthesis in plants. (A) The classical cytosolic acetate/mevalonate pathway. (AACT = acetoacetyl-CoA thiolase, HMGS = hydroxymethyl-glutaryl-CoA (HMG-CoA) synthase, HMGR = HMG-CoA reductase. (B) The newly discovered plastidic DOXP/MEP pathway of IPP formation. Greytone: C_2 unit derived from pyruvate. (DXS = deoxyxylulose-5-phosphate synthase, DXR = deoxyxylulose-5-phosphate reductoisomerase).

3. THE DEOXYXYLULOSE PHOSPHATE/ METHYLERYTHRITOL PHOSPHATE PATHWAY

After the discovery of a mevalonate-independent pathway for IPP-biosynthesis in eubacteria (Rohmer et al. 1993), this 1-deoxy-D-xylulose 5-phosphate pathway was also established in algae (Lichtenthaler et al. 1995; Schwender et al. 1995, 1996) and higher plants (Lichtenthaler et al. 1997a, b). It is absent in fungi and yeasts (Disch and Rohmer 1998) and also in archea and mammals, all of which lack the genes of the DOXP/MEP pathway (Lichtenthaler et al. 2000, Lichtenthaler 2000). The present knowledge on the non-mevalonate DOXP pathway of IPP formation of plants is reviewed by Lichtenthaler (1999, 2000) and the references cited therein give access to the contributions of various other groups to this new pathway.

3.1   The DOXP synthase

The plant DOXP pathway of IPP formation starts with a transketolase-like reaction between pyruvate and glyceraldehyde-3-phosphate (GAP) involving a decarboxylation of pyruvate after binding to thiamine pyrophosphate (TPP). This enzymic reaction yields 1-deoxy-D-xylulose-5-phosphate (DOXP, Fig. 3B). Cloning of the gene (*dxs*) of the DOXP synthase has been reported from different sources: *E. coli* (Sprenger et al. 1997; Lois et al. 1998), *Mentha x piperita* (Lange et al. 1998), *Capsicum annuum* L. (Bouvier et al. 1998), *Chlamydomonas reinhardtii* (see Lichtenthaler 1999). Further *dxs* sequences from plants (e.g. rice), photosynthetic (*Rhodobacter, Synechocystis*) and non-photosynthetic bacteria (e.g. *Bacillus subtilis*) are found in databases. This distribution pattern fits into the hypothesis of the endosymbiontic origin of plastids from cyanobacteria-like ancestors. DOXP synthase is transcriptionally upregulated when large amounts of isoprenoids are needed: e.g. prior to the peak of monoterpene biosynthesis in *Mentha x piperita* (Lange et al. 1998), during chloroplast to chromoplast transition in *Capsicum annuum* L. (Bouvier et al. 1998) and during chloroplast development in the light in *Arabidopsis* (Mandel et al. 1996). All these results suggest that DOXP synthase has an important control function in isoprenoid biosynthesis. DOXP marks a metabolic branchpoint because it is not only a precursor for IPP but also for thiamine (vitamin B_1) and pyridoxine (vitamin B_6) (cf. Fig. 6 in Lichtenthaler 1999).

3.2   The DOXP reductoisomerase

2-*C*-methyl-D-erythritol-4-phosphate (MEP, Fig. 2B) is the next intermediate of the DOXP pathway in plants and bacteria. The second enzyme of the DOXP pathway, the DOXP reductoisomerase (DXR), yields MEP by an intramolecular C-skeleton rearrangement of DOXP and a NADPH-dependent reduction. The gene of DXR (*dxr*) has been cloned and characterized from *E. coli* (Takahashi et al. 1998), *Mentha x piperita* (Lange and Croteau 1999) and *Arabidopsis thaliana* (Schwender et al. 1999). The enzymatic reaction of DXR can specifically be blocked by the antibiotic fosmidomycin and related substances (Fig. 3B) in plants (Lichtenthaler et al. 1999; Schwender 1999; Schwender et al. 1999) and in bacteria (Kuzuyama et al. 1998). Fosmidomycin blocks the accumulation of phytol and carotenoids in barley, duckweed and ripening tomato, and also inhibits the

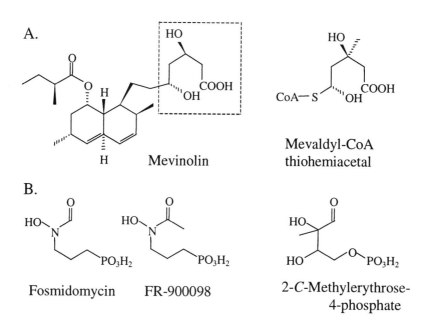

A.

Mevinolin

Mevaldyl-CoA
thiohemiacetal

B.

Fosmidomycin   FR-900098

2-*C*-Methylerythrose-
4-phosphate

Figure 3. Inhibitors of cytosolic and plastidic isoprenoid biosynthesis in plants: (A) Mevinolin, an inhibitor of hydroxymethylglutaryl-CoA reductase (HMGR), the committed enzyme of the classical acetate/mevalonate pathway of IPP formation. A structural similarity to mevaldyl-CoA thiohemiacetal, the proposed intermediate of the HMGR reaction, is shown. (B) Fosmidomycin and FR-900098 are inhibitors of DOXP reductoisomerase (DXR), the committed enzyme of the DOXP/MEP pathway of IPP formation. 2-C-methylerythrose-4-phosphate is the supposed intermediate of the DXR reaction and structurally related to the inhibitors.

emission of isoprene (Zeidler et al. 1998) and methylbutenol (Zeidler and Lichtenthaler 2001), as well as the incorporation of radioactive DOXP into β-carotene and geranylgeraniol (Fellermeier et al. 1999). The malaria parasite *Plasmodium falciparum* obviously also contains the DOXP pathway in its plastid-like organelle, the apicoplast. Fosmidomycin inhibited *in vitro* growth as well as the recombinant DXR enzyme from *Plasmodium* and cured mice infected with *Plasmodium vinckei* (Jomaa et al. 1999).

3.3 The steps between methylerythritol phosphate and IPP/DMAPP

Three further intermediates of the DOXP/MEP pathway between MEP and IPP have been identified. These are CDP-methylerythritol, CDP-methylerythritol 2-phosphate and methylerythritol 2,4-cyclodiphosphate (Rohdich et al. 1999, 2000; Kuzuyama et al. 2000a, b, Lüttgen et al. 2000, Herz et al. 2000, Takagi et al. 2000). The further enzymatic steps are not yet known, but at least two reductases and dehydratases can be anticipated. The steps from MEP to IPP apparently require 3 ATP and 2 NADPH (Fig. 2B). An essential question in this context was if either IPP or its isomer DMAPP are formed as first C_5-unit in this DOXP/MEP pathway. In higher plants IPP is apparently the end product of the DOXP/MEP pathway as suggested by recent experiments in peppermint (McCaskill and Croteau 1999) and *Catharanthus roseus* (Arigoni et al. 1999). IPP can then be transformed to DMAPP by IPP isomerase, which is found in chloroplasts. Details on IPP isomerases are given by Ramos-Valdivia et al. (1997). In bacteria (*E. coli*, *Synechocystis*) there is, however, a branching point in the DOXP/MEP pathway, possibly at the ME cyclodiphosphate, one branch leading to IPP, the other to DMAPP (Ershov et al. 2000; Rodriguez-Conception et al. 2000). It is to be expected that this branching point may also exist at least in some plants.

4. TERPENES MADE VIA THE DOXP/MEP PATHWAY

After the discovery of the DOXP/MEP pathway of IPP biosynthesis the biosynthetic origin of several secondary terpenoids has been re-examined, mostly by analysis of ^{13}C-labeling patterns after feeding [1-^{13}C]-glucose or labeled deoxyxylulose (DOX). The general view at present is that the DOXP/MEP pathway is located in the plastids and is involved in the formation of the volatile hemiterpenes isoprene and methylbutenol, monoterpenes (e.g. geraniol, menthol), diterpenes (e.g. phytol, geranylgeraniol) and tetraterpenes (e.g. carotenoids, secondary carotenoids). The cytosolic acetate/MVA pathway, in turn, produces IPP for the biosynthesis of sesqui- and triterpenes (Fig. 4).

The contributions of the two IPP forming pathways to the biosynthesis of secondary terpenoids seem to differ somewhat depending on the plant species, the type of terpenoid and the experimental or physiological conditions of the plants. In agreement with the scheme, shown in Fig. 4, is the biosynthesis of the following secondary terpenoids: Isoprene is synthesized in *Chelidonium majus* L., *Salix viminalis* L., *Populus nigra* L. (Schwender et al. 1997; Zeidler et al. 1997) *Platanus x acerifolia* L., *Quercus robur* L. and *Robinia pseudoacacia* L. (Zeidler and Lichtenthaler 1998) via the DOXP/MEP pathway. Also the volatile isoprenoid 2-methyl-3-buten-2-ol (MBO), emitted from several pine species from Western United States in a light and temperature dependent manner (Harley et al. 1998; Baker et al. 1999; Schade et al. 2000), is synthesized via the DOXP/MEP pathway (Zeidler and Lichtenthaler 2001). This also applies to monoterpenes (menthone, thymol, geraniol and pulegone) from *Mentha x piperita*, *Pelargonium graveolens*, *Thymus vulgaris* and *Mentha pulegium* (Eisenreich et al. 1997; Sagner et al. 1998) and a monoterpenoid indole alkaloid precursor (secologanin) from *Catharanthus* cell cultures (Contin et al. 1998). Also the monoterpenes borneol and bornyl acetate as well as the diterpene phytol in the liverworts *Ricciocarpos natans* and *Conocephalum conicum* were made via the DOXP/MEP pathway (Adam et al. 1998; Thiel et al. 1997). In contrast, sesquiterpenes and stigmasterol were built up via the MVA pathway which is in agreement with the scheme shown in Figure 4. The diterpenes marubiin (Knöss et al. 1997) and taxol (Eisenreich et al. 1996) are made via the DOXP/MEP pathway of IPP formation.

Only 50% of the terminal isoprenic units of the sesquiterpenes bisabololoxide A and chamazulene of chamomile (*Matricaria recutita*) were derived from MVA, whereas the rest of IPP came from the DOXP/MEP pathway (Adam and Zapp 1998) indicating some cooperation between both IPP forming pathways. Piel et al. (1998) have found varying but mostly high incorporation ratios of deuterated DOX into volatile mono-, sesqui-, and diterpenes from different plant species. Also the sesquiterpenoid blumenol C from mycorrhizal barley roots is made via the DOXP/MEP pathway (Maier et al. 1997). The authors state that this could be explained, if blumenol C were derived from carotenoids by degradation, as has been shown for abscisic acid, which is made via the DOXP/MEP pathway (Milborrow and Lee 1998). All plant carotenoids have been found to be synthesized principally via the DOXP/MEP pathway (Lichtenthaler et al. 1995, 1997b; Schwender et al. 1996; Arigoni et al. 1997; Disch et al. 1998; Milborrow and

Lee 1998). The isoprenoid chain of shikonin from *Lithospermum*, however, is derived from mevalonate (Li et al. 1998) providing the first monoterpenoid plant product that is not made via the DOXP/MEP pathway.

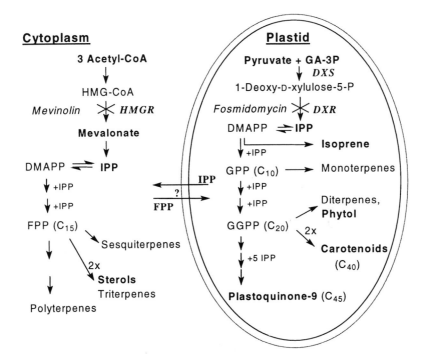

Figure 4. Compartmentation of isoprenoid biosynthesis in higher plants: The cytosolic acetate/mevalonate pathway and the plastidic DOXP/MEP pathway deliver IPP for the biosynthesis of terpenoids in the cytosol or in the plastid, respectively. The inhibition of both pathways by antibiotics, either Mevinolin or Fosmidomycin, is indicated. (HMGR = HMG-CoA reductase, DXS = DOXP synthase, DXR = DOXP reductoisomerase).

In summary, the results mentioned here and reviewed in more detail elsewhere (Lichtenthaler 1999, 2000) demonstrate the wide spread occurrence of the DOXP/MEP pathway in higher plants and photosynthetic algae. A possible exchange of intermediates between the cytosolic and plastidic isoprenoid pathways, which appears to be possible to some extent under certain physiological conditions, is not yet fully understood. The biosynthesis of the primary functional isoprenoids of the plant cell such as sterols, carotenoids, phytol (chlorophylls) and plastoquinone-9 proceeds either via the MVA or the DOXP/MEP pathway as indicated in the scheme shown in Figure 4. The hemiterpene isoprene, and many mono- and diterpenes are predominantly or exclusively made via the plastidic DOXP/MEP pathway and the sesquiterpenes primarily via the cytosolic acetate/MVA pathway of IPP formation. In contrast, some secondary plant isoprenoids may show a diverse behaviour in possibly using both pathways of IPP biosynthesis. The plastidic DOXP/MEP pathway plays an essential role in providing the carbon skeletons not only for natural plant products that are of human interest but also for the emission of several hundred million tonnes of carbon per year as isoprene and terpenes (see Sharkey et al. 1991).

5.   ISOPRENE EMISSION AND BIOSYNTHESIS

2-Methyl-1,3-butadiene or isoprene is the dominant volatile hydrocarbon emitted from various plants and in particular from forests. It was first identified as a plant product by Sanadze (1957) and different aspects of isoprene emission have recently been reviewed by Sharkey 1996, Lerdau et al. 1997, and Kreuzwieser et al. 1999. Harley et al. (1999) compared the taxonomic distribution of isoprene emission based on a list of isoprene and monoterpene emitting plants made out of over 100 publications available on the World Wide Web. A selection of the highest emitters of volatile isoprene from this list is given in Table 1. The primary parameters causing isoprene emission are high light and high temperatures of >30°C. Isoprene emission is directly linked to photosynthetic CO_2 assimilation (Delwiche and Sharkey 1993), and the fraction of recently fixed carbon lost as isoprene can be very high e.g. in water stressed kudzu leaves (*Pueraria lobata*) with a loss of 67% of the fixed carbon at 35°C (Sharkey and Loreto 1993). An isoprene emission of even 170% of the photosynthetic rates has been reported in the palm *Thrinax morrisii* at an irradiance of 1500 μmol m^{-2} s^{-1} and a temperature above 35°C (Lerdau and Keller 1997). This indicates that plants under certain conditions can synthesize isoprene also from stored carbon sources such as starch. The physiological function of isoprene emission is not really known. It could function in the protection of the photosynthetic photosystems (pigment apparatus) against photooxidation and heat damage (Singsaas et al. 1997) and could also serve as a valve for the removal of excess ATP and reducing equivalents.

Isoprene is synthesized via the DOXP/MEP pathway (Zeidler et al. 1997; Zeidler and Lichtenthaler 1998). Its precursor substance is DMAPP from which a diphosphate is eliminated in a reaction catalysed by isoprene synthase. Isoprene synthase has been purified to near homogeneity from aspen (*Populus tremuloides*) (Silver and Fall 1995). The enzyme has a molecular mass of 98–137 kDa with 58 kDa and 62 kDa subunits, an isoelectric point of 4.7, an apparent K_m for DMAPP of 8mM, a pH optimum of 8 and a requirement for Mg^{2+} or Mn^{2+}. Similar characteristics are shown for an isoprene synthase from *Quercus petraea* (Schnitzler et al. 1996).

Isoprene emission of *Quercus robur* was shown to vary during the season parallel to differential amounts of extractable isoprene synthase activity (Schnitzler et al. 1997). Isoprene synthase was also found in willow (*Salix discolor* L.) (Wildermuth and Fall 1998) as a thylakoid-bound enzyme together with a soluble form in the stroma.

Table 1. Isoprene emission rates of various green plants calculated from the original data for standard conditions of 30°C and a photon flux density of 1000 μmol m^{-2} s^{-1} (based on Hewitt et al. 1997).

Species	Common name	Isoprene emission (μg g^{-1} dw h^{-1})
Acacia nigrescens	African akazia	110
Arundo donax	Giant reed	39–174
Capparis cynophollophora	Jamaican Caper Tree	268
Eucalyptus globulus	Blue gum	13–63
Liquidambar styraciflua	Sweet gum	16–86
Mucuna pruriens	Velvet bean	317
Myrtus communis	Common myrtle	25–137
Populus nigra	Black poplar	29–76
Pueraria spp.	Kudzu	97
Quercus coccinea	Scarlet oak	20–130
Quercus robur	European oak	40–77
Salix babylonica	Weeping willow	23–233

Another volatile hydrocarbon is *2-methyl-3-buten-2-ol* (MBO) which is emitted from several pines (*P. ponderosa*, *P. contesta* and *P. sabiniana*) under similar light and temperature dependent conditions as isoprene from deciduous trees (Harley et al. 1998; Baker et al. 1999; Schade et al. 2000). These pines do not emit isoprene, but its chemically related product MBO. Recently an MBO synthase activity has been found in these pines (Fisher et al. 2000). Like isoprene, MBO seems to have a significant impact on the oxidative capacity of the atmosphere through the consumption of hydroxy radicals. Its biosynthesis proceeds via the DOXP/MEP pathway and can be inhibited by fosmidomycin (Zeidler and Lichtenthaler 2001).

6.   BIOSYNTHESIS OF MONO- AND SESQUITERPENES

Besides isoprene (primarily emitted by deciduous trees) and of MBO from ponderosa pines, the most significant non-methane plant hydrocarbons involved in atmospheric chemistry are monoterpenes (emitted mainly by coniferous trees). 128 to 147 million tons of monoterpenes are emitted throughout the world per year (Helas et al. 1997; and refs. therein). Their biosynthesis apparently proceeds via the DOXP/MEP pathway, but this assumption would need confirmation for each individual monoterpene. Volatile terpenes are not restricted to coniferous trees. Various angiosperms contain essential oils which consist predominantly of mono- and sesquiterpenes. Such terpenoids are usually stored in special structures such as resin ducts of conifers, glandular trichomes or flower petals but some of them can also be emitted to the atmosphere.

6.1   GPP and FPP synthases

Prior to actual mono-, sesqui- and diterpene biosynthesis the prenyl diphosphate precursors GPP, FPP and GGPP have to be formed (Fig. 1) by prenyltransferases which possess different chain length specificity (McGarvey and Croteau 1995; Kellogg and Poulter 1997; Ohnuma et al. 1997; Ogura and Koyama 1998). GPP synthase and GGPP synthase are principally localized in the plastids where GPP and GGPP are further processed into mono- and diterpenes (Bohlmann et al. 1998). GGPP also leads to the tetraterpenoid carotenoids by tail-to-tail condensation. In contrast, FPP synthase resides in the cytosol forming the precursors of sesquiterpenes and, again by tail-to-tail dimerization, squalene, the precursor of the cytosolic steroids and triterpenes.

An exception from this "regular compartmentation rule" is given in the case of *Lithospermum*. There GPP for shikonin biosynthesis is delivered by a cytosolic GPP synthase (Sommer et al. 1995) which is in agreement with the cytosolic MVA pathway as carbon source for the shikonin isoprenoid chain (Li et al. 1998). *Arabidopsis* contains two genes *fps1* and *fps2* for cytosolic FPP synthases. Recently, an additional mitochondrial FPP synthase isoform deriving from the differentially transcribed *fps1* gene has been described in *Arabidopsis thaliana* (Cunilera et al. 1997). In contrast, the rice genome might contain only one FPP synthase gene (Sanmiya et al. 1997).

6.2   Terpene cyclases

GPP and FPP are transformed into mono- and sesquiterpenes, respectively, by various mono- and sesquiterpene cyclases or synthases (for reviews see e.g. McGarvey and Croteau 1995; Bohlmann et al. 1998).

Monoterpene cyclases: GPP itself cannot be cyclized and therefore all monoterpene cyclases catalyse a rearrangement of GPP to the tertiary allylic isomer linalyl diphosphate which is cyclized to the α-terpinyl cation. This universal precursor of cyclic monoterpenes, may undergo various kinds of reactions such as hydride shifts, C-skeleton rearrangements or further cyclizations. Finally a deprotonation or addition of water or pyrophosphate yields an uncharged monoterpene. Other monoterpene synthases produce acyclic molecules such as myrcene and linalool. Monoterpene synthases are soluble, plastidic enzymes of 50-100 kDa mass with a monomeric or homodimeric quartery structure, a requirement for Mg^{2+} or Mn^{2+} and a pH optimum of 6-8. Differences exist between angiosperm and gymnosperm monoterpene synthases: The latter depend on an additional monovalent cation (K^+), prefer Mn^{2+} or Fe^{2+} over Mg^{2+} and have a higher pH optimum. Several monoterpene synthases are able to form different products, e.g. (+)-pinene cyclase from *Salvia officinalis* generates a mixture of (+)-α-pinene, (+)-camphene, (+)-limonene, myrcene, ocimene and terpinolene (Arigoni et al. 1993). Recent insights into the mechanism of monoterpene synthases indicate a role for the highly conserved N-terminal arginin pair in the diphosphate rearrangement involved in the formation of the intermediate linalyl diphosphate preceding the cyclization of mutagenized (-)-4S-limonene synthase from spearmint (Williams et al. 1998). Linalool synthase employs the same mechanism as limonene synthase but its gene has also sequence motifs indicative of terpene synthases with a different, proton-promoted cyclization mechanism (e.g. the diterpene synthase copalyl diphosphate synthase) (Cseke et al. 1998). Monoterpenes in conifers act as solvents for diterpenoid resin acids in oleoresin and deter herbivores. Their synthesis can be induced by wounding (Litvak and Monson 1998; Steele et al. 1998) and their volatilization can contribute significantly to VOC emissions during resin tapping in forests (Pio and Valente 1998). In the evergreen mediterranean oak *Quercus ilex* L. monoterpene emissions seem to play a different role: They show a marked short-term temperature and light dependence (Staudt and Bertin 1998) similar to isoprene emission found e.g. in other oak species (Loreto et al. 1998a). Similar to isoprene, monoterpenes from *Quercus ilex* L. are not stored, quickly labeled by $^{13}CO_2$ (Loreto et al. 1996a and b) and may confer heat resistance to photosynthetic membranes (Loreto et al. 1998b).

Sesquiterpene cyclases: Sesquiterpenes are derived from FPP and display a greater structural diversity than the monoterpenes, because the additional isoprene unit with its additional double bond allows more different cyclizations leading to ring sizes from 6 to 11 C-atoms. The formation of sesquiterpenes with a 6-membered ring requires, as with the monoterpenes, a preliminary isomerization of FPP to nerolidyl diphosphate, the sesquiterpenoid analogue of linalyl diphosphate, before ionization and cyclization can take place. Internal additions, rearrangements and other reactions lead to a broad range of structures. Thousands of oxidized or otherwise modified sesquiterpenes with more than 300 distinct carbon skeletons have been identified to date (Cane 1998). Reaction mechanisms and properties of the cytosolic sesquiterpene synthases are similar to those of the plastidic monoterpene synthases. Recently crystal structures of 5-epi-aristolochene synthase and pentalenene synthase have provided new insights into the mechanisms of sesquiterpene cyclases (Lesburg et al. 1997; Starks et al. 1997).

7.   CONCLUSIONS

One of the major discoveries in plant biochemistry in recent years was the detection of the non-mevalonate pathway for IPP and isoprenoid biosynthesis in plants. This DOXP/MEP pathway explains many of the former inconsistencies found in the biosynthesis of primary and secondary plant isoprenoids and terpenoids. Although the DOXP/MEP pathway answers many questions, it opens a wide field for new research and new questions. The fact, that higher plants possess two separate and independent biochemical pathways for IPP and isoprenoid formation, requires a reinvestigation also of the biogenetic origin of secondary plant terpenoids. The new non-mevalonate pathway, which is linked to the photosynthetic chloroplast and plastidic carbohydrate metabolism, has to be considered in mechanistic models of isoprene, methylbutenol and monoterpene emission. The biosynthetic steps from IPP to the huge variety of plant terpenoids provide an essentially infinite field for future research.

REFERENCES

Adam K-P & Zapp J (1998) Biosynthesis of the isoprene units of chamomile sesquiterpenes. Phytochemistry 48: 953-959

Adam K-P, Thiel R, Zapp J & Becker H (1998) Involvement of the mevalonic acid pathway and the glyceraldehyde-pyruvate pathway in terpenoid biosynthesis of the liverworts *Ricciocarpos natans* and *Conocephalum conicum*. Arch Biochem Biophys 354: 181-187

Arigoni D, Cane DE, Shim JH, Croteau Rm & Wagschal K (1993) Monoterpene cyclization mechanisms and the use of natural abundance deuterium NMR – short cut or primrose path? Phytochemistry 32: 623-631

Arigoni D, Eisenreich W, Latzel C, Sagner S, Radykewicz T, Zenk MH & Bacher A (1999). Dimethylallyl pyrophosphate is not the committed precursor of isopentenyl pyrophosphate during terpenoid biosynthesis from 1-deoxy-D-xylulose in higher plants. Proc Natl Acad Sci USA 96: 1309-1314

Arigoni D, Sagner S, Latzel C, Eisenreich W, Bacher A & Zenk MH (1997). Terpenoid biosynthesis from 1-deoxy-D-xylulose in higher plants by intramolecular skeletal rearrangement. Proc Natl Acad Sci USA 94: 10600-10605

Bach TJ (1995) Some aspects of isoprenoid biosynthesis in plants - a review. Lipids 30: 191-202

Bach TJ & Lichtenthaler HK (1982) Mevinolin a highly specific inhibitor of microsomal 3-hydroxy-3-methyl-glutaryl-coenzyme A reductase of radish plants. Z Naturforsch 37c: 46-50

Bach TJ & Lichtenthaler HK (1983) Inhibition of plant growth, sterol formation and pigment accumulation. Physiol Plant 59: 50-60

Bach TJ, Boronat A, Campos N, Ferrer A & Vollack KU (1997) Mevalonate biosynthesis in plants. In: Bach TJ, Boronat A, Campos N, Ferrer A & Vollack KU (eds) Biochemistry and Function of Sterols, pp 135-150. CRC Press, Boca Raton, U.S.A.

Baker B, Guenther A, Greenberg J, Goldstein A & Fall R (1999) Canopy fluxes of 2-methyl-3-buten-2-ol over a ponderosa pine forest by relaxed eddy accumulation: field data and model comparison. J Geophys Res 104: 26107-26114

Bohlmann J, Meyer-Gauen G & Croteau R (1998) Plant terpenoid synthases: molecular biology and phylogenetic analysis. Proc Natl Acad Sci USA 95: 4126-4133

Bouvier F, d'Harlingue A, Suire C, Backhaus RA & Camara B (1998) Dedicated roles of plastid transketolases during the early onset of isoprenoid biogenesis in pepper fruits. Plant Physiol 117: 1423-1431

Cane DE (1998) Comprehensive Natural Product Chemistry Vol. 2 Isoprenoids including Carotenoids and Steroids. Pergamon, Oxford

Chaykin S, Law J, Philipps AH & Bloch K (1958) Phosphorylated intermediates in the synthesis of squalene. Proc Natl Acad Sci USA 44: 998-1004

Chappell J (1995) Biochemistry and molecular biology of the isoprenoid biosynthetic pathway in plants. Ann Rev Plant Physiol Plant Mol Biol 46: 521-547

Contin A, van der Heijden R, Lefeber AW & Verpoorte R (1998) The iridoid glucoside secologanin is derived from the novel triose phosphate/pyruvate pathway in a *Catharanthus roseus* cell culture. FEBS Lett 434: 413-416

Cseke L, Dudareva N & Pichersky E (1998) Structure and evolution of linalool synthase. Mol Biol Evol 15: 1491-1498

Cunilera N, Boronat A & Ferrer A (1997) The *Arabidopsis thaliana* Fps1 gene generates a novel messenger-RNA that encodes a mitochondrial farnesyl diphosphate synthase isoform. J Biol Chem 272: 15381-15388

Delwiche CF & Sharkey TD (1993) Rapid appearance of ^{13}C in biogenic isoprene when $^{13}CO_2$ is fed to intact leaves. Plant Cell Environ 16: 587-591

Dewick PM (1999). The biosynthesis of C_5-C_{25} terpenoid compounds. Nat Prod Rep 16: 97

Disch A & Rohmer M (1998) On the absence of the glyceraldehyde 3-phosphate/pyruvate pathway for isoprenoid biosynthesis in fungi and yeasts. FEMS Microbiol Lett 168: 201-208

Disch A, Schwender J, Müller C, Lichtenthaler HK & Rohmer M (1998) Distribution of the mevalonate and glyceraldehyde phosphate/pyruvate pathways for isoprenoid biosynthesis in unicellular algae and the cyanobacterium *Synechocystis* PCC 6714. Biochem J 333: 381-388

Eisenreich W, Menhard B, Hylands PJ, Zenk MH & Bacher A (1996) Studies on the biosynthesis of taxol: The taxane carbon skeleton is not of mevalonoid origin. Proc Natl Acad Sci USA 93: 6431-6436

EisenreichW, Sagner S, Zenk MH & Bacher A (1997) Monoterpenoid essential oils are not of mevalonoid origin. Tetrahedron Lett 38: 3889-3892

Ershov Y, Gantt RR, Cunningham FX & Gantt E (2000) Isopentenyl diphosphate isomerase deficiency in *Synechocystis* sp. strain PCC6803. FEBS Lett. 473: 337-340

Eschenmoser A, Ruzicka L, Jeger O & Arigoni D (1955) Zur Kenntnis der Triterpene. Eine stereochemische Interpretation der biogenetischen Isoprenregel bei den Triterpenen. Helv Chim Acta 38: 1890-1904

Fellermeier M, Kis K, Sagner S, Maier U, Bacher A & Zenk M (1999) Cell-free conversion of 1-deoxy-D-xylulose 5-phosphate and 2-C-methyl-D-erythritol 4-phosphate into β-carotene in higher plants and its inhibition by fosmidomycin. Tetrahedron Lett 40: 2743-2746

Fisher AJ, Bahr BM, Greenberg JP & Fall R (2000) Enzymatic synthesis of methylbutenol from dimethylallyl diphosphate in needles of *Pinus sabiniana*. Arch Biochem Biophys 383: 128-134

Gershenzon J & Croteau RB (1993) Terpenoid biosynthesis: The basic pathway and formation of monoterpenes, sesquiterpenes and diterpenes. In: Gershenzon J & Croteau RB (eds) Lipid Metabolism in Plants, pp 339-388. CRC Press, Boca Raton, U.S.A.

Goodwin T W (1965) Regulation of terpenoid biosynthesis in higher plants. In: Pridham JB & Swain T (eds) Biosynthetic Pathways in Higher Plants, pp 57-71. Academic Press, London, U.K.

Harley P, Fridd-Stroud V, Greenberg J, Guenther A & Vasconcellos P (1998) Emission of 2-methyl-3-buten-2-ol by pines: a potentially large natural source of reactive carbon to the atmosphere. J Geophys Res D 103: 25479-25486

Harley PC, Monson RK & Lerdau MT (1999) Ecological, evolutionary aspects of isoprene emission from plants. Oecologia 118: 109-123

Helas G, Slanina J & Steinbrecher R. (1997) Biogenic Volatile Organic Compounds in the Atmosphere. SPB Academic Publishing, Amsterdam, The Netherlands

Herz S, Wungsintaweekul J, Schuhr CA, Hecht S, Lüttgen H, Sagner S, Fellermeier M, Eisenreich M, Zenk MH, Bacher A & Rohdich F (2000) Biosynthesis of terpenoids: YgbB protein converts 4-diphosphocytidyl-2C-methyl-D-erythritol 2-phosphate to 2C-methyl-D-erythritol 2,4-cyclodiphosphate. Proc Natl Acad Sci USA 97: 2486-2490

Hewitt CN, Stewart H, Street RA & Scholefield PA (1997) Isoprene, monoterpene – emitting species survey 1997, http://www.es.lancs.ac.uk/cnhgroup/download.html

Kellogg BA & Poulter CD (1997) Chain elongation in the isoprenoid biosynthetic pathway. Curr Opin Chem Biol 1: 570-578

Kesselmeier J & Staudt M (1999) Biogenic volatile organic compounds (VOC): An overview on emission, physiology, ecology. J Atmos Chem 33: 23-88

Knöss W, Reuter B & Zapp J (1997) Biosynthesis of the labdane diterpene marrubiin in *Marubium vulgare* via a non-mevalonate pathway. Biochem J 326: 449-454

Kreuzwieser J, Schnitzler J-P & Steinbrecher R (1999) Biosynthesis of organic compounds emitted by plants. Plant Biol 1: 149-159

Kuzuyama T, Shimizu T, Takahashi S & Seto H (1998a) Fosmidomycin, a specific inhibitor of 1-deoxy-D-xylulose 5-phosphate reductoisomerase in the nonmevalonate pathway for terpenoid synthesis. Tetrahedron Lett 39: 7913-7916

Kuzuyama T, Takagi M, Kaneda K, Dairi T & Seto H (2000a) Formation of 4-(cytidine 5'-diphospho)-2-C-methyl-D-erythritol from 2-C-methyl-D-erythritol 4-phosphate by 2-C-methyl-D-erythritol 4-phosphate cytidylyltransferase, a new enzyme in the nonmevalonate pathway. Tetrahedron Lett 41: 703-706

Kuzuyama T, Takagi M, Kaneda K, Watanabe H, Dairi T & Seto H (2000b) Studies on the nonmevalonate pathway: conversion of 4-(cytidine 5'-diphospho)-2-C-methyl-d-erythritol to its 2-phospho derivative by 4-(cytidine 5'-diphospho)-2-C-methyl-d-erythritol kinase. Tetrahedron Lett 41: 2925-2928

Lange BM, Wildung MR, McCaskill D & Croteau R (1998) A family of transketolases that directs isoprenoid biosynthesis via a mevalonate-independent pathway. Proc Natl Acad Sci USA 95: 2100-2104

Lange, B.M. & Croteau, R. (1999) Isoprenoid biosynthesis via a mevalonate-independent pathway in plants: cloning, heterologous expression of 1-deoxy-D-xylulose-5-phosphate reductoisomerase from peppermint. Arch Biochem Biophys 365: 170-174

Lerdau M & Keller M (1997) Controls on isoprene emission from trees in a subtropical dry forest. Plant Cell Environ 20: 569-578

Lerdau M, Guenther A & Monson R (1997) Plant production, emission of volatile organic compounds. BioScience 47: 373-383

Lesburg CA, Zhai GZ, Cane DE & Christianson DW (1997). Crystal-structure of pentalenene synthase – mechanistic insights on terpenoid cyclization reactions in biology. Science 277: 1820-1824

Li S-M, Hennig S & Heide L (1998) Shikonin: A geranyl diphosphate-derived plant hemiterpenoid formed via the mevalonate pathway. Tetrahedron Lett 39: 2721-2724

Lichtenthaler HK (1999) The 1-deoxy-D-xylulose-5-phosphate pathway of isoprenoid biosynthesis in plants. Ann Rev Plant Physiol Plant Molec Biol 50: 47-65

Lichtenthaler HK (2000) The non-mevalonate isoprenoid biosynthesis: enzymes, genes, inhibitors. Biochem Soc Trans 28: 794-795

Lichtenthaler HK, Schwender J, Seeman M & Rohmer M (1995). Carotenoid biosynthesis in green algae proceeds via a novel biosynthetic pathway. In: Mathis P (ed) Photosynthesis: from Light to Biosphere, pp 115-118. Kluwer Academic Publishers, Amsterdam, The Netherlands

Lichtenthaler HK, Rohmer M & Schwender J (1997a) Two independent biochemical pathways for isopentenyl diphosphate, isoprenoid biosynthesis in higher plants. Physiol Plant 101: 643-652

Lichtenthaler HK, Schwender J, Disch A & Rohmer M (1997b) Biosynthesis of isoprenoids in higher plant chloroplast proceeds via a mevalonate independent pathway. FEBS Lett 400: 271-274

Lichtenthaler HK, Müller C, Schwender J, Jomaa H & Zeidler J (1999) Specific inhibition of isoprenoid (carotene, chlorophyll, isoprene) biosynthesis in plants by fosmidomycin. Plant Physiol 105: Abstract No. 263

Lichtenthaler HK, Zeidler J, Schwender J & Müller C (2000) The non-mevalonate isoprenoid biosynthesis of plants as a test-system for new herbicides, drugs against pathogenic bacteria, the malaria parasite. Z. Naturforsch 55c: 305-313

Litvak ME & Monson R (1998) Patterns of induced, constitutive monoterpene production in conifer needles in relation to insect herbivory. Oecologia 114: 531-540

Lois LM, Campos N, Rosa-Putra S, Danielsen K, Rohmer M & Boronat A (1998) Cloning, characterization of a gene from *Escherichia coli* encoding a tranketolase-like enzyme that catalyzes the synthesis of D-1-deoxyxylulose 5-phosphate, a common precursor for isoprenoid, thiamin, pyridoxol biosynthesis. Proc Natl Acad Sci USA 95: 2105-2110

Loreto F, Ciccioli P, Brancaleoni E, Cecinato A, Frattoni M & Sharkey TD (1996a) Different sources of reduced carbon contribute to form three classes of terpenoid emitted by *Quercus ilex* L. leaves. Proc Natl Acad Sci USA 93: 9966-9969

Loreto F, Ciccioli P, Cecinato A, Brancaleoni E, Frattoni M, Fabozzi C & Tricoli D (1996b) Evidence of the photosynthetic origin of monoterpenes emitted by *Quercus ilex* L. leaves by ^{13}C labeling. Plant Physiol 110: 1317-1322

Loreto F, Ciccioli P, Brancaleoni E, Valentini R, De Lillis M, Csiky O & Seufert G (1998a) A
    hypothesis on the evolution of isoprenoid emission by oaks based on the correlation
    between emission type, *Quercus* taxonomy. Oecologia 115: 302-305

Loreto F, Förster A, Dürr M, Csiky O & Seufert G (1998b) On the monoterpene emission
    under heat stress, on the increased thermotolerance of leaves of *Quercus ilex* L. fumigated
    with selected monoterpenes. Plant Cell Environ 21: 101-107

Lüttgen H, Rohdich F, Herz S, Wungsintaweekul J, Hecht S, Schuhr C A, Fellermeier M,
    Sagner S, Zenk MH, Bacher A & Eisenreich W (2000) Biosynthesis of terpenoids: YchB
    protein of *Escherichia coli* phosphorylates the 2-hydroxy group of 4-diphosphocytidyl-2C-
    methyl-D-erythritol. Proc Natl Acad Sci USA 97: 1062-1067

Lynen F, Eggerer H, Henning U & Kessel I (1958) Farnesylpyrophosphat, 3-Methyl-Δ^3-
    butenyl-1-pyrophosphat, die biologischen Vorstufen des Squalens. Angew. Chem. 70:
    738-742

Maier W, Schneider B & Strack D (1997) Biosynthesis of sesquiterpenoid cyclohexenone
    derivatives in mycorrhizal barley roots proceeds via the glyceraldehyde 3-
    phosphate/pyruvate pathway. Tetrahedron Lett 39: 521-524

Mandel MA, Feldmann KA, Herrera-Estrella L, Rocha-Sosa M & León P (1996) CLA1, a
    novel gene required for chloroplast development, is highly conserved in evolution. Plant J
    9: 649-658

McCaskill D & Croteau R (1998) Some caveats for bioengeneering terpenoid metabolism in
    plants. Trends Biotechnol 16: 349-355

McCaskill D & Croteau R (1999). Isopentenyl diphosphate is the terminal product of the
    deoxyxylulose-5-phosphate pathway for terpenoid biosynthesis in plants. Tetrahedron Lett
    40: 653-656

McGarvey DJ & Croteau R (1995) Terpenoid metabolism. Plant Cell 7: 1015-1026

Milborrow BV & Lee H-S (1998) Endogenous biosynthetic precursors of (+)-abscisic acid.
    VI. Carotenoids, ABA are formed by the 'non-mevalonate' triose-pyruvate pathway in
    chloroplasts. Austral J Plant Physiol 25: 507-512

Ogura K & Koyama T (1998) Enzymatic aspects of isoprenoid chain elongation. Chem Rev
    98: 1263-1276

Ohnuma S, Hirooka K, Ohto C & Nishino T (1997) Conversion from archaeal geranylgeranyl
    diphosphate synthase to farnesyl diphosphate synthase – 2 amino-acids before the first
    aspartate-rich motif solely determine eukaryotic farnesyl diphosphate synthase activity. J
    Biol Chem 272: 5192-5198

Piel J, Donath J, Bandemer K & Boland W (1998) Mevalonate-independent biosynthesis of
    terpenoid volatiles in plants: Induced, constitutive emission of volatiles. Angew Chem Int
    Ed 37: 2478-2480

Pio CA & Valente AA (1998) Atmospheric fluxes, concentrations of monoterpenes in resin-
    tapped pine forests. Atmos Environ 32: 683-692

Ramos-Valdivia A-C, van der Heijden R & Verpoorte R (1997). Isopentenyl diphosphate
    isomerase: a core enzyme in isoprenoid biosynthesis. A review of its biochemistry,
    function. Nat Prod Rep 14: 591-604

Rodríguez-Concepción M, Campos N, Lois LM, Maldonado C, Hoeffler JF, Grosdemange-
    Billiard C, Rohmer M & Boronat A (2000) Genetic evidence of branching in the
    isoprenoid pathway for the production of isopentenyl diphosphate, dimethylallyl
    diphosphate in *Escherichia coli*. FEBS Lett. 473: 328-332

Rohdich F, Wungsintaweekul J, Fellermeier M, Sagner S, Herz S, Kis K, Eisenreich W, Bacher A & Zenk MH (1999) Cytidine 5'-triphosphate-dependent biosynthesis of isoprenoids: YgbP protein of *Escherichia coli* catalyzes the formation of 4-diphosphocytidyl-2-C-methylerythritol. Proc Natl Acad Sci USA 96: 11758-11763

Rohdich F, Wungsintaweekul J, Eisenreich W, Richter G, Schuhr CA, Hecht S, Zenk MH & Bacher A (2000) Biosynthesis of terpenoids: 4-Diphosphocytidyl-2C-methyl-D-erythritol synthase of *Arabidopsis thaliana*. Proc Natl Acad Sci USA 97: 6451-6456

Rohmer M, Knani M, Simonin P, Sutter B & Sahm H (1993) Isoprenoid biosynthesis in bacteria: a novel pathway for early steps leading to isopentenyl diphosphate. Biochem J 295: 517-524

Ruzicka L, Eschenmoser A & Heusser H (1953) The isoprene rule, the biogenesis of terpenoic compounds. Experientia 9: 357-396

Sagner S, Latzel C, Eisenreich W, Bacher A & Zenk MH (1998) Differential incorporation of 1-deoxy-D-xylulose into monoterpenes, carotenoids in higher plants. Chem Commun 221-222

Sanadze JA (1957) Emission of organic matters by leaves of *Robinia pseudoacacia* L. Soobsh Acad Nauk GSSR 19: 83

Sanmiya K, Iwasaki T, Matsuoka M, Miyao M & Yamamoto N (1997) Cloning of a cDNA that encodes farnesyl diphosphate synthase, the blue-light induced expression of the corresponding gene in the leaves of rice plants. Biochim Biophys Acta 1350: 240-246

Schade GW, Goldstein AH, Gray DW & Lerdan MT (2000) Canopy, leaf level 2-methyl-3-butenol-2-ol fluxes from a ponderosa pine plantation. Atmos Environ 34: 3535-3544

Schnitzler J-P, Arenz R, Steinbrecher R & Lehning A (1996). Characterization of an isoprene synthase from leaves of *Quercus Petraea* (Matuschka) Liebl. Bot Acta 109: 216-221

Schnitzler J-P, Lehning A & Steinbrecher R (1997) Seasonal pattern of isoprene synthase activity in *Quercus robur* leaves, its significance for modelling isoprene emission rates. Bot Acta 110: 240-243

Schwender J, Lichtenthaler HK, Seeman, M & Rohmer M (1995) Biosynthesis of isoprenoid chains of chlorophylls, plastoquinone in *Scenedesmus* by a novel pathway. In: Mathis P (ed), Photosynthesis: from Light to Biosphere pp 1001-1004. Kluwer Academic Publishers, Amsterdam, The Netherlands

Schwender J, Seemann M, Lichtenthaler HK & Rohmer M (1996) Biosynthesis of isoprenoids (carotenoids, sterols, prenyl side-chains of chlorophyll, plastoquinone) via a novel pyruvate/glycero-aldehyde-3-phosphate non-mevalonate pathway in the green alga *Scenedesmus.* Biochem J 316: 73-80

Schwender J, Zeidler J, Gröner R, Müller C, Focke M, Braun S, Lichtenthaler FW & Lichtenthaler HK (1997) Incorporation of 1-deoxy-D-xylulose into isoprene, phytol by higher plants, algae. FEBS Lett 414: 129-134

Schwender J (1999) Die Mevalonat-unabhängige Isoprenoid-Biosynthese und deren Verbreitung in Pflanzen. Karlsr. Contrib. Plant Physiol. 35: 1-115

Schwender J, Müller C, Zeidler J & Lichtenthaler HK (1999) Cloning, heterologous expression of a cDNA encoding 1-deoxy-D-xylulose-5-phosphate reductoisomerase of *Arabidopsis thaliana*. FEBS Lett. 455, 140-144

Sharkey TD (1996) Isoprene synthesis by plants, animals. Endeavour 20: 74-78

Sharkey TD & Loreto F (1993) Water stress, temperature,, light effects on the capacity for isoprene emission, photosynthesis of kudzu leaves. Oecologia 95: 328-333

Sharkey TD & Singsaas EL (1995) Why plants emit isoprene. Nature 374: 769

Sharkey TD, Holland EA & Mooney HA (1991) Trace Gas Emissions by Plants. Academic Press, Inc., San Diego, California, U.S.A.

Silver GM & Fall R (1995) Characterization of aspen isoprene synthase, an enzyme responsible for leaf isoprene emission to the atmosphere. J Biol Chem 270: 13010-13016

Singsaas EL, Lerdau M, Winter K & Sharkey, TD (1997) Isoprene increases thermotolerance of isoprene-emitting leaves. Plant Physiol 115: 1413-1420

Sommer S, Severin K, Camara B & Heide L (1995) Intracellular localization of geranylpyrophosphate synthase from cell cultures of *Lithospermum erythrorhizon*. Phytochemistry 38: 623-627

Sprenger GA, Schörken U, Wiegert T, Grolle S, de Graaf AA, Taylor SV, Begley TP, Bringer-Meyer S & Sahm H (1997) Identification of a thiamin-dependent synthase in *Escherichia coli* required for the formation of 1-deoxy-D-xylulose-5-phosphate precursor to isoprenoids, thiamin,, pyridoxol. Proc Natl Acad Sci USA 94: 12857-12862

Spurgeon SL & Porter JW (1981) Biosynthesis of isoprenoid compounds. John Wiley & Sons, New York, NY

Starks CM, Back KW, Chapell J & Noel JP (1997) Structural basis for cyclic terpene biosynthesis by tobacco 5-epi-aristolochene synthase. Science 277: 1815-1820

Staudt M & Bertin N (1998) Light, temperature dependence of the emission of cyclic, acyclic monoterpenes from holm oak (*Quercus ilex* L.) leaves. Plant Cell Environ 21: 385-395

Steele CL, Katoh S, Bohlmann J & Croteau R (1998). Regulation of oleoresinosos in grand fir (*Abies grandis*) – differential transcriptional control of monoterpene, sesquiterpene, diterpene synthase genes in response to wounding. Plant Physiol 116: 1497-1504

Takagi M, Kuzuyama T, Kaneda K, Watanabe H, Dairi T & Seto H (2000) Studies on the nonmevalonate pathway: formation of 2-C-methyl-d-erythritol 2,4-cyclodiphosphate from 2-phospho-4-(cytidine 5'-diphospho)-2-C-methyl-d-erythritol. Tetrahedron Lett 41: 3395-3398

Takahashi S, Kuzuyama T, Watanabe H & Seto H (1998) A 1-deoxy-D-xylulose 5-phosphate reductoisomerase catalyzing the formation of 2-*C*-methyl-D-erythritol 4-phosphate in an alternative nonmevalonate pathway for terpenoid biosynthesis. Proc Natl Acad Sci USA 95: 9879-9884

Thiel R, Adam KP, Zapp J & Becker H (1997) Isopentenyl diphosphate biosynthesis in liverworts. Pharm Pharmacol Lett 7: 103-105

Trainer M, Williams EJ, Parrish DD, Buhr MP, Allwine EJ, Westberg HH, Fehsenfeld FC & Liu SC (1987) Models, observations of the impact of natural hydrocarbons on rural ozone. Nature 329: 705-707

Wallach O (1885) Zur Kenntnis der Terpene und ätherischen Öle. Liebigs Annal Chem 227: 277-302

Wildermuth MC &, Fall R (1998) Biochemical characterization of stromal, thylakoid-bound isoforms of isoprene synthase in willow leaves. Plant Physiol 116: 1111-1124

Williams DC, McGarvey DJ, Katahira EJ & Croteau R (1998) Truncation of limonene synthase preprotein provides a fully active pseudomature form of this monoterpene cyclase, reveals the function of the amino-terminal arginine pair. Biochemistry 37: 12213-12220

Zeidler JG, Lichtenthaler HK, May HU & Lichtenthaler FW (1997) Is isoprene emitted by plants synthesized via the novel isopentenyl pyrophosphate pathway? Z Naturforsch 52c: 15-23

Zeidler JG & Lichtenthaler HK (1998) Two simple methods for measuring isoprene emission of leaves by UV-spectroscopy, GC-MS. Z Naturforsch 53c: 1087-1089

Zeidler J, Schwender J, Müller C, Wiesner J, Weidemeyer C, Beck E, Jomaa H & Lichtenthaler HK (1998) Inhibition of the non-mevalonate 1-deoxy-D-xylulose-5-phosphate pathway of plant isoprenoid biosynthesis by fosmidomycin. Z Naturforsch 53c: 980-986

Zeidler J & Lichtenthaler HK (2001) Biosynthesis of 2-methyl-3-buten-2-ol emitted from needles of *Pinus ponderosa* via the non-mevalonate DOXP/MEP pathway of isoprenoid formation. Planta 213: 323-326

Chapter 1.5

Biosynthesis of aldehydes and organic acids

Jürgen Kreuzwieser
Institut für Forstbotanik und Baumphysiologie, Professur für Baumphysiologie, Georges-Köhler-Allee Geb. 053/054, D-79110 Freiburg i. Br., Germany

1. INTRODUCTION

Trees emit a wide range of VOCs including oxygenated species such as aldehydes and organic acids into the atmosphere. Depending on their reactivity the calculated lifetimes of these compounds in the troposphere vary in the range from only a few (e.g. formaldehyde 3.6 h) up to some hours (e.g. acetaldehyde 20 h) which is in the same range as isoprene (c. 8 h) or monoterpenes (0.5 to 7 h) (Kotzias et al. 1997). Photochemical destruction and oxidation of aldehydes and organic acids lead to the generation of free radicals which participate in numerous atmospheric reactions and control many atmospheric chemical processes (Kotzias et al. 1997). These reactions strongly affect HO_X-radical chemistry in the troposphere thereby leading to the production of HO_2-radicals. In the presence of NO and NO_2 in ambient air, these reactions cause a net generation of ozone. Moreover, the destruction of acetaldehyde, for example, may be connected to the production of peroxyacylnitrates (PANs) which are known for their adverse effects on human health and plant growth (Sakaki 1998; Kotzias et al. 1997). Formic and acetic acids in the atmosphere strongly participate in the acidification of rainwater. Especially in rural regions the contribution of these acids to rainwater acidity is significant and amounts to around 50 to 64 % (see Galloway et al. 1982; Andreae et al. 1988; Bode et al. 1997). To date, it is assumed that the main portion of formic and acetic acids in the atmosphere is derived from biogenic sources (see Chapter 3.4 this issue) and that other higher molecular weight organic

R. Gasche et al. (eds.), Trace Gas Exchange in Forest Ecosystems, 101–114.
© 2002 *Kluwer Academic Publishers. Printed in the Netherlands.*

acids and aldehydes are also emitted from vegetation in considerable amounts. Although in recent years much progress was achieved in identifying compounds which are emitted by plants, the knowledge on biogenic pathways and environmental conditions controlling these pathways is still incomplete. This review provides an overview of biosynthetic pathways in plants which are possibly involved in the production of the most important aldehydes and organic acids found in the atmosphere and considered to be emitted by vegetation.

2.   OTHER SOURCES OF ATMOSPHERIC ALDEHYDES AND ORGANIC ACIDS THAN PLANTS

2.1   Indirect sources

Although it is difficult to give a proper estimate of the VOC amount which is generated in the atmosphere, indirect VOC production is thought to contribute significantly to ambient VOC concentrations. A considerable portion of organic acids and aldehydes in the atmosphere is produced by photo-oxidation of non-methane-hydrocarbons (NMHC; Altshuller 1993; Kotzias et al. 1997). Acetic and formic acid, for example, are known to be synthesized in the atmosphere by the reaction of ozone with ethene and propene which are emitted from anthropogenic sources (Neeb et al. 1994). Oxidation of isoprene, the quantitatively most important volatile emitted by plants, results in significant generation of formic and pyruvic acid (Jacob and Wofsky 1988) and aldehydes (Altshuller 1993). The formation of acetaldehyde and formaldehyde from the photo-oxidation of isoprene in the troposphere has been shown by Arnts *et al.* (1981). Moreover, the typical "leaf alcohol" (Z)-3-hexene-1-ol or (Z)-3-hexenylacetate which is released by the leaves of numerous plant species can also be converted to aldehydes such as propanal, formaldehyde and acetaldehyde (Atkinson 1990).

2.2   Anthropogenic sources

The finding of considerably higher concentrations of aldehydes in urban than in rural areas (Puxbaum et al. 1988; Granby et al. 1997) indicates the significant contribution of anthropogenic emission sources for the total budget of these substances. Most important direct sources of aldehydes and organic acids are industrial emissions (Carlier et al. 1990) and automobile exhaust (Anderson et al. 1996). For example, around 30% of the formaldehyde in the urban atmosphere is probably derived from the combustion of fossil fuels by automobile traffic (see Kotzias et al. 1997). Further important direct emission sources, preferentially at rural sites, are biomass burning, particularly of forests (Lipari et al. 1984; Hurst et al. 1994) which especially promote the production of acetic acid.

2.3 Other biogenic sources than plants

Besides the leaves of plants, the physiological activity of microorganisms in the soil (Sanhueza and Andreae 1991), insects such as formic acid producing ants (Johnson and Dawson 1993) and animal faeces (Nethea and Narayan 1972) may also contribute to the total budget of oxygenated hydrocarbons in the atmosphere. The strength of these sources is presently uncertain.

3. BIOSYNTHESIS OF ALDEHYDES AND ORGANIC ACIDS IN PLANTS

3.1 Formaldehyde and formic acid

Formaldehyde and formate production in plants is assumed to be closely connected to the occurrence of methanol which is synthesized during seed development (Obendorf et al. 1990), cell expansion, cell wall degradation (MacCann and Roberts 1991; Levy and Staehelin 1992; Nemecek-Marshall et al. 1995; Fall and Benson 1996) and senescence processes in plants (see Harriman et al. 1991 and Fig. 1). The enzymes involved in methanol biosynthesis during these processes have been identified as pectin methylesterases (PMEs; EC 3.1.1.11). PMEs are cell wall enzymes which catalyze the cleavage of pectin thereby generating carboxylated pectin as well as methanol and a proton (Ricard and Noat 1986). PME activity has been observed in all higher plants examined (Harriman et al. 1991). However, so far neither the role of different PME isoforms in plant growth and development is well understood nor the genetics of PMEs (Gaffe et al. 1994) which may either be encoded by different genes or represent post-translational modifications of the same PME protein (Gaffe et al. 1994).

Beside pectin demethylation by PMEs, methanol in plants is also derived from the action of protein methyltransferases and protein repair reactions (Mudgett and Clarke 1993), as well as from tetrahydrofolate metabolism (Cossins 1987) and the degradation of lignin in plant secondary cell walls by fungi (Ander et al. 1985) (see Fig. 1).

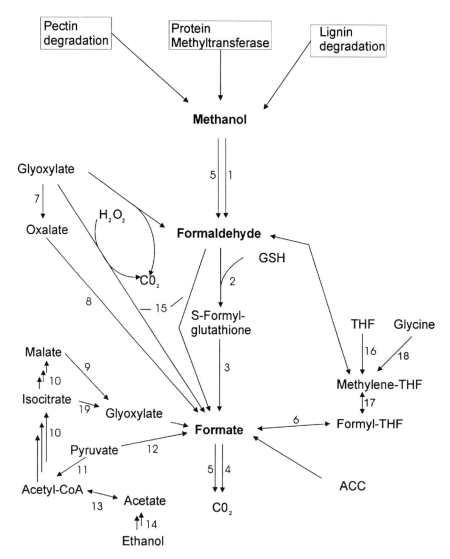

Figure 1. Biosynthesis of methanol, formaldehyde and formate in trees. 1: methanol oxidase;
2: formaldehyde dehydrogenase; 3: S-formyl-hydrolase; 4: formate dehydrogenate; 5:
catalase; 6: 10-formyl-THF synthetase; 7: glycolate oxidase; 8: oxalate decarboxylase;
9: malate synthase; 10: reactions of TCA cycle; 11: pyruvate dehydrogenase; 12: pyruvate-
formate-lyase; 13: acetyl-CoA synthetase; 14: alcohol dehydrogenase and aldehyde
dehydrogenase; 15: non-enzymic; 16: serine-hydroxymethyl transferase; 17: methenyl-THF -
cyclohydrolase and methenyl-THF-dehydrogenase; 18: glycine decarboxylase, 19: isocitrate
lyase.

Since methanol is a cytotoxic compound, plants have developed
methanol degradation pathways in order to avoid self-poisoning. From these

detoxification reactions, formaldehyde and formate originate as intermediates which usually are metabolized to CO_2. Due to their low vapor pressures traces of formate and formaldehyde may escape from tissues into the environment. Methanol which is present in the cytoplasm is converted into formaldehyde probably by the action of methanol oxidase (Cossins 1964). Although this enzyme has so far only been found in microorganisms and not in plants, its occurrence in plants may be assumed because several studies have shown that methanol indeed is converted to formate (Cossins 1964; Nonomura and Benson 1992; Devlin et al. 1994; Rowe et al. 1994; Hourton-Cabassa et al. 1998). In a subsequent reaction catalyzed by NAD^+ / glutathione-dependent formaldehyde dehydrogenase (EC 1.2.1.1.), the tripeptide glutathione in its reduced form (GSH) is bound to formaldehyde, which results in the generation of S-formyl-glutathione (Uotila and Koivusalo 1979, Giese et al. 1994; Fliegmann and Sandermann Jr. 1997).

Formaldehyde dehydrogenase is considered a conservative enzyme whith significance for the detoxification of endogenous and external formaldehyde (Fliegmann and Sandermann Jr. 1997; Wippermann et al. 1999). Suzuki et al. (1998) demonstrated that the expression of formaldehyde dehydrogenases is increased under conditions of iron deficiency but also during anaerobiosis. The authors concluded that formaldehyde dehydrogenase induction might be caused by the anoxia induced by iron deficiency in spite of normal oxygen availability. Iron deficiency could lead to reduced energy metabolism due to inhibited heme synthesis in mitochondria which finally causes effects similar to anaerobiosis (Suzuki et al. 1998). It may therefore be concluded that formaldehyde dehydrogenase activity may contribute to significant formate production under such stress conditions (Suzuki et al. 1998). For complete metabolisation, S-formyl-glutathione is converted into formate by S-formyl hydrolase. In a further reaction formate dehydrogenase (FDH; EC 1.2.1.2.) cleaves formate thereby releasing CO_2. In higher plants FDHs are mitochondrial, NAD^+-dependent enzymes (Oliver 1981; Colas des Francs-Small et al. 1993; Hourton-Cabassa et al. 1998). FDHs seem to be more abundant in mitochondria of non-green tissues as compared to leaf mitochondria (Colas de Francs-Small et al. 1993; Hourton-Cabassa et al. 1998), suggesting distinct differences in formate metabolism between these tissues. Consistent with the findings of formaldehyde dehydrogenase induction, several stresses such as hypoxia, drought, chilling and wounding led also to significantly increased FDH expression in potato leaves suggesting an increased production of formaldehyde and formate in response to these environmental factors (Hourton-Cabassa et al. 1998).

Besides methanol oxidase, catalase (EC 1.11.1.6) in the peroxisomes may also contribute to the metabolization of methanol and the subsequent production of formaldehyde. Since H_2O_2 is necessary for the formation of the active form of catalase, the oxidation of methanol with catalase depends on a steady availability of H_2O_2 (Halliwell and Gutteridge 1989). Because the permanent formation of H_2O_2 is ensured in peroxisomes by photo-respiration, this reaction may indeed be of significance for methanol metabolism and, therefore, also for formaldehyde and formate generation. In addition to methanol oxidation, catalase contributes to the degradation of formate (Halliwell and Gutteridge 1989). Formate catabolism in peroxisomes could be significant because, in addition to the cytosolic production of formate described above, considerable amounts of formate may also arise from glyoxylate catabolism in the peroxisomes. Glyoxylate produced during photo-respiration is degraded to a minor extent either non-enzymatically or by the sequential action of glycollate oxidase and oxalate decarboxylase, to CO_2 and formate (Halliwell and Butt 1974). This pathway seems to be stimulated when normal metabolization of glyoxylate is inhibited (Amory and Cresswell 1986). The formate synthesized is then oxidatively decarboxylated by the action of 10-formyl-THF synthetase resulting in a generation of 5,10-methylene-THF, which in two subsequent steps, catalyzed by 5,10-methenyl-THF dehydrogenase and 5,10-methenyl-THF cyclohydrolase, is converted into methylene-THF (Wingler et al. 1999). Hourton-Cabassa et al. (1998) suggested that formate may also be derived from serine under drought stress which would require the enzymes of C1 metabolism as well as the action of 10-formyl-THF synthetase. Formate is also produced if glycolysis is stimulated which, for example, occurs under anaerobic conditions or during times of darkness (Colas des Francs-Small et al. 1993; Hourton-Cabassa et al. 1998; Suzuki et al. 1998). As inducing compounds ethanol, pyruvate and acetate are discussed under these conditions (Hourton-Cabassa et al. 1998). When these substances act as direct formate precursors, formate may be synthesized from pyruvate either via malate, isocitrate and glyoxylate which includes part of the TCA cycle and isocitrate lyase, or it is produced by the action of pyruvate-formate lyase which has been found in bacteria (Ferry 1990) and unicellular algae (Kreuzberg 1984) but so far not in higher plants (Hourton-Cabassa et al. 1998). Beside these pathways formate is also generated during the production of the gaseous phyto-hormone ethylene. In the last step of ethylene production formate may be synthesized by the oxidation of 1-aminocyclopropane-1-carboxylate (ACC).

3.2   Acetaldehyde and acetic acid

Just as formaldehyde and formate production is connected to methanol synthesis, acetaldehyde generation in plants has to be seen in correlation with the production of ethanol (Fig. 2). Ethanol synthesis in plant cells takes place if oxygen availability limits oxidative respiration. The only known pathway for ethanol synthesis in higher plants so far is alcoholic fermentation. The biochemical reactions take place in the cytoplasm and include the sequential action of pyruvate decarboxylase (PDC; EC 4.1.1.1) and alcohol dehydrogenase (ADH; EC 1.1.1.1). The PDC reaction is the rate limiting step in alcoholic fermentation which is necessary to avoid accumulation of phytotoxic acetaldehyde in the cells; accordingly *in vitro* activity of ADH is usually much higher than PDC activity (e.g. Kimmerer 1987). The different isoforms of ADHs identified include inducable species and constitutively expressed forms. The physiological significance of anaerobically inducable ADHs is the maintenance of glycolysis by regeneration of NAD$^+$. Alcoholic fermentation gathers ecological importance if the oxygen supply of roots is limited, for example as a response of submergence. However, also compressed soil may lead to oxygen shortage of plant cells and may induce alcoholic fermentation. Beside roots, alcoholic fermentation is known to take place also in the cambium of trees (MacDonald and Kimmerer 1991) and in fruits especially during ripening (see Beaulieu et al. 1997; Zuckerman et al. 1997), or in pollen (Tadege and Kuhlemeier 1997). The significance of the constitutively expressed ADH isoforms often found in the leaves of trees is so far not fully understood. Since the leaves of plants are in an aerobic environment and even produce oxygen during photosynthesis, the constitutive expression of ADH activity does not appear to be necessary. These isoforms may be of significance when ethanol, produced either in hypoxic parts of the cambium or in the roots (MacDonald and Kimmerer 1991; Kreuzwieser et al. 1999), is transported to the leaves by the transpiration stream. This assumption is supported by the fact that ethanol is often found in the xylem sap of trees (Kimmerer and Stringer 1988; Crawford and Finegan 1989; MacDonald and Kimmerer 1991). Considerably increased amounts of ethanol (mM range) in the xylem sap can be detected if roots are exposed to hypoxia (MacDonald and Kimmerer 1991; Kreuzwieser et al. 1999). The major portion of ethanol transported to the leaves by the transpiration stream is oxidized in leaf cells (MacDonald and Kimmerer 1993) by ADH (Kreuzwieser et al. 2001), thereby releasing acetaldehyde from which minor portions are emitted into the atmosphere (Kreuzwieser et al. 1999). Acetaldehyde is further converted into acetate by the action of aldehyde dehydrogenase (ALDH; EC 1.2.1.5) (Kreuzwieser et al. 2001).

The acetate synthesized may further be converted into acetyl-CoA by acetyl-CoA synthetase which enables acetate to enter general metabolic pathways like the TCA cycle or lipid synthesis (MacDonald and Kimmerer 1993). Acetate in plants usually is found in its activated form as acetyl-CoA. This compound plays a central role as a precursor of several biochemical pathways and is synthesized from a variety of biochemical processes like degradation of carbohydrates and fatty acids. Acetate in its free form acts as a transport form between mitochondria and chloroplasts or the cytoplasm. Uptake, activation and metabolization of free acetate has been demonstrated in several studies (e.g. Groeneveld et al. 1991).

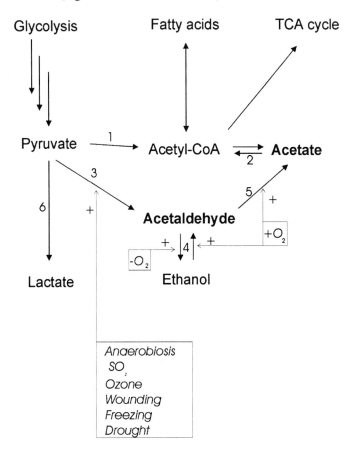

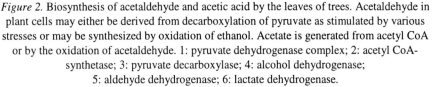

Figure 2. Biosynthesis of acetaldehyde and acetic acid by the leaves of trees. Acetaldehyde in plant cells may either be derived from decarboxylation of pyruvate as stimulated by various stresses or may be synthesized by oxidation of ethanol. Acetate is generated from acetyl CoA or by the oxidation of acetaldehyde. 1: pyruvate dehydrogenase complex; 2: acetyl CoA-synthetase; 3: pyruvate decarboxylase; 4: alcohol dehydrogenase; 5: aldehyde dehydrogenase; 6: lactate dehydrogenase.

The origin of acetaldehyde emitted by flooded trees may be explained by the oxidation of ethanol produced through alcoholic fermentation in the roots. However, it remains to be elucidated what metabolic pathways lead to acetaldehyde generation and emission of non-flooded plants. The relatively low rates of acetaldehyde emission frequently observed in non-flooded plants may also be caused by oxidation of ethanol which is generally present in distinct concentrations in the xylem sap of a wide range of tree species (Barta 1984; Kimmerer and Stringer 1988; MacDonald and Kimmerer 1991). However, other pathways of acetaldehyde synthesis cannot be excluded in non-flooded trees. An explanation for acetaldehyde generation during daytime has been provided by Kimmerer and Kozlowski (1982). The authors found that oxidative stress, e.g. caused by fumigating of plants with ozone or SO_2, led to a significant production and emission of ethanol and acetaldehyde by the leaves of birch and pine trees. They proposed that the stress probably caused an oxygen shortage even in an oxygen rich environment and, as a consequence, PDC activity was stimulated at the expense of pyruvate dehydrogenase (Fig. 2). For this reason the plants synthesized the products of alcoholic fermentation, ethanol and acetaldehyde, under normoxic conditions. The release of acetaldehyde and ethanol due to oxidative stress can also be explained by the oxidation of membrane lipids by radicals as proposed by Halliwell and Gutteridge (1989).

3.3 Higher molecular weight aldehydes and organic acids

At a quantitative level formaldehyde, acetaldehyde as well as formic and acetic acid are certainly the most important oxygenated hydrocarbons which are emitted by vegetation; however, higher molecular weight aldehydes and organic acids are emitted in considerable amounts as well. For example, pyruvic acid has also been found in significant amounts in the atmosphere (Talbot et al. 1990). Such organic acids are probably derived from general metabolic pathways such as TCA cycle. Isidorov et al. (1985) reported the emission of a variety of higher chained aldehydes from tree species, shrubs and herbaceous plants, such as propanal (sorb, fern), butanal (fern), isobutanal (Northern white cedar), isobutenal (aspen, poplar, larch) and crotonaldehyde (Chinese arbor vitae). Other studies (Buttery et al. 1982, 1985; Arey et al. 1991; König et al. 1995) identified the emission of 2-methyl-2-pentenal, hexanal, (E)-2-hexenal, (Z)-3-hexenal, nonanal and benzaldehyde from vegetation. Most of these compounds are odorous substances and some are discussed to be involved in plant defense mechanisms against insects, fungi and bacteria.

It is not possible to describe all of the biosynthetic pathways of these aldehydes in this article; more detailed information is provided on the synthesis of the C_6-aldehydes, hexanal and hexenals ((3Z)-hexenal, (2E)-hexenal, (3E)-hexenal, n-hexenal) which, together with their corresponding alcohols, mediate the typical odor and flavor of leafs, fruits and flowers

(Riley and Thompson 1998). It is assumed that these C_6-aldehydes are derived from polyunsaturated fatty acid precursors which are generated by the action of lipolytic acyl hydrolase (LAH) on membrane lipids (Riley and Thompson 1998). Hexanal and hexenal are produced from linoleic (18:2) and α-linolenic acid (18:3), respectively (Fig. 3). Two enzymes are involved in these reactions: Lipoxygenase adds stereoselectively O_2 to unsaturated fatty acids with a (1Z, 4Z)-pentadiene moiety (e.g. α-linolenic and linoleic acids) to synthesize 13-(S)-hydroperoxides (Hatanaka 1993) which in a subsequent reaction catalyzed by hydroperoxide lyase are cleaved between C-12 and C-13 resulting in the release of the C_6-aldehydes. Both enzymes, LOX and HPL, are bound to the thylakoid membrane of the chloroplasts of green leaves (Hatanaka 1993).

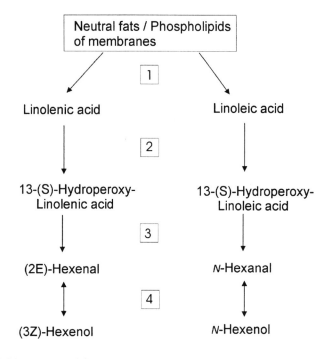

Figure 3. Biosynthesis of C_6-aldehydes in the leaves of plants. Most of the C_6-aldehydes emitted by plants are probaly produced via oxidation of membrane components. 1: lipolytic acyl hydrolase; 2: lipoxygenase; 3: hydroperoxide lyase; 4: isomerization (simplified after Hatanaka, 1993).

4.   FUTURE OBJECTIVES

In recent years several aldehydes and organic acids present in ambient air were found to be derived from vegetation. Using cuvette systems emission rates of the different compounds have been determined in order to get information on their source strength for the global budget of these VOCs. In order to predict VOC emission from vegetation in the natural environment more precisely, it is necessary to know the biosynthetic pathways of the individual compounds. To date, knowledge about these pathways in plants is still incomplete. Although some mechanisms leading to the production of C_1-compounds are known, for most of the other compounds emitted neither biochemical pathways which lead to their synthesis nor environmental conditions which induce or inhibit these pathways are identified. Future work should combine the determination of emission rates with the elucidation of biosynthetic pathways responsible for the production of the compounds emitted and should also focus on the identification of environmental conditions controlling these pathways.

REFERENCES

Altshuller AP (1993) Production of aldehydes as primary emissions and secondary atmospheric reaction of alkanes and alkenes during night and early morning hours. Atmos Environ 27: 21-31

Amory AM & Cresswell CF (1986) Role of formate in the photorespiratory metabolism of *Themada triandra* Forssk. J Plant Physiol 124: 247-255

Ander P, Eriksson MER & Eriksson KE (1985) Methanol production from lignin-related substances by *Phanerochaete chrysosporium*. Physiol Plant 65: 317-321

Andreae MO, Talbot RW, Andreae TW & Harriss RC (1988) Formic and acetic acid over the Central Amazon region, Brazil. 1. Dry Season. J Geophys Res 93: 1616-1624

Arey J, Winer AM, Atkinson R, Aschmann SM, Long WD & Morrison CL (1991) The emission of (Z)-3-hexen-1-ol, (Z)-3-hexenylacetate and other oxygenated hydrocarbons from agricultural plant species. Atmos Environ 25A: 1063-1075

Arnts RR, Gay BW & Bufalini JJ (1981) Photochemical oxidant potential of the biogenic hydrocarbons. In: Bufakini JJ & Arnts RR (eds) Atmospheric biogenic hydrocarbons (Vol 2), pp 117-133. Ann Arbor Science Publishers Inc., Michigan, U.S.A.

Atkinson R (1990) Gas-phase tropospheric chemistry of organic compounds: a review. Atmos Environ 24A: 1-41

Beaulieu JC, Peiser G & Saltveit ME (1997) Acetaldehyde is a causal agent responsible for ethanol-induced ripening inhibition in tomato fruit. Plant Physiol 113: 431-439

Bode K, Helas G & Kesselmeier J (1997) Biogenic contribution to atmospheric organic acids. In: Helas G, Slanina J & Steinbrecher R (eds) Biogenic volatile organic compounds in the atmosphere, pp 157-170. SPB Academic Publishing, Amsterdam, The Netherlands

Buttery RG, Ling LC & Wellso SG (1982) Oat leaf volatiles: possible insect attractants. J Agr Food Chem 30: 791-792

Buttery RG, Xu C & Ling LC (1985) Volatile components of wheat leaves (and stems): possible insect attractants. J Agr Food Chem 33: 115-117

Colas des Francs-Small C, Ambard-Bretteville F, Small ID & Remy R (1993) Identification of a major soluble protein in mitochondria from nonphotosynthetic tissues as NAD-dependent formate dehydrogenase. Plant Physiol 102: 1171-1177

Cossins EA (1964) The utilization of carbon-1 compounds by plants. I. The metabolism of methanol-C14 and its role in amino acid biosynthesis. Can J Botany 42: 1793-1802

Cossins EA (1987) Folate biochemistry and the metabolism of one-carbon units. In: Davies DD (ed) The Biochemistry of Plants, Vol 11, pp 317-353, Academic Press, San Diego, U.S.A.

Crawford RMM & Finegan DM (1989) Removal of ethanol from lodgepole pine roots. Tree Physiol 5: 53-61

Devlin RM, Bhowmik PC & Karczmarczyk SJ (1994) Influence of methanol on plant growth. Plant Growth Regul 22: 102-108

Fall R & Benson AA (1996) Leaf methanol: the simpliest natural product from plants. Trends Plant Sci 1: 296-301

Ferry JG (1990) Formate dehydrogenase. FEMS Microbiol Rev 87: 377-382

Fliegmann J & Sandermann H Jr. (1997) Maize glutathione-dependent formaldehyde dehydrogenase cDNA: a novel plant gene of detoxification. Plant Mol Biol 34: 843-854

Gaffe J, Tieman DM & Handa AK (1994) Pectin methylesterase isoforms in tomato (*Lycopersicon esculentum*) tissues. Effects of expression of a pectin methylesterase antisense gene. Plant Physiol 105: 199-203

Giese M, Bauer-Doranth U, Langebartels C & Sandermann H (1994) Detoxification of formaldehyde by the spider plant (*Chlorophytum comosum* L.) and by soybean (*Glycine max* L.) cell-suspension cultures. Plant Physiol 104: 1301-1309

Granby K, Christensen CS & Lohse C (1997) Urban and semi-rural observations of carboxylic acids and carbonyls. Atmos Environ 31: 1403-1415

Groeneveld HW, Binnekamp A & Sekens D (1991) Cardenolide biosynthesis from acetate in *Asclepias urassavica*. Phytochemistry 30: 2577-2585

Halliwell B & Butt, VS (1974) Oxidative decarboxylation of glycollate and glyoxylate by leaf peroxisomes. Biochem J 138: 271-224

Halliwell B & Gutteridge JMC (1989) Free radicals in biology and medicine. Second Edition. Oxford University Press, Oxford, U.K.

Harriman RW, Tieman DM & Hanada AK (1991) Molecular cloning of tomato pectin methylesterase gene and its expression in Rutgers, ripening inhibitor, non-ripening, and never ripe tomato fruits. Plant Physiol 97: 80-87

Hatanaka A (1993) The biogeneration of green odour by green leaves. Phytochemistry 34: 1201-1218

Hourton-Cabassa C, Ambard-Bretteville F, Moreau F, Davy de Virville J, Rémy R & Colas des Francs C (1998) Stress induction of mitochondrial formate dehydrogenase in potato leaves. Plant Physiol 116: 627-635

Isidorov VA, Zenkevich IG & Ioffe BV (1985) Volatile organic compounds in the atmosphere of forests. Atmos Environ 19: 1-8

Johnson BJ & Dawson GA (1993) A preliminary study of the carbon-isotopic content of ambient formic acid and two selected sources: automobile exhaust and formicine ants. J Atmos Chem 17: 123-140

Kimmerer TW (1987) Alcohol dehydrogenase and pyruvate decarboxylase activity in leaves and roots of Eastern cottonwood (*Populus deltoides* Bartr.) and soybean (*Glycine max* L.). Plant Physiol 84: 1210-1213

Kimmerer TW & Kozlowski TT (1982) Ethylene, ethane, acetaldehyde, and ethanol production by plants under stress. Plant Physiol 69: 840-847

Kimmerer TW & MacDonald RC (1987) Acetaldehyde and ethanol biosynthesis in leaves of plants. Plant Physiol 84: 1204-1209

Kimmerer TW & Stringer MA (1988) Alcohol dehydrogenase and ethanol in the stems of trees. Plant Physiol 87: 693-697

König G, Brunda M, Puxbaum H, Hewitt CN, Duckham SC & Rudolph J (1995) Relative contribution of oxygenated hydrocarbons to the total biogenic VOC emissions of selected mid-European agricultural and natural plant species. Atmos Environ 29: 861-874

Kotzias D, Kondiari C & Sparta C (1997) Carbonyl compounds of biogenic origin - emission and concentrations in the atmosphere. In: Helas G, Slanina J & Steinbrecher R (eds) Biogenic volatile compounds in the atmosphere, pp 317-353. SPB Academic Publishing, Amsterdam, The Netherlands

Kreuzberg K (1984) Starch fermentation via formate producing pathway *Chlamydomonas reinhardii, Chlorogonium elongatum*, and *Chlorella fusca*. Physiol Plant 61: 87-94

Kreuzwieser J, Scheerer U & Rennenberg H (1999) Metabolic origin of acetaldehyde emitted by trees. J Exp Bot 50: 757-765

Kreuzwieser J, Harren FJM, Laarhoven LJJ, Boamfa I, te Lintel-Hekkert S, Scheerer U, Hüglin C & Rennenberg H (2001) Acetaldehyde emission by the leaves of trees-correlation with physiological and environmental parameters. Physiol Plant 113: 41

Levy S & Staehelin LA (1992) Synthesis, assembly and function of plant cell wall macromolecules. Curr Opin Cell Biol 4: 856-862

MacCann MC & Roberts K (1991) Architecture of the primary cell wall. In: Lloyd CW (ed) The Cytoskeletal Basis of Plant Growth and Form. pp. 109-129. Academic Press, San Diego, U.S.A.

MacDonald RC & Kimmerer TW (1991) Ethanol in the stems of trees. Physiol Plant 82: 582-588

MacDonald RC & Kimmerer, TW (1993) Metabolism of transpired ethanol by eastern cottonwood (*Populus deltoides* Bartr). Plant Physiol 102: 173-179

Mudgett MB & Clarke S (1993) Characterization of plant L-isoaspartyl methyltransferases that may be involved in seed survival. Purification, characterization and sequence analysis of the wheat germ enzyme. Biochemistry 32: 11100-11111

Nemecek-Marshall M, MacDonald RC, Franzen JJ, Wojciechowski CL & Fall R (1995) Methanol emission from leaves. Plant Physiol 108: 1359-1368

Nonomura AM & Benson AA (1992) The path of carbon in photosynthesis: improved crop yields with methanol. P Natl Acad Sci USA 89: 9794-9798

Obendorf RL, Koch JL, Gorecki RJ, Amable RA & Aveni MT (1990) Methanol accumulation in maturing seeds. J Exp Bot 41: 489-495

Oliver DJ (1981) Formate oxidation and oxygen reduction by leaf mitochondria. Plant Physiol 68: 703-705

Puxbaum H, Rosenberg C, Gregori M, Lanzerstorfer C, Ober E & Winiwarter W (1988) Atmospheric concentrations of formic and acetic acid and related compounds in eastern and northern Austria. Atmos Environ 22: 2841-2850

Ricard J & Noat G (1986) Electrostatic effects and the dynamics of enzyme reactions at the surface of plant cells. I. A theory of the ionic control of a complex multi-enzyme system. Eur J Biochem 155: 199-202

Riley JC & Thompson JE (1998) Ripening-induced acceleration of volatile aldehyde generation following tissue disruption in tomato fruit. Physiol Plant 104: 571-576

Rowe RN, Farr DJ & Richards BAJ (1994) Effects of foliar and root applications of methanol or ethanol on the growth of tomato plants (*Lycopersicon esculentum* Mill.) New Zeal J Crop Hort 22: 335-337

Sanhueza E & Andreae MO (1991) Emission of formic and acetic acid from tropical savanna soils. Geophys Res Lett 18: 1707-1710

Suzuki K, Itai R, Suzuki K, Nakashiani H, Nishizawa N-K, Yoshimura E & Mori S (1998) Formate dehydrogenase, an enzyme of anaerobic metabolism, is induced by iron deficiency in barley roots. Plant Physiol 116: 725-732

Tadege M & Kuhlemeier C (1997) Aerobic fermentation during tobacco pollen development. Plant Mol Biol 35: 343-354

Talbot RW, Andreae MO, Berresheim H, Jacob DJ & Beecher KM (1990) Sources and sinks of formic, acetic and pyruvic acids over Central Amazonias. 2. Wet season. J Geophys Res 95: 16799-16811

Wingler A, Lea PJ & Leegood RC (1999) Photorespiratory metabolsim of glyoxylate and formate in glycine-accumulating mutants of barley and *Amaranthus edulis*. Planta 207: 518-526

Wippermann U, Fliegmann J, Bauw G, Langebartels C, Maier K & Sandermann H Jr. (1999) Maize glutathione-dependent formaldehyde dehydrogenase: protein sequence and caralytic properties. Planta 208: 12-18

Zuckerman H, Harren FJM, Reuss J & Parker DH (1997) Dynamics of acetaldehyde production during anoxia and post-anoxia in Red Bell Pepper studied by photoacoustic techniques. Plant Physiol 113: 925-932

EXCHANGE OF TRACE GASES AT THE SOIL-ATMOSPHERE INTERFACE

Chapter 2.1

NO, NO$_2$ and N$_2$O

Rainer Gasche and Hans Papen
Fraunhofer Institut für Atmosphärische Umweltforschung, Abteilung für Bodenmikrobiologie, Kreuzeckbahnstrasse 19, D-82467 Garmisch-Partenkirchen, Germany

1.   INTRODUCTION

The soil of forest ecosystems can function as both, a source and a sink for atmospheric N$_2$O, NO, and even NO$_2$ (e.g. Butterbach-Bahl et al. 1998; Papen and Butterbach-Bahl 1999; Gasche and Papen 1999). The fluxes of NO and N$_2$O observed at the soil/atmosphere interface are the result of simultaneous dynamic production and consumption processes taking place in soils predominantly by microorganisms. The most important microbial processes in soils contributing to fluxes of NO and N$_2$O are nitrification and denitrification, whereas at least for NO$_2$ deposition (consumption) nitrification must be considered to be potentially involved (for details see Chapter 1.1). Furthermore, there is general agreement that besides microbial processes especially for NO physico-chemical processes (e.g. chemodenitrification) may be at least under certain environmental conditions and/or site characteristics of significant importance for the observed NO production and emission (e.g. Conrad et al. 1996; Davidson 1992). The contribution of natural ecosystems like forests to the global budgets of these N-trace gases is still highly uncertain. The reasons for this uncertainty are multiple. The estimates

– are still based on a limited number of data sets from field observations comprising in most cases only sporadic measurements not covering complete seasons and/or years;
– do not differentiate between the contribution of temperate coniferous and deciduous forests to the global N$_2$O and NO emission;

R. Gasche et al. (eds.), Trace Gas Exchange in Forest Ecosystems, 117–140.
© 2002 *Kluwer Academic Publishers. Printed in the Netherlands.*

– do not consider direct and indirect anthropogenic factors on N-cycling in forest ecosystems like atmospheric N- input into forest ecosystems and countermeasures against soil acidification as liming of forest soils.

Nevertheless, our understanding about the fluctuation of N-trace gas fluxes from forest soils in space and time and about the abiotic and biotic factors modulating the exchange of N_2O, NO and NO_2 between forest soils and the atmosphere has improved significantly during the last decade. Especially, the results obtained from intensive studies on the process level have successfully been used for the development of mechanistic models simulating N-trace gas exchange at the forest soil/atmosphere interface. These process oriented models have been applied for estimating the N_2O and NO source strength of different forest ecosystems and represent a powerful tool to come up with more reliable estimates of the source strength of forest ecosystems on a local, regional and global scale.

This review summarizes the current knowledge about temporal and spatial variability of N-trace gas fluxes and the abiotic and biotic factors modulating these fluxes from forest soils of different climatic zones.

2. SPATIAL AND TEMPORAL VARIATIONS OF FLUX RATES

2.1 Spatial variations

Chamber measurements used in most studies for the determination of N_2O, NO and NO_2 fluxes from forest soils have revealed pronounced variations of fluxes in space and time (e.g. Mosier 1989; Sitaula and Bakken 1993; Papen and Butterbach-Bahl 1999; Gasche and Papen 1999; Groffman et al. 2000).

With respect to spatial variability of NO emission rates it was shown for a forest ecosystem in southeastern USA that lowest and highest fluxes from individual chambers differed by a factor of 3 (Valente and Thornton 1993).

In accordance with these findings it was demonstrated for a spruce and beech forest in Denmark that mean spatial variability of N_2O emissions was 384% and 285%, respectively (Ambus and Christensen 1995). These marked differences were due to the high spatial variability of key variables (e.g. soil moisture, soil temperature) or the combination of the key variables regulating N-trace gas flux.

Most of the investigations about spatial variations of N-trace gas fluxes at the soil/atmosphere interface of forest ecosystems were performed with regard to topogradients. It was found that forests in downslope position revealed higher N_2O emissions as compared to forests in midslope and tophill position (Ambus and Christensen 1995; Bowden et al. 1992). These differences in magnitude of N_2O emissions could be explained most likely by differences in soil hydrological conditions and differences in soil temperature. However, for spruce forests in Scotland N_2O emissions were higher at high altitude stands as compared to low altitude stands (Macdonald et al. 1997). This unexpected result was most likely due to the fact that the high altitude stands were exposed to higher amounts of atmospheric N input as compared to the low altitude stands, thus N-deposition becoming a stronger modulating factor than soil temperature and soil moisture Macdonald et al. 1997).

For tropical forest ecosystems spatial variability of N-trace gas fluxes seems to be not as pronounced as found for temperate forests. Spatial variability of N_2O emission rates varied between 94%-196% (Vitousek et al. 1989), 100%-118% (Verchot et al. 1999), and 14%-132% (Breuer et al. 2000). Nevertheless – as mentioned for temperate forest in case of N_2O emissions – the variability of N_2O and NO emission rates from tropical forest soils was also related to topography, i.e. higher flux rates at downslope as compared to hilltop positions (Le Roux et al. 1995; Reiners et al. 1998), primarily due to higher soil moisture content and nutrient availability at the downslope areas. Besides topographical position differences in water filled pore space and C/N ratio were found to be important modulators of N_2O emission from tropical rain forest soils (Breuer et al. 2000).

More recent investigations revealed that considering soil areas in direct vicinity to tree trunks is of crucial importance to come up with a more reliable estimate about the source strengths of forest soils for atmospheric N trace gases on an ecosystem scale. Neglecting these areas especially of deciduous forest soils was shown to result in a significant underestimation of the source strength of the forest soil for atmospheric N_2O and NO by approx. 20% and 40%, respectively (Butterbach-Bahl et. al. 2002, Gasche and Papen 2002). Also Pilegaard et al. (1999) reported for a N-affected Norway spruce forest in Denmark that measuring plots near tree trunks exhibited markedly higher NO emissions as compared to plots in distance to tree trunks. The observed higher N-trace gas fluxes from soil areas in direct vicinity to tree trunks are most likely due to canopy induced higher water inputs (i.e. for deciduous trees) as well as nutrient inputs (for both, deciduous and coniferous trees) into the soil areas close to the stems (Pilegaard et al. 1999; Gasche and Papen 2002) resulting in enhanced microbial N-turnover rates (Chang and Matzner 2000). Furthermore, recently published results from an intense study in pine forests of the Northeastern Lowlands of Germany have revealed that micro-scale spatial variability of N_2O and NO flux rates from the soils was highly dependent on the micro-scale patterns of ground vegetation as well as on the canopy structure above the measuring plots (Jenssen et al. 2000, 2002).

2.2   Temporal variations

Though most studies performed in temperate forest soils reported diel and daily variations of N_2O flux rates (e.g. Bowden et al. 1990, 1993; Brumme and Beese 1992) as well as of NO flux rates (e.g. Williams and Fehsenfeld 1991; Pilegaard et al. 1999), a clear diel pattern of NO and N_2O emissions could not be demonstrated to exist. The latter finding did also apply for NO fluxes in an Amazon rain forest in Brazil (Bakwin et al. 1990). However, for a Miombo forest in Southern Africa a diel pattern of NO emissions was demonstrated to exist and followed the daily variation of soil temperature as long as soil moisture was not limiting (Meixner et al. 1997).

Besides diel and daily fluctuations of N_2O, NO and NO_2 flux rates, many forest ecosystems have been demonstrated to exhibit pronounced seasonal as well as interannual variations of N-trace gas flux rates (e.g. Macdonald et al. 1997; Papen and Butterbach-Bahl 1999; Gasche and Papen 1999). Seasonal variations in N_2O and NO emissions from temperate forest soils are mainly controlled by seasonal changes in soil water and temperature (Pilegaard et al. 1999; Skiba et al. 1999). Most studies performed in temperate forests revealed that N_2O as well as NO and NO_2 fluxes from the soils were highest during spring and summer and lowest during late autumn and winter (Brumme and Beese 1992; Valente and Thornton 1993; Thornton et al. 1997; Henrich and Haselwandter 1997; Corre et al. 1999; Gasche and Papen 1999; Papen and Butterbach-Bahl 1999; Brumme et al. 1999; van Dijk and Duyzer 1999; Hahn et al. 2000; Butterbach-Bahl et al. 2002a; Gasche and Papen 2002). On the other hand, temperate forests which are affected by summer droughts were reported to show lowest N_2O emission rates during summer months or did even function as a sink for atmospheric N_2O during such periods (Schmidt et al. 1988; Bowden et al. 1990; Ambus and Christensen 1995). However, Brumme et al. (1999) and Bowden et al. (1993) have described temperate forest ecosystems not exhibiting any seasonal pattern of N_2O emissions at all. The reason, why there are forest ecosystems exhibiting a pronounced seasonal pattern of N_2O emission and others lacking such a pattern remains to be clarified.

Episodes of high N_2O emissions were observed during extended frost periods and thawing of frozen forest soils in early spring contributing 50% or even > 70% to the total annual N_2O release from the soils (Goodroad and Keeney 1984; Schmidt et al. 1988; Tietema et al. 1991; Papen et al. 1993; Brooks et al. 1997; Papen and Butterbach-Bahl 1999; Teepe et al. 2000). The magnitude of N₂O release during such periods seems to be dependent on the duration and intensity of the frost period (Papen and Butterbach-Bahl 1999; Butterbach-Bahl et al. 2002a). Both microbial activity (Christensen and Tiedje 1990; Christensen and Christensen 1991, Papen et al. 1993, Flessa et al. 1995; Papen and Butterbach-Bahl 1999) as well as release of entrapped N₂O produced in unfrozen parts of the soil are discussed (Tietema et al. 1991; Papen et al. 1993; Burton and Beauchamp 1994) to be responsible for the peak emissions observed during frost periods. For the Höglwald Forest (near Munich, Germany) it was demonstrated by Papen and Butterbach-Bahl (1999) that the enormous N₂O emissions observed during the frost period were almost exclusively due to actual microbial activity (tight coupling of ammonification, nitrification and denitrification) within the upper organic layer of the soils (spruce and beech) taking place even at temperatures below the freezing point and were not primarily due to physical release of trapped N₂O after melting of the ice barrier as had been suggested before (Bremner et al. 1980; Goodroad and Keeney 1984). The sources for this enormous N₂O production and release were the easily decomposable substrates derived from the dead microbial biomass produced during long lasting frost periods, which was used by the surviving microbes for N₂O production. It is noteworthy that such dramatic frost effects on N₂O emissions from forest soils could not be demonstrated for NO (Gasche and Papen 1999). Since during frost periods both nitrification as well as denitrification were proven to be active (Papen and Butterbach-Bahl 1999), it is most likely that the NO that had been produced via these processes was immediately consumed by denitrifiers and eventually released in form of N₂O from the organic layer of the soil into the atmosphere.

Long term continuous measurements over multiple years of N-trace gas fluxes from soils of a temperate spruce and beech forest ecosystem (The Höglwald experimental site, Bavaria, Germany) have revealed that enormous interannual variations of N₂O (factor up to 7.3) and NO (factor up to 1.6) emissions did occur (Papen and Butterbach-Bahl 1999; Gasche and Papen 1999, 2002; Butterbach-Bahl et al. 2002a). Brumme and Beese (1992) and Brumme (1995) reported for a beech stand at the Solling, Germany, variations in annual mean N₂O emissions between different observation years up to a factor of 11. From these results it was concluded that several years of data and year-round measurements are required to produce more reliable estimates of annual N₂O flux (Groffman et al. 2000).

With respect to tropical forest ecosystems it was generally found that N₂O emission rates during the wet season were markedly higher as compared to the dry season (Vitousek et al. 1989; Garcia Méndez et al. 1991; Steudler et al. 1991; Davidson et al. 1993; Keller and Reiners 1994; Donoso et al. 1993; Verchot et al. 1999; Breuer et al. 2000). However, highest N$_2$O emission rates were observed during the transition period from the wet season to the dry season at tropical rain forests of the Atherton Tablelands in Australia (Breuer et al. 2000). As for N$_2$O fluxes, most researchers reported highest NO emissions during the wet season (Davidson et al. 1991, 1993; Serça et al. 1994, 1998), though there are also reports about markedly higher NO emission rates during the dry rather than the wet season (Meixner et al. 1997; Verchot et al. 1999).

3.   ABIOTIC FACTORS INFLUENCING N-TRACE GAS EXCHANGE

The N$_2$O and NO emissions observed are the result of biological and physico-chemical processes in the soils (see Chapter 1.1) which themselves are regulated by important abiotic environmental factors like e.g. soil temperature, soil moisture, pH value as well as N-status of the soil (e.g. N-concentration, N-input, C/N ratio).

3.1   Soil temperature and soil moisture

For both temperate and tropical forest soils most researchers found a strong positive relationship between soil temperature and magnitude of N$_2$O as well as NO emission rates (Johansson 1984; Williams et al. 1988; Sitaula and Bakken 1993; Macdonald et al. 1997; Thornton et al. 1997; Gasche and Papen 1999; Papen and Butterbach-Bahl 1999; Pilegaard et al. 1999; van Dijk and Duyzer 1999). However, Bowden et al. (1991), Kaplan et al. (1988) and Johansson et al. (1988) reported that N trace gas emissions were independent from changes in soil temperature indicating that other environmental factors, e.g. soil moisture, were stronger modulators for N$_2$O and NO emission rates than soil temperature. For example Pilegaard et al. (1999) could not detect any correlation between soil temperature and NO emission rates during a very dry year, in which soil moisture was the limiting factor for NO production and NO emission. Interestingly, a very pronounced temperature dependency was observed in the following year at the same site, when rainfall was more evenly distributed as compared to the dry year. Also Papen and Butterbach-Bahl (1999) and Gasche and Papen (1999) have demonstrated that the effects of soil temperature and soil moisture can not be seen independently of each other. These authors have shown that best correlations between soil temperatures and N$_2$O and NO emissions were found when water filled pore space (WFPS) was in an optimal range. Temperature simulation experiments in a Northern Hardwood forest of American Beech and sugar maple, in which the soil was artificially heated in the field, revealed that soil water content was a stronger modulator of N$_2$O flux than soil temperature (McHale et al. 1998).

In accordance with these results, most researchers found that N$_2$O flux in temperate, boreal and tropical forest ecosystems is positively correlated to precipitation and WFPS (Goodroad and Keeney 1984; Sitaula and Bakken 1993; Keller and Reiners 1994; Riley and Vitousek 1995; Corre et al. 1999; Verchot et al. 1999). However, marked differences seem to exist between different forest types: while at a beech stand up to 58% of variation in N$_2$O flux could be explained by variations in WFPS, at a spruce stand only 4.7% of this variation could be related to changes in WFPS (Papen and Butterbach-Bahl 1999). On the other hand, there are some reports both for tropical forest ecosystems (Matson et al. 1990) as well as temperate forests (Bowden et al. 1991) according to which no obvious relationship between soil moisture and magnitude of N$_2$O emissions could be detected indicating that other factors besides temperature and soil moisture, as e.g. N-limitation and, in consequence, restricted nitrification/denitrification activity could be responsible for lack of variation in N$_2$O flux observed.

Many authors have reported enormous pulses of NO and N$_2$O emissions from tropical forest soils after rewetting of the soil at the transition period between dry season to wet season with the onset of rainfall indicating that lack of water limited microbial production of NO and N$_2$O (Davidson et al. 1991, 1993; Meixner et al. 1997; Serça et al. 1998).

With regard to the N$_2$O/NO emission ratio from temperate as well as tropical forest soils most results published in the literature are in excellent agreement with the results reported by Davidson (1993) according to which the production ratio of N$_2$O to NO increases with increasing soil water content: while at values of WFPS < 60% NO was the predominant N-trace gas produced, at WFPS values > 70% N$_2$O production dominates (Riley and Vitousek 1995; Gasche and Papen 1999; Papen and Butterbach-Bahl 1999; van Dijk and Duyzer 1999; Verchot et al. 1999).

3.2   Soil pH value

There are remarkable conflicting results reported in the literature regarding the effect of soil pH on magnitude of both in situ N$_2$O as well as in situ NO emission rates from forest soils. These differences seem to be related to whether the pH value of the soil investigated reflects the natural development of the soil or the soil pH is changed by anthropogenic manipulations as e.g. by surface application of lime to acid forest soils.

For different mixed coniferous forests in the San Bernardino Mountains in Southern California, USA, a negative correlation was found between soil pH value and magnitude of N_2O and NO emission, i.e. the lower the pH value of the forest soil the higher the NO and N_2O emission rates observed (Fenn et al. 1996). A comparable negative correlation was described by Sitaula and Bakken (1993) for a spruce forest soil in Europe. In accordance with these results, Schmidt et al. (1988) described for different deciduous forests in Germany that sites characterized by higher pH values showed significantly lower mean N_2O emission rates as compared to those sites characterized by lower pH values. Also Struwe and Kjøller (1994) studying different beech forest soils in Denmark with different pH values reported highest N_2O emission rates at pH 5.5 and lowest rates at pH 7.2. These results can be interpreted that with increasing soil pH nitrous oxide reductase, which is sensitive to low pH values, is more active and, in consequence, less N_2O is released via denitrification because of enhanced N_2O reduction to molecular dinitrogen (N_2).

However, totally different results were reported for forest soils, if the pH value in the uppermost organic layer was increased artificially by application of lime to the soil surface, a widely used forest management practice at least within Europe to counteract the problem of increasing acidification of forest soils due to acid rain and increasing atmospheric N-input. Papen and Butterbach-Bahl (1999) did observe a 1.4-fold increase of N_2O emission (long year annual mean increase) at the Höglwald experimental site from soil under spruce which had received lime (surface application of 4 t dolomite ha^{-1}) as compared to a untreated spruce site in direct vicinity to the limed site. In accordance with these results Borken and Brumme (1997) observed a trend to enhanced N_2O emissions after surface application of lime to a spruce forest using 6 lime ha^{-1}. However, a reduction in N_2O emission was observed by these authors from soil under spruce if extremely high doses of lime were applied (> 40 t ha^{-1}) and also for a beech stand that had received 30 t lime ha^{-1} indicating that different amounts of lime applied will lead to different effects on N_2O emission from the soil.

Moreover, in contrast to N_2O emission at the Höglwald Forest (see above) a 1.9-fold decrease in NO emission (long year annual mean) was observed at the limed spruce site as compared to the untreated spruce control site (Gasche and Papen 1999). For NO_2, no significant differences in magnitude of deposition rates were detectable between both sites (Gasche and Papen 1999). In an earlier study at the Höglwald Forest, Papke and Papen (1998) also have described a significant reduction of NO emission after liming of a spruce forest soil as compared to a spruce soil that had not received lime. These results indicate that liming of a coniferous forest soil with a dose typically used within Europe (4 t ha^{-1}) will most likely result in significantly enhanced N_2O emission, but simultaneously in significantly reduced NO emission rates from soil under spruce. Since at the Höglwald experimental site liming of the soil did not change the relative contribution of nitrification and denitrification for both NO as well as N_2O emission, Gasche and Papen (1999) concluded that the ratio of NO/N_2O produced during both nitrification and denitrification was shifted in favour of N_2O.

3.3   Inorganic nitrogen

3.3.1   Soil ammonium and nitrate concentration

Since ammonium and nitrate are the key substrates for the soil microbial processes nitrification and denitrification responsible for N_2O and NO production and emission, a strong positive correlation between soil ammonium and nitrate concentrations and magnitude of N-trace gas emission from the soil should be expected. However, though most reports describe such a relationship (Williams et al. 1988; Matson et al. 1990; Keller and Reiners 1994; Macdonald et al. 1997; Skiba et al. 1998a, b; Corredor et al. 1999; Speir et al. 1999; Hahn et al. 2000; Erickson et al. 2001), there are other reports who could not detect any relationship between these parameters (Bowden et al. 1991; Corre et al. 1999). It is noteworthy that in many cases in which a positive relationship between soil inorganic nitrogen concentrations and N-trace gas fluxes was detected, this relationship, though significant, remained only weak (e.g. Matson and Vitousek 1990; Bowden et al. 1991; Ambus and Christensen 1995; Brumme 1995; Henrich and Haselwandter 1997; Gasche and Papen 1999; Papen and Butterbach-Bahl 1999). These results indicate that actual soil pools of ammonium and nitrate might be of minor significance for the prediction of N-trace gas fluxes (Skiba et al. 1998b) and that actual microbial N-turnover rates might be better predictors of in situ N-trace gas flux (Gasche and Papen 1999; Papen and Butterbach-Bahl 1999; see also Section 4.1).

3.3.2   Atmospheric N-input

Pristine forest ecosystems did retrieve inorganic nitrogen in the form of ammonium and nitrate almost exclusively from litter and soil decomposition i.e. via the microbial processes of mineralization and nitrification. As far as temperate forests are concerned, those forest ecosystems are primarily limited in growth by the availability of inorganic nitrogen. There is growing evidence that such N-limited forests are only weak sources of both N_2O as well as NO because of strong competition between soil microbes and tree roots for the available soil inorganic nitrogen for growth and, thus, leaving less inorganic nitrogen available for gaseous losses in form of N_2O and NO via nitrification and denitrification (e.g. Castro et al. 1993; Rennenberg et al.

1998, 2001; Papen et al. 2001; Butterbach-Bahl et al. 1998). Such forest soils can function – at least seasonally – even as net sinks for atmospheric N₂O (Bowden et al. 1990, 1991, 1993; Castro et al. 1993; Butterbach-Bahl et al. 1998; Skiba et al. 1999; Papen et al. 2001). However, this natural situation has changed dramatically due to anthropogenic activities, i.e. fossil fuel combustion and intensive agriculture, leading to enhanced atmospheric N-deposition into these historically N-limited ecosystems and thus, changing these systems from N-limited into N-saturated and, in consequence, besides NO₃⁻ leaching, into systems of increasing N₂O and NO release from soils (e.g. Aber et al. 1989, 1998; Fenn et al. 1996, 1998).

The effect of elevated levels of N-deposition into forests on magnitude of N-trace gas emissions from the soils has been studied in the past mostly by simulation experiments, i.e. artificial application of inorganic nitrogen fertilizers onto the soil. Most of these studies revealed a significant positive effect of N-application on the magnitude of both N₂O as well as NO released from the soil (Johansson 1984; Bowden et al. 1991; Brumme and Beese 1992; Matson et al. 1992; Sitaula et al. 1995; Rennenberg et al. 1998; van Dijk and Duyzer 1999; Papen et al. 2001). The results from these studies strongly indicated that chronical atmospheric N-input – simulated in these studies by N-application – into naturally N-limited ecosystems like forests will lead in the long run to a change of these ecosystems from N-limited to N-saturated ecosystems and, thus, into systems

– of weakened competition between vegetation and soil microbes for N-substrates,
– of increased availability of N for nitrification and denitrification and, as a consequence
– of increasing N₂O and NO production and emission into the atmosphere.

However, Skiba et al. (1999) have pointed out that "the history of N inputs to a site will precondition the response to subsequent N inputs. Experimental applications of elevated atmospheric N to a 'pristine' soil may in the short term (several years) underestimate the likely impact of N-deposition on trace gas emissions". Thus, the simulation experiments might not an appropriate approach for prediciting the effect of long term N-deposition on magnitude of N-trace gas emissions.

The existence of a close positive relationship between amount of N-deposition into forests and magnitude of N₂O and NO release has also been demonstrated in in situ studies. E.g. at the Höglwald experimental site, Butterbach-Bahl et al. (1997), Papen and Butterbach-Bahl (1999) and Gasche and Papen (1999) have unequivocally demonstrated for the first time that there was a strong direct positive correlation between the amount of actual N-input into the forest soil via throughfall and the magnitude of actual in situ N₂O and NO emission rates. Furthermore, in a study in which different pine forest ecosystems in the Northeastern Lowlands of Germany exposed for decades to different levels of atmospheric N-input were compared, a linear positive relationship between amount of in situ N-input and magnitude of in situ N_2O and NO emission could be demonstrated to exist (Butterbach-Bahl et al. 2002b). As an example in Figure 1 the positive linear relationship ($r^2 = 0.85$; $p < 0.01$) between long-year annual mean in situ N-input and magnitude of in situ NO emission rates for different coniferous and deciduous sites in Germany is shown (compilation of data from: Gasche and Papen 1999; Butterbach-Bahl et al. 2002b). A linear positive relationship was also demonstrated to exist for different forests in Scotland exposed to elevated levels of atmospheric N-input from a poultry farm (Skiba et al., 1998a). Skiba et al (1999) found for a woodland in

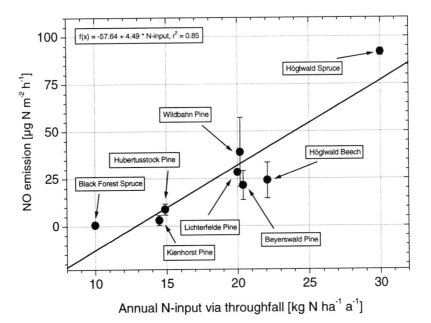

Figure 1. Correlation analysis between magnitude of in situ NO emission rates from different coniferous and deciduous forest ecosystems in Germany and long year annual mean in situ atmospheric N-input by wet deposition (measured in the throughfall) into the forest soils.

Scotland that in situ nitrogen deposition did lead to increased N_2O as well as NO emissions. It is noteworthy that in this study it was found that 6% and 14% of the ammonia deposited to the soil was re-emitted in the form of N_2O-N and NO-N, respectively. These results are in good agreement with those described by Papen and Butterbach-Bahl (1999) who reported that 0.5% (spruce) and 10% (beech) from the actual in situ N-input was released again from the soil into the atmosphere in the form of N_2O-N; for NO-N these figures were 17% (spruce) and 7% (beech), respectively (Gasche and Papen 1999).

The possible consequences of increasing atmospheric N-input into tropical forest ecosystems on nitrogen cycling and magnitude of N_2O and NO emission is discussed in detail in Chapter 6.1.

4.   BIOTIC FACTORS INFLUENCING N-TRACE GAS EXCHANGE

In this section we concentrate on biotic factors affecting the magnitude of N_2O and NO emission from forest soils. Besides the soil microbial processes of nitrification and denitrification, most recently results obtained about additional biotic factors affecting magnitude of N-trace gas emissions, like forest type (tree species) and ground vegetation, are especially highlighted.

4.1   Denitrification rates/nitrification rates

The most important microbial processes of N_2O and NO production, nitrification and denitrification, are discussed in detail in Chapter 1.1. The present section focuses on the importance/contribution of these processes in different forest ecosystems for the actual N_2O and NO release from the soils into the atmosphere.

Skiba and Smith (2000) and Goodroad and Keeney (1985) have pointed out that in most forest soils the prevalence of nitrification or denitrification as the main source of N_2O and NO is not static and can switch very rapidly, as the soil aeration state within the biologically active site changes due to e.g. rainfall or increased O_2 demand caused by the presence of easily mineralizable organic matter. Though denitrification has been demonstrated to be a very important source for N_2O and NO from many forest soils (e.g. Kaplan et al. 1988; Keller et al. 1988; Bakwin et al. 1990; Davidson et al. 1993; Keller and Reiners 1994; Serça et al. 1994), there is increasing evidence that nitrification is the prevailing process of N_2O and NO emission at least from well aerated temperate forest soils during most parts of the year (e.g. Castro et al. 1993; Sitaula and Bakken 1993; Fenn et al. 1996). In a long term study at the Höglwald experimental site Gasche and Papen (1999) and Papen and Butterbach-Bahl (1999) found that the annual mean contribution of nitrification to actual NO and N_2O emission from soils of a spruce and beech stand was 65% and 70%, respectively, whereas the mean contribution of denitrification was 35% and 30%, respectively. At these sites a strong positive correlation between in situ net nitrification rates and in situ NO and N_2O emission rates was demonstrated to exist. In Figure 2 the correlation analysis between in situ NO and N_2O emission rates and in situ net nitrification rates for the soil under beech at the Höglwald experimental site (Bavaria, Germany) is shown. Variations in net nitrification rates could explain between 68% and 62% of variations in in situ NO and N_2O emission rates.

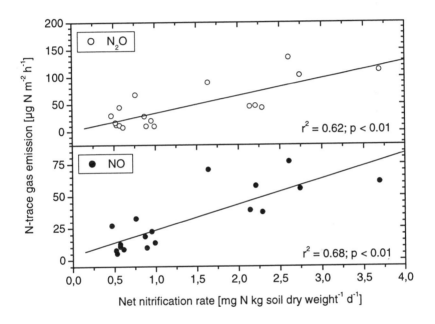

Figure 2. Correlation analysis between in situ N_2O and NO emission rates and in situ net nitrification rates in the uppermost organic layer of the soil at the beech site of the Höglwald Forest for September 1994 to July 1996 (Bavaria, Germany).

Only during frost and thawing periods of the soil, the enormous N_2O releases from the soils into the atmosphere were almost exclusively due to denitrification activity (Papen and Butterbach-Bahl 1999; see Section 2.2)

Also for a tropical forest ecosystems in Puerto Rico 71% of variation of NO and N_2O emission rates could be explained by variations in net nitrification rates (Erickson et al. 2001). For a tropical rain forest in Brazil this value was 70% (Matson et al. 1990). Riley and Vitousek (1995) have reported for a Hawaiian montane forest that the sum of N_2O plus NO emissions were almost linearly related to gross nitrification rates and that there was a strong positive relationship between N₂O emission rates and net nitrification rates.

Serça et al. (1994) reported for a tropical forest in Africa that the dominant source for NO emission was nitrification during the dry season, whereas it was denitrification during the wet season. With respect to N₂O the authors assumed that the primary source for N₂O emissions was denitrification. Similar results were obtained by Keller and Reiners (1994) for a wet tropical forest in the Lowlands of Costa Rica and by Davidson et al. (1993) for a deciduous forest in Mexico.

4.2   Tree species and forest type

It has been demonstrated by different authors (e.g. Slemr et al. 1984; Conrad and Seiler 1985; Goodroad and Keeney 1985; Schmidt et al. 1988; Skiba et al. 1993; Borken and Brumme 1997; Dong et al. 1998; Papke and Papen 1998; Gasche and Papen 1999; Papen and Butterbach-Bahl 1999) that the soil horizon contributing most to the actual N₂O and NO emission from temperate coniferous forest soils into the atmosphere is the uppermost organic layer. Papen and Butterbach-Bahl (1999) mentioned that approx. 70% of the actual N₂O emission from the soil of a spruce stand originated from the organic layer, whereas the contribution of the mineral soil was approx. 30%. These findings are in line with results presented by Borken and Brumme (1997), who estimated the contribution of the organic layer of a spruce stand to the actual N₂O flux at 73%. In contrast, for a deciduous forest (beech forest) the contribution of the mineral soil to the actual N₂O emission was approx. 78%, while the organic layer contributed only 22% (Papen and Butterbach-Bahl 1999). Removal of the humus layer in forest stands dominated by beech and oak and by pure beech, respectively, did lead to a reduction of approx. 50% in N₂O emission indicating that the organic layer and the mineral soil contributed to the same extent to the flux into the atmosphere (Dong et al. 1998; Borken and Brumme 1997). Also for tropical rain forests in Australia the mineral soil was identified to represent the most important soil layer for N₂O release into the atmosphere (Breuer et al. 2000).

These findings indicate that forest type itself has a huge impact on the soil layer, contributing most to the actual N₂O flux from the soil into the atmosphere: while in coniferous forests the most important soil horizon is the organic layer; in deciduous forests the mineral soil is at least as important or even more important than the organic layer for the in situ N₂O flux into the atmosphere. While for deciduous forests data are lacking, the most important soil layer for NO released into the atmosphere from temperate coniferous forest soils (up to 86%) was identified to be the uppermost organic layer (Skiba et al. 1993; Papke and Papen 1998; Gasche and Papen 1999).

There is growing evidence that the tree species itself has not only a huge impact on the soil layer the N-trace gases stem from, but also on the N-trace gas species which is predominantly emitted from the soil into the atmosphere (Butterbach-Bahl et al. 1997; Gasche and Papen 1999, 2002; Papen and Butterbach-Bahl 1999; Butterbach-Bahl et al. 2002a). At the Höglwald experimental site, where beech and spruce stock on identical soil type and are exposed to identical climatic conditions, it was found that N_2O emissions from soils under beech were dramatically higher than N_2O emissions from soils under spruce (Papen and Butterbach-Bahl 1999; Butterbach-Bahl et al. 2002a). For NO emissions at the same sites the opposite applied: emissions under spruce were dramatically higher as compared to those under beech (Gasche and Papen, 1999, 2002). Thus, the tree species influences the predominant N-trace gas emitted from the soil. As a consequence, forest type (coniferous or deciduous) or tree species has a strong differentiating impact on the ratio of NO to N_2O emitted from the soil. This ratio was found to be >> 1.0 for spruce, 0.9 for pine and << 1.0 for beech (Butterbach-Bahl et al. 1997). These results are in excellent agreement with reports by van Dijk and Duyzer (1999) who also observed significantly lower emission rates (by 60%) from soil under beech as compared to soil under Douglas fir.

These results strongly indicate that the different litter quality produced by different trees (i.e. C/N ratio of the litter) is a strong modulator for the magnitude of N_2O and NO production and emission (Papen and Butterbach-Bahl 1999). This conclusion is supported by the results obtained by other reseachers: e.g. Erickson et al. (2001) could demonstrate for a humid lower montane subtropical forests in Puerto Rico that the magnitude of N_2O and NO emissions observed increased with decreasing C/N ratio of the litter layer. Also Menyailo and Huwe (1999) reported for Siberian forest soils that the C/N ratio was a key parameter that could explain best differences in magnitude of N_2O emission between soils planted to different tree species. The C/N ratio increased and, in consequence, the magnitude of N_2O emission decreased in the following sequence of tree species studied:

birch > aspen > cedar > spruce > pine > larch.

Also forest sites in the tropics dominated by leguminuous trees showed 100-fold higher NO and N_2O emission rates than sites lacking those nitrogen fixing tree species (Erickson et al. 2001). Comparable results were obtained in the temperate zone for an alder plantations compared with a birch plantation stocking on the same soil: N_2O emissions from the soil under alder were more than twice as large as those from soil under birch (Skiba et

al. 1998b) indicating that nitrogen fixation in alder enhances the production
and emission of N_2O.

4.3   Ground vegetation

Besides tree species and forest type, the ground vegetation in forests has
to be considered as an additional biotic factor modulating the magnitude of
N_2O and NO emission from forest soil as well as NO_2 deposition to the
forest floor. However, it has to be mentioned that only a few data are
available on the impact of ground vegetation on the magnitude of NO_2
deposition to the forest floor. Hanson et al. (1989) found that the litter itself
had no or only at least little conductance for NO_2, while the conductance to
bare soil was as great as to the intact forest floor. Papke and Papen (1998)
have shown that the ground vegetation at the Höglwald spruce forest site
(Bavaria, Germany) dominated by *Oxalis acetosella* and mosses did not
contribute to the in situ NO_2 deposition to the forest floor. These results
further support the early findings of Abeles et al. (1971) and Ghiorse and
Alexander (1976) who could not attribute NO_2 deposition to soils to any
biological activity (plants and microorganisms).

Pilegaard et al. (1999) found for a N-affected Norway spruce forest in
Denmark that NO emissions decreased with increasing percentage of moss
cover of the soil. In an early study Johansson (1984) reported for Scots pine
forests in rural Sweden marked differences in magnitude of NO emission
rates between an N-fertilized and N-unfertilized site and concluded that
differences in ground vegetation were responsible for these differences,
since other environmental factors as e.g. soil temperature remained
unchanged between both sites.

Jenssen et al. (2000) and Jenssen et al. (2002) have demonstrated that
ground vegetation type could be used as an indicator for the magnitude of
NO and N_2O emission rates from pine forest soils of the Northeastern
German Lowlands. These authors reported that e.g. with increasing presence
of raspberry – an indicator for increased nitrogen availability – in different
Scots pine forest ecosystems of the Northeastern German Lowlands, there
was a significant increase in magnitude of NO emission rates from the soils.
Furthermore, at these sites typical N_2O and NO emission patterns could be
identified and attributed to specific mosaic-like composition of the ground
vegetation, information that may be useful in future studies for upscaling of
N_2O and NO emission on a regional scale.

5. CONTRIBUTION OF FOREST ECOSYSTEMS OF DIFFERENT CLIMATIC ZONES TO THE GLOBAL SOURCE STRENGTH OF FOREST SOILS FOR ATMOSPHERIC N₂O AND NO

There is general agreement that tropical forest soils are on a global scale very important N_2O sources. A recently published re-estimation of the contribution of tropical forest ecosystems to the global budget of N_2O was 3.55 Tg N_2O-N yr^{-1} (Breuer et al. 2000), which is c. 50% higher as compared to an earlier published estimation of approx. 2.40 Tg N_2O-N yr^{-1} (Matson and Vitousek 1990). This new estimation was based on a much broader data set from field measurements in both tropical forests of the Neotropics as well as outside the Neotropics (for a detailed compilation of individual data sets used for the calculation see: Breuer et al. 2000). As compared to the data sets available for N_2O emissions from tropical forest soils, those for NO emissions from this source remain extremely scarce. In consequence, the published estimations have a very high degree of uncertainty. According to a published global inventory of nitric oxide emissions from soils, the source strength of tropical forest soils (evergreen, deciduous and montane forests) was roughly estimated to be 1.3 Tg NO-N yr^{-1} based on a total of nine estimates (Davidson and Kingerlee 1997). According to Lee et al. (1997), considering a canopy reduction factor, the net release of NO-N from tropical forests (above canopy emission) into the atmosphere is as low as c. 0.5 Tg NO-N yr^{-1}.

Temperate and boreal forests have been considered in the past to represent only weak sources for N_2O as well as NO. The contribution of these forests to the global budget of N_2O was estimated to be approx. 0.5 Tg N_2O-N yr^{-1} (Potter et al. 1996). However, these authors stressed that in this estimation the effect of nitrogen deposition into these forests on the magnitude of N_2O and NO release from the soil had not been considered. Furthermore, the effects of different forest types (deciduous, coniferous) and of frost periods and freezing/thawing cycles on the magnitude of N_2O release from the soils had also not been considered in these estimates. Considering these effects Papen and Butterbach-Bahl (1999) and Gasche and Papen (1999) estimated that the contribution of these forest soils to the global budgets of N_2O and NO is most likely > 1.0 Tg N_2O-N yr^{-1} and 0.3 Tg NO-N yr^{-1}, respectively, the latter estimate being in excellent agreement with that published by Davidson and Kingerlee (1997) for N-affected temperate forests.

6.   CONCLUDING REMARKS

There has been considerable progress within the last decade in our understanding of the biological processes involved in N₂O and NO production and emission from forest soils and about the different abiotic environmental factors modulating these processes and, in consequence, the magnitude of N₂O and NO emission. Multiple regression analysis approaches have been used for identification of the sequence of importance of key parameters influencing the magnitude of N-trace gas flux at individual forest sites and for prediction of N-trace gas flux from these sites. However, such regression models developed for a specific site often failed to precisely predict N-trace gas fluxes if applied to other forest sites. In consequence, process oriented models, which strictly consider the kinetics of the relevant biochemical reactions, are the most promising approach to precisely predict N-trace gas fluxes. Such strictly process driven models are applicable across ecosystems, soil types and eventually climate zones for a more accurate prediction of N-trace gas fluxes and for the establishment of future more reliable estimates of N₂O and NO emissions from forest soils on a local, regional or even global scale. Such models have already been developed (e.g. Parton et al. 1996; Potter et al. 1996; Li et al. 2000) and successfully been used to predict with high accuracy both N₂O as well as NO emissions from different temperate forest soils on a site (Stange et al. 2000) as well as on a regional scale (Butterbach-Bahl et al. 2001).

REFERENCES

Abeles FB, Craker LE, Forrence LE & Leather GR (1971) Fate of air pollutants: removal of ethylene, sulfur dioxide, and nitrogen dioxide by soil. Science 173: 914-916

Aber J, McDowell W, Nadelhoffer K, Magill A, Berntson G, Kamakea M, McNulty S, Currie W, Rustad L & Fernandez I (1998) Nitrogen saturation in temperate forest ecosystems - Hypotheses revisited. BioScience 48: 921-934

Aber JD, Nadelhoffer KJ, Steudler P & Melillo JM (1989) Nitrogen saturation in northern forest ecosystems; excess nitrogen from fossil fuel combustion may stress the biosphere. BioScience 39: 378-386

Ambus P & Christensen S (1995) Spatial and seasonal nitrous oxide and methane fluxes in Danish forest-, grassland-, and agroecosystems. J Environ Qual 24: 993-1001

Bakwin PS, Wofsy SC, Fan S-M, Keller M, Trumbore SE & Da Costa JM (1990) Emission of Nitric Oxide (NO) From Tropical Forest Soils and Exchange of NO Between the Forest Canopy and Atmospheric Boundary Layers. J Geophys Res 95: 755-764

Borken W & Brumme R (1997) Liming practice in temperate forest ecosystems and the effects on CO₂, N₂O and CH₄ fluxes. Soil Use Manage 13: 251-257

Bowden RD, Melillo JM, Steudler PA & Aber JD (1991) Effects of nitrogen additions on annual nitrous oxide fluxes from temperate forest soils in the Northeastern United States. J Geophys Res 96: 9321-9328

Bowden RD, Castro MS, Melillo JM, Steudler PA & Aber JD (1993) Fluxes of greenhouse gases between soils and the atmosphere in a temperate forest following a simulated hurricane blowdown. Biogeochem 21: 61-71

Bowden RD, Steudler PA, Melillo JM & Aber JD (1990) Annual nitrous oxide fluxes from temperate forest soils in the Northeastern United States. J Geophys Res 95: 13997-14005

Bowden WB, MxDowell WH, Asbury CE & Finley AM (1992) Riparian nitrogen dynamics in two geomorphologically distinct tropical rain forest watersheds: nitrous oxide fluxes. Biogeochem 18: 77-99

Bremner JM, Robbins SG & Blackmer AM (1980) Seasonal variability in emission of nitrous oxide from soil. Geophys Res Lett 7: 641-644

Breuer L, Papen H & Butterbach-Bahl K (2000) N₂O emission from tropical forest soils of Australia. J Geophys Res 105: 26353-26367

Brooks PD, Schmidt SK & Williams MW (1997) Winter production of CO₂ and N₂O from alpine tundra: environmental controls and relationship to inter-system C and N fluxes. Oecologia 110: 403-413

Brumme R & Beese F (1992) Effects of liming and nitrogen fertilization on emissions of CO₂ and N₂O from a temperate forest. J Geophys Res 97: 851-858

Brumme R (1995) Mechanisms of carbon and nutrient release and retention in beech forest gaps: III. Environmental regulation of soil respiration and nitrous oxide emissions along a microclimatic gradient. Plant Soil 168-169: 593-600

Brumme R, Borken W & Finke S (1999) Hierarchical control on nitrous oxide emission in forest ecosystems. Glob Biogeochem Cycle 13: 1137-1148

Burton DL & Beauchamp EG (1994) Profile nitrous oxide and carbon dioxide concentrations in a soil subject to freezing. Soil Sci Soc Am J 58: 115-122

Butterbach-Bahl K, Gasche R, Breuer L & Papen H (1997) Fluxes of NO and N₂O from temperate forest soils: impact of forest type, N deposition and of liming on the NO and N₂O emissions. Nutr Cycl Agroecosyst 48: 79-90

Butterbach-Bahl K, Gasche R, Huber CH, Kreutzer K & Papen H (1998) Impact of N-input by wet deposition on N-trace gas fluxes and CH₄-oxidation in spruce forest ecosystems of the temperate zone in Europe. Atmos Environ 32: 559-564

Butterbach-Bahl K, Rothe A & Papen H (2002a) Effect of tree distance on N₂O and CH₄-fluxes from soils in temperate forest ecosystems. Plant Soil (in press)

Butterbach-Bahl K, Breuer L, Gasche R, Willibald G & Papen H (2002b) Exchange of trace gases between soils and the atmosphere in Scots pine forest ecosystems of the North Eastern German Lowlands 1. Fluxes of N₂O, NO/NO₂ and CH₄ at forest sites with different N-deposition. For Ecol Manage (in press)

Castro MS, Steudler PA, Melillo JM, Aber JD & Millham S (1993) Exchange of N₂O and CH₄ between the atmosphere and soils in spruce-fir forests in the northeastern United-States. Biogeochem 18: 119-135

Chang SC & Matzner E (2000) Soil nitrogen turnover in proximal and distal stem areas of European beech trees. Plant Soil 218: 117-125

Christensen S & Tiedje JM (1990) Brief and vigorous N₂O production by soil at spring thaw. J Soil Sci 41: 1-4

Christensen S & Christensen BT (1991) Organic matter available for denitrification in different soil fractions: effect of freeze/thaw cycles and straw disposal. J Soil Sci 42: 637-647

Conrad R (1996) Metabolism of nitric oxide in soil and soil microorganisms and regulation of flux in to the atmosphere. In: Murrell JC & Kelly DP (eds) Microbiology of Atmospheric Trace Gases: Sources, Sinks and Global Processes. pp. 167-203, Springer Verlag, Berlin, Germany

Conrad R & Seiler W (1985) Localization of microbial activities relevant to the emission of nitrous oxide from the soil into the atmosphere. Soil Biol Biochem 17: 893-895

Corre MD, Pennock DJ, Van Kessel C & Elliott DK (1999) Estimation of annual nitrous oxide emissions from a transitional grassland-forest region in Saskatchewan, Canada. Biogeochem 44: 29-49

Corredor JE, Morell JM & Bauza J (1999) Atmospheric nitrous oxide fluxes from mangrove sediments. Marine Pollut Bull 38: 473-478

Davidson EA, Vitousek PM, Matson PA, Riley R, García-Méndez G & Maass JM (1991) Soil emissions of nitric oxide in a seasonally dry tropical forest of México. J Geophys Res 96: 15439-15445

Davidson EA, Matson PA, Vitousek PM, Riley R, Dunkin K, García-Méndez G & Maass JM (1993) Processes regulating soil emissions of NO and N_2O in a seasonally dry tropical forest. Ecology 74: 130-139

Davidson EA (1992) Sources of nitric oxide and nitrous oxide following wetting of dry soil. Soil Sci Soc Am J 56: 95-112

Davidson EA (1993) Soil water content and the ratio of nitrous oxide to nitric oxide emitted from soil. In: Oremland RS (ed) Biogeochemistry of Global Change: Radiatively Active Trace Gases, pp 369-386. Chapman & Hall, New York, U.S.A.

Davidson EA & Kingerlee W (1997) A global inventory of nitric oxide emissions from soils. Nutr Cycl Agroecosyst 48: 37-50

Dong Y, Scharffe D, Lobert JM, Crutzen PJ & Sanhueza E (1998) Fluxes of CO_2, CH_4 and N_2O from a temperate forest soil: the effects of leaves and humus layers. Tellus 50B: 243-252

Donoso L, Santana R & Sanhueza E (1993) Seasonal variation of N_2O fluxes at a tropical savannah site: soil consumption of N_2O during the dry season. Geophys Res Lett 20: 1379-1382

Erickson H, Keller M & Davidson EA (2001) Nitrogen oxide fluxes and nitrogen cycling during postagricultural succession and forest fertilization in the humid tropics. Ecosystems 4: 67-84

Fenn ME, Poth MA & Johnson DW (1996) Evidence for nitrogen saturation in the San Bernardino Mountains in southern California. For Ecol Manage 82: 211-230

Fenn ME, Poth MA, Aber JD, Baron JS, Bormann BT, Johnson DW, Lemly AD, McNulty SG, Ryan DE & Stottlemyer R (1998) Nitrogen excess in North American ecosystems: Predisposing factors, ecosystem responses, and management strategies. Ecol Appl 8: 706-733

Flessa H, Dörsch P & Beese F (1995) Seasonal variation of N_2O and CH_4 fluxes in differently managed arable soils in southern Germany. J Geophys Res 100: 23115-23124

García-Méndez G, Maass JM, Matson PA & Vitousek PM (1991) Nitrogen transformations and nitrous oxide flux in a tropical deciduous forest in México. Oecologia 88: 362-366

Gasche R & Papen H (1999) A 3-year continuous record of nitrogen trace gas fluxes from untreated and limed soil of a N-saturated spruce and beech forest ecosystem in Germany 2. NO and NO_2 fluxes. J Geophys Res 104: 18505-18520

Gasche R & Papen H (2002) Spatial variability of NO and NO_2 flux rates from soil of spruce and beech forest ecosystems. Plant Soil (in press)

Ghiorse WC & Alexander M (1976) Effect of microorganisms on the sorption and fate of sulfur dioxide and nitrogen dioxide in soil. J Environ Qual 5: 227-230

Goodroad LL & Keeney DR (1984) Nitrous oxide emission from forest, marsh, and prairie ecosystems. J Environ Qual 13: 448-452

Goodroad LL & Keeney DR (1985) Site of nitrous oxide production in field soils. Biol Fertil Soils 1: 3-7

Groffman PM, Brumme R, Butterbach-Bahl K, Dobbie KE, Mosier AR, Ojima D, Papen H, Parton WJ, Smith KA & Wagner-Riddle C (2000) Evaluating annual nitrous oxide fluxes at the ecosystem scale. Glob Biogeochem Cycle 14: 1061-1070

Hahn M, Gartner K & Zechmeister-Boltenstern S (2000) Greenhouse gas emissions (N$_2$O, CO$_2$ and CH$_4$) from three forest soils near Vienna (Austria) with different water and nitrogen regimes. Bodenkultur 51: 115-125

Hanson PJ, Rott K, Taylor Jr. GE, Gunderson CA, Lindberg SE & Ross-Todd BM (1989) NO$_2$ deposition to elements representative of a forest landscape. Atmos Environ 23: 1783-1794

Henrich M & Haselwandter K (1997) Denitrification and gaseous nitrogen losses from an acid spruce forest soil. Soil Biol Biochem 29: 1529-1537

Jenssen M, Butterbach-Bahl K, Hofmann G & Papen H (2000) Modellierung von Beziehungen zwischen Emission von N-Spurengasen aus Waldböden und den Vegetationsstrukturen in Kiefernökosystemen des nordostdeutschen Tieflandes. Beiträge der Forstwissenschaft und Landschaftsökologie 34: 121-127

Jenssen M, Butterbach-Bahl K, Hofmann G & Papen H (2002) Exchange of trace gases between soils and the atmosphere in Scots pine forest ecosystems of the North Eastern German Lowlands 2. A novel approach to scale up N$_2$O- and NO-fluxes from forest soils by modeling their relationship to vegetation structure. For Ecol Manage (in press)

Johansson C (1984) Field measurements of emission of nitric oxide from fertilized and unfertilized forest soil in Sweden. J Atmos Chem 1: 429-442

Johansson C, Rodhe H & Sanhueza E (1988) Emission of NO in a tropical savannah and a cloud forest during the dry season. J Geophys Res 93: 7180-7192

Kaplan WA, Wofsy SC, Keller M & Da Costa JM (1988) Emission of NO and deposition of O$_3$ in a tropical forest system. J Geophys Res 93: 1389-1395

Keller M, Kaplan WA, Wofsy SC & Da Costa JM (1988) Emissions of N$_2$O from tropical forest soils: Response to fertilization with NH$_4^+$, NO$_3^-$, and PO$_4^{3-}$. J Geophys Res 93: 1600-1604

Keller M & Reiners WA (1994) Soil-atmosphere exchange of nitrous oxide, nitric oxide, and methane under secondary succession of pasture to forest in the Atlantic lowlands of Costa Rica. Glob Biogeochem Cycle 8: 399-409

Le Roux X, Abbadie L, Lensi R & Serça D (1995) Emission of nitrogen monoxide from African tropical ecosystems - control of emission by soil characteristics in humid and dry savannahs of West Africa. J Geophys Res 100: 23133-23142

Lee DS, Köhler I, Grobler E, Rohrer F, Sausen R, Gallardo-Klenner L, Olivier JGJ, Dentener FJ & Bouwman AF (1997) Estimations of global NO$_X$ emissions and their uncertainties. Atmos Environ 31: 1735-1749

Li CS, Aber J, Stange F, Butterbach-Bahl K & Papen H (2000) A process-oriented model of N$_2$O and NO emissions from forest soils: 1. Model development. J Geophys Res 105: 4369-4384

Macdonald JA, Skiba U, Sheppard LJ, Ball B, Roberts JD, Smith KA & Fowler D (1997) The effect of nitrogen deposition and seasonal variability on methane oxidation and nitrous oxide emission rates in an upland spruce plantation and moorland. Atmos Environ 31: 3693-3706

Matson PA, Vitousek PM, Livingston GP & Swanberg NA (1990) Sources of variation in nitrous oxide flux from Amazonian ecosystems. J Geophys Res 95: 16789-16798

Matson PA & Vitousek PM (1990) Ecosystem approach to a global nitrous oxide budget. BioScience 40: 667-672

Matson PA, Gower ST, Volkmann C, Billow C & Grier CC (1992) Soil nitrogen cycling and nitrous oxide flux in a Rocky Mountain Douglas-fir forest: effects of fertilization, irrigation and carbon addition. Biogeochem 18: 101-117

McHale PJ, Mitchell MJ & Bowles FP (1998) Soil warming in a northern hardwood forest: trace gas fluxes and leaf litter decomposition. Can J For Res 28: 1365-1372

Meixner FX, Fickinger T, Marufu L, Serça D, Nathaus FJ, Makina E, Mukurumbira L & Andreae MO (1997) Preliminary results on nitric oxide emission from a southern African savannah ecosystem. Nutr Cycl Agroecosyst 48: 123-138

Menyailo OV & Huwe B (1999) Activity of denitrification and dynamics of N_2O release in soils under six tree species and grassland in central Siberia. J Plant Nutr Soil Sci 162: 533-538

Mosier AR (1989) Chamber and isotope techniques. In: Andreae MO & Schimel DS (eds) Exchange of Trace Gases between Terrestrial Ecosystems and the Atmosphere, pp 175-187. John Wiley & Sons Ltd., Chichester, New York, U.S.A.

Papen H & Butterbach-Bahl K (1999) A 3-year continuous record of nitrogen trace gas fluxes from untreated and limed soil of a N-saturated spruce and beech forest ecosystem in Germany 1. N_2O emissions. J Geophys Res 104: 18487-18503

Papen H, Hellmann B, Papke H & Rennenberg H (1993) Emission of N-oxides from acid irrigated and limed soils of a coniferous forest in Bavaria. In: Oremland RS (ed) Biogeochemistry of Global Change: Radiatively Active Trace Gases, pp 245-260. Chapman & Hall Inc., New York, U.S.A.

Papen H, Daum M, Steinkamp R & Butterbach-Bahl K (2001) N_2O- and CH_4-fluxes from soils of a N-limited and N-fertilized spruce forest ecosystem of the temperate zone. J Appl Bot 75: 159-163

Papke H & Papen H (1998) Influence of acid rain and liming on fluxes of NO and NO_2 from forest soil. Plant Soil 199: 131-139

Parton WJ, Mosier AR, Ojima DS, Valentine DW, Schimel DS, Weier K & Kulmala AE (1996) Generalized model for N_2 and N_2O production from nitrification and denitrification. Glob Biogeochem Cycle 10: 401-412

Pilegaard K, Hummelshoj P & Jensen NO (1999) Nitric oxide emission from a Norway spruce forest floor. J Geophys Res 104: 3433-3445

Potter CS, Matson PA, Vitousek PM & Davidson EA (1996) Process modeling of controls on nitrogen trace gas emissions from soils worldwide. J Geophys Res 101: 1361-1377

Reiners WA, Keller M & Gerow KG (1998) Estimating rainy season nitrous oxide and methane fluxes across forest and pasture landscapes in Costa Rica. Water, Air, and Soil Pollut 105: 117-130

Rennenberg H, Kreutzer K, Papen H & Weber P (1998) Consequences of high loads of nitrogen for spruce (*Picea abies*) and beech (*Fagus sylvatica*) forests. New Phytol 139: 71-86

Rennenberg H, Stoermer H, Weber P, Daum M & Papen H (2001) Competition of spruce trees for substrates of microbial N_2O-production and -emission in a forest ecosystem. J Appl Bot 75: 101-106

Riley RH & Vitousek PM (1995) Nutrient dynamics and nitrogen trace gas flux during ecosystem development in montane rain forest. Ecology 76: 292-304

Schmidt J, Seiler W & Conrad R (1988) Emission of nitrous oxide from temperate forest soils into the atmosphere. J Atmos Chem 6: 95-115

Serça D, Delmas R, Jambert C & Labroue L (1994) Emissions of nitrogen oxides from equatorial rain forest in central Africa: origin and regulation of NO emission from soils. Tellus 46B: 243-254

Serça D, Delmas R, Le Roux X, Parsons DAB, Scholes MC, Abbadie L, Lensi R, Ronce O & Labroue L (1998) Comparison of nitrogen monoxide emissions from several African tropical ecosystems and influence of season and fire. Glob Biogeochem Cycle 12: 637-651

Sitaula BK & Bakken LR (1993) Nitrous oxide release from spruce forest soil: relationships with nitrification, methane uptake, temperature, moisture and fertilization. Soil Biol Biochem 25: 1415-1421

Sitaula BK, Bakken LR & Abrahamsen G (1995) N-fertilization and soil acidification effects on N_2O and CO_2 emission from temperate pine forest soil. Soil Biol Biochem 27: 1401-1408

Skiba U, Smith KA & Fowler D (1993) Nitrification and denitrification as sources of nitric oxide and nitrous oxide in a sandy loam soil. Soil Biol Biochem 25: 1527-1536

Skiba U, Sheppard LJ, Pitcairn CER, Leith I, Crossley A, Van Dijk S, Kennedy VH & Fowler D (1998) Soil nitrous oxide and nitric oxide emissions as indicators of the exceedance of critical loads of atmospheric N deposition in seminatural ecosystems. Environ Pollut 102: 457-461

Skiba U, Sheppard LJ, Pitcairn CER, Van Dijk S & Rossall MJ (1999) The effect of N deposition on nitrous oxide and nitric oxide emissions from temperate forest soils. Water Air Soil Pollut 116: 89-98

Skiba UM, Sheppard LJ, MacDonald J & Fowler D (1998) Some key environmental variables controlling nitrous oxide emissions from agricultural and semi-natural soils in Scotland. Atmos Environ 32: 3311-3320

Skiba U & Smith KA (2000) The control of nitrous oxide emissions from agricultural and natural soils. Chemosphere-Global Change Science 2: 379-386

Slemr F & Seiler W (1984) Field measurements of NO and NO_2 emissions from fertilized and unfertilized Soils. J Atmos Chem 2: 1-24

Speir TW, Townsend JA, More RD & Hill LF (1999) Short-lived isotopic method to measure nitrous oxide emissions from a soil under four low-fertility management systems. Soil Biol Biochem 31: 1413-1421

Stange F, Butterbach-Bahl K, Papen H, Zechmeister-Boltenstern S, Li CS & Aber J (2000) A process-oriented model of N_2O and NO emissions from forest soils 2. Sensitivity analysis and validation. J Geophys Res 105: 4385-4398

Steudler PA (1991) The effects of natural and human disturbances on soil nitrogen dynamics and trace gas fluxes in a Puerto Rican wet forest. Biotropica 4a: 356-363

Struwe S & Kjøller A (1994) Potential for N_2O production from bech (*Fagus sylvatica*) forest soil with varying pH. Soil Biol Biochem 26: 1003-1009

Teepe R, Brumme R & Beese F (2000) Nitrous oxide emissions from frozen soils under agricultural, fallow and forest land. Soil Biol Biochem 32: 1807-1810

Thornton FC, Pier PA & Valente RJ (1997) NO emissions from soils in the southeastern United States. J Geophys Res 102: 21189-21195

Tietemja A, Bouten W & Wartenbergh PE (1991) Nitrous oxide dynamics in an oak-beech forest ecosystem in the Netherlands. For Ecol Manage 44: 53-61

Valente RJ & Thornton FC (1993) Emissions of NO from soil at a rural site in Central Tennessee. J Geophys Res 98: 16745-16753

van Dijk SM & Duyzer JH (1999) Nitric oxide emissions from forest soils. J Geophys Res 104: 15955-15961

Verchot LV, Davidson EA, Cattanio JH, Ackerman IL, Erickson HE & Keller M (1999) Land use change and biogeochemical controls of nitrogen oxide emissions from soils in eastern Amazonia. Glob Biogeochem Cycle 13: 31-46

Vitousek P, Matson P, Volkmann C, Manuel JM & Garcia G (1989) Nitrous oxide flux from dry tropical forests. Glob Biogeochem Cycle 3: 375-382

Williams EJ & Fehsenfeld FC (1991) Measurement of soil nitrogen oxide emissions at three North American ecosystems. J Geophys Res 96: 1033-1042

Williams EJ, Parrish DD, Buhr MP, Fehsenfeld FC & Fall R (1988) Measurement of soil NO_x emissions in Central Pennsylvania. J Geophys Res 93: 9539-9546

Chapter 2.2

CH$_4$

Klaus Butterbach-Bahl
Fraunhofer Institut für Atmosphärische Umweltforschung, Kreuzeckbahnstrasse 19, D-82467 Garmisch-Partenkirchen, Germany

1.   INTRODUCTION

Soils of forest ecosystems can function as net sinks or as net sources for atmospheric methane. The exchange of CH$_4$ between soils and the atmosphere is the net-result of simultaneous production of CH$_4$ in predominantly anaerobic zones of the soil and oxidation of CH$_4$ in predominantly oxic zones of the soil, i.e.:

CH$_4$-flux = CH$_4$-production − CH$_4$-consumption (± CH$_4$-storage)

Production and consumption of CH$_4$ within the soil profile is controlled by several environmental factors, from which O$_2$ and substrate availability, soil properties and climate are the most important.

So far, forest ecosystems in different climates of the world have not been evaluated in total with regard to the magnitude of fluxes of CH$_4$ between soils and the atmosphere. Therefore, a global estimate for the source/ sink strength of forest ecosystems for atmospheric CH$_4$ is not available. This chapter summarizes the knowledge about the processes involved in production, consumption and emission of CH$_4$ with regard to forest soils and gives an overview about published values of CH$_4$-exchange between forest ecosystems and the atmosphere for different climate zones of the world.

R. Gasche et al. (eds.), Trace Gas Exchange in Forest Ecosystems, 141–156.
© 2002 *Kluwer Academic Publishers. Printed in the Netherlands.*

2.   FOREST SOILS AS SOURCES FOR ATMOSPHERIC CH$_4$

The main primarily control if forest soils are sinks or sources for atmospheric CH$_4$ is the oxygen availability in the soil profile. Forest soils which are temporarily or permanently water logged and therefore become anaerobic at least in certain periods of time are net sources for atmospheric CH$_4$ since CH$_4$-production in anoxic parts of the soil exceeds CH$_4$-consumption in oxic parts of the soil. In waterlogged soils CH$_4$ production – the final step of anaerobic matter decomposition – is primarily controlled by substrate availability and soil redox potential (Schimel et al. 1993; Conrad 1996). In nature the dominating substrates for methanogenesis are acetate and CO$_2$-H$_2$ (Knowles 1993). The availability of these substrates and the magnitude of CH$_4$ production rates is suggested to be estimated by the net primary production in an ecosystem (Aselmann and Crutzen 1988). Further controllers of CH$_4$ production are soil pH (most methanogenes showed optimum growth at pH 6–8, Knowles 1993) and temperature (Conrad 1996).

Since methane is mainly produced in a certain depth in wetland soils it must be transported to the soil surface prior its emission to the atmosphere. During this process CH$_4$ can be oxidized by methanotrophic bacteria. Several studies in wetland ecosystems have shown that 40–95% of total CH$_4$ produced in the soil (e.g. Galchenko et al. 1989; Butterbach-Bahl et al. 1997) is oxidized either in the surface aerobic soil layer, or in oxigenized zones around plant roots (e.g. Gilbert and Frenzel 1995) or even within the aerenchyma of plants (Gilbert and Frenzel 1998) before it is released to the atmosphere. In principal three different emission pathways for CH$_4$ from the production sites within the soil to the atmosphere must be considered: diffusion, ebullition or plant mediated transport. At least for rice paddies diffusion of CH$_4$ through water saturated sediments have been shown to be of a minor importance as compared to the other emission pathways (Butterbach-Bahl et al. 1997), and it can be hypothesized that this is also true for flooded forest ecosystems (see e.g. Wassmann et al. 1992) since diffusion of gases in water is approx. a factor of 10000 lower than in air.

In consequence of CH$_4$ production and CH$_4$ over-saturation in soil pore water also gas bubbles will form in the water saturated soil sediment. If the buoyancy force of the bubble is greater than the sum of hydrostatic and atmospheric pressures and the pressure required to move soil particles the gas bubble will move to the sediment surface before it finally will be released to the atmosphere. Especially in ecosystems were plant mediated transport is negligible this emission mechanism is of significant importance. This may also be true for flooded forest ecosystems. Data from Bartlett et al. (1988) and Wassmann et al. (1992) indicate that the main mechanism of CH$_4$

emission from flooded forests in the central Amazon region was ebullition (> 60%) rather than diffusion, though the importance of plant mediated transport was not investigated in these studies.

As a morphological adaptation to flooding conditions, many wetland trees, like e.g. black alder (Schröder 1989), willows (Gill 1970), bald cypress (Kludze et al. 1994), mangroves (Scholander et al. 1955) or even flood tolerant pines (Topa and McLeod 1986), develop an aerenchyma system which improves the O_2 availability in the submerged roots (Crawford 1982), but which may also be a means for the exchange of gases between the soil and the atmosphere (Rusch and Rennenberg 1998). Rusch and Rennenberg (1998) showed that CH_4-emission from the bark of a 3 year old black alder tree increased to up to 60 mg CH_4 m^{-2} h^{-1} if the alders were grown for 40 days in flooded soils. This mechanism of plant mediated CH_4-emissions from the soil to the atmosphere is apparently different from the observation that methanogenesis, i.e. CH_4-production, may occur also in the trunk of trees (Heyer 1990). This is mainly due to the fact that in decaying heartwood of living trees methanogenic bacteria can be found (Zeikus and Ward 1974; Zeikus and Henning 1975). This may lead to high pressures of CH_4 in the stem of trees, which can finally result in an outburst of CH_4 to the atmosphere. So far, it is not well established to which extend plant mediated transport of CH_4 from the soil to the atmosphere via the aerenchyma or the CH_4-production in woody trees contributes to the CH_4-exchange between forest ecosystems and the atmosphere. But for other wetland ecosystems it has been shown that plant mediated transport is the main mechanism by which CH_4 is emitted from the soil to the atmosphere (e.g. Schütz et al. 1991; Whiting and Chanton 1992; Schimel et al.1993; Butterbach-Bahl et al. 1997) and it can be hypothesized that this may be true also for some wetland forest ecosystems like e.g. flooded alder forests.

3. FOREST SOILS AS SINKS FOR ATMOSPHERIC CH₄

It is well established that well aerated forest soils are significant sinks for atmospheric CH_4 (e.g. Keller et al. 1983; Yavitt et al. 1990a, b; Crill 1991; Adamsen and King 1993; Castro et al. 1995). Though from a holistic point of view well aerated forest soils are sinks, the balance between production and consumption of CH_4 could still strongly vary depending on soil depth, indicating that at a certain depth either CH_4-consumption or CH_4 production is the dominating process. Figure 1 shows results of measure-

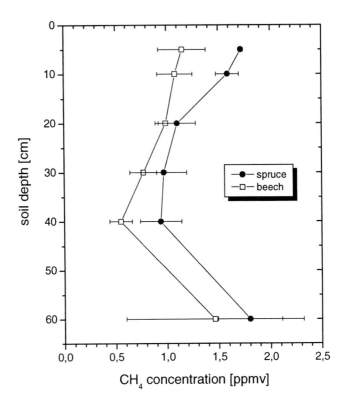

ments of CH_4 concentrations in different soil depths at two forests sites (age >90yr) at the Hoeglwald (Germany), which visualize the change of soil zones with predominantly consumption or production of CH_4 with soil depth by referring to measurements of CH_4-concentrations in the soil profile. The CH_4 profiles (Fig. 1) show that in the first 30-40 cm of the mineral soil CH_4 oxidation is higher than CH_4-production, since concentrations of methane in these soil zones are below atmospheric concentration. But in deeper soil layers an increase in CH_4-concentration even above atmospheric concentrations was observed, clearly demonstrating that in these soil zones CH_4-production is present. Such CH_4-concentration patterns, which were also observed for other forest sites (e.g. Klemedtsson and Klemedtsson 1997), can easily explain observations of different investigators that sometimes well aerated forest soils are turning to weak sources for atmospheric CH₄, if CH₄ oxidation in the uppermost mineral soil layers is reduced due to e.g. low temperatures or high soil moisture (e.g. Yavitt et al. 1995). An example of the change between emission and deposition of CH₄ at the forest floor/atmosphere interface is shown in Figure 2 for winter measurements of CH₄-fluxes at the spruce control site at the Höglwald Forest. The figure shows that until the end of march the spruce forest soil was a weak source for atmospheric CH₄, but turned to a significant sink after temperatures increased and soil moisture somewhat decreased at the end of winter.

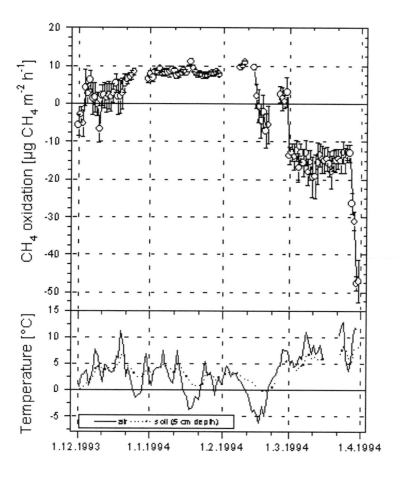

Figure 2. Fluxes of CH₄ at the spruce control site at the Höglwald Forest in winter 1993/1994. Shown are daily averages of CH₄ fluxes ± SE of 10 chambers and of temperatures of air and soil (5 cm soil depth). For detailed site and method descriptions see Papen and Butterbach-Bahl (1999) and Butterbach-Bahl et al. (1998).

Main uptake activity of atmospheric CH_4 has been shown to be concentrated in the uppermost mineral horizon of forest soils (e.g. Adamsen and King 1993; Bender and Conrad 1994; Saari et al. 1998; Brumme and Borken 1999, Steinkamp et al. 1999), whereas the organic layer has low CH_4-oxidation activity or may even show low rates of net-production of CH_4 (Sexstone et al. 1990; Adamsen and King 1993). The reasons for this widespread observation that the main CH_4 oxidation activity of forest soils is located in the uppermost mineral soil layer and not in the organic layer which is directly exposed to the atmosphere are still uncertain. Some authors assumed that higher NH_4^+ content in the organic layer as compared to the uppermost mineral layer inhibit CH_4-oxidation (Bender and Conrad 1994; Schnell and King 1994; Conrad 1996), since atmospheric CH_4-oxidation has been demonstrated to be extremely sensitive to increased inorganic N-concentrations in the soil (Steudler et al. 1989; Sitaula et al. 1995; Castro et al. 1995; Macdonald et al. 1997). Laboratory experiments indicate that the inhibitory effect of increased inorganic N-contents on CH_4-oxidation activity may be due to a competitive inhibition of the enzyme methane monooxygenase by NH_3 (Adamsen and King 1993; King and Schnell 1994a; Dunfield and Knowles 1995) or the production of toxic products (NO_2^- and NH_2OH) produced during enhanced NH_4^+ oxidation (King and Schnell 1994b). Furthermore, the organic layer may contain further compounds that may control CH_4 oxidation activity as was recently shown by Amaral and Knowles (1998) who found that CH_4 oxidation rates can be inhibited by increasing concentrations of monoterpenes. These authors suggested that soil monoterpene distribution may be responsible for the stratification of CH_4 oxidation activity in soils.

Besides the availability of inorganic N in forest soils, environmental factors like e.g. soil temperature, soil moisture and soil structure are major controllers of CH_4-oxidation activity.

Atmospheric CH_4-uptake by soils is largely controlled by gas diffusion resistance within the soil. Therefore, the structure of the organic layer (e.g. Dong et al. 1998; Brumme and Borken 1999) and the texture of the mineral soil (e.g. Dörr et al. 1993; Boeckx et al. 1997) have a huge impact on the magnitude of rates of CH_4-oxidation in forest soils. Soil moisture directly influence the gas permeability of soils and therefore also affect the exchange of gases between soils and the atmosphere. Consequently, an increase in soil moisture will reduce CH_4 oxidation activity by limitation of substrate availability (e.g. Castro et al. 1995; Gulledge and Schimel 1998; Brumme and Borken 1999). Furthermore, high soil moisture contents will also decrease O_2-availability in the soil profile, thereby inhibiting the process of CH_4-oxidation, since for the oxidation of methane to methanol by the enzyme methane monooxygenase, molecular oxygen is a prerequisite

(Knowles 1993; Gulledge and Schimel 1998). That rates of atmospheric CH_4-oxidation in forest soils and soil moisture are negatively correlated was confirmed in several field studies (e.g. Adamsen and King 1993; Czepiel et al. 1995; Steinkamp et al. 2001). On the other hand, if soil moisture is too low the activity of microbial CH_4-oxidation will be restricted due to physiological water stress (Conrad 1996; Schnell and King 1996; Gulledge and Schimel 1998).

As for every biological process also microbial CH_4-oxidation will increase with increasing temperature since an optimum has been reached. Several studies have shown that soil temperature is only a major controller of CH_4-oxidation in forest soils if temperatures are below approx. 10°C (e.g. Crill 1991; Castro et al. 1995; Steinkamp et al. 2001). At temperatures > 10°C other factors, like e.g. soil moisture, have been demonstrated to be of higher importance for the magnitude of CH_4-uptake than soil temperature (Castro et al. 1995; Steinkamp et al. 2001). Reported Q_{10} values for microbial oxidation of atmospheric CH_4 by forest soils are rather low and often < 1.5 (e.g. Born et al. 1990; Crill 1991; King and Adamsen 1992; Primé and Christensen 1997; Dong et al. 1998; Steinkamp et al. 2001).

CH_4-uptake rates of forest soils are strongly influenced by human activities. On a long term increasing rates of atmospheric N-deposition is discussed to lead on a long term to N-saturation of naturally N-limited forest ecosystems (e.g. Aber et al. 1989, Aber 1992) and it must be hypothesized that due to the sensitivity of microbial CH_4-oxidation to increased N-availability in soils this also will affect rates of atmospheric CH_4-oxidation. This hypothesis is strongly supported by results of N-fertilization experiments (Steudler et al. 1989; Castro et al. 1995; Sitaula et al. 1995; Macdonald et al. 1996) as well as by results of correlation analysis between N-input by wet deposition and actual CH_4-oxidation rates (Butterbach-Bahl et al. 1998), which have clearly shown that CH_4-oxidation in different forest ecosystem will be strongly reduced by increasing N-availability. CH_4-oxidation in soils has been shown to be extremely sensitive to physical disturbance of the soil structure as may occur if land use is changed from forestry to agriculture. Conversion of forests to agricultural land has been shown to reduce CH_4-uptake rates in temperate regions by at least 60% (Dobbie et al. 1996; Priemé et al. 1997), whereas for soils of the wet tropics results of Keller and Reiners (1994) suggest, that conversion of forests to pasture may even turn soils from sinks to sources of atmospheric CH_4. In contrast to the short-term effects of land use change from forestry to agriculture on CH_4-fluxes, it may take decades after conversion of arable land to forests to regain CH_4-uptake rates which are comparable to undisturbed forest sites (Priemé et al. 1997; Keller and Reiners 1994). The reason for these discrepancies is still poorly understood, but may be associated with major changes in the physical structure of soils, e.g. compaction of the soil, associated with agricultural use (Ojima et al. 1993; Dobbie et al. 1996).

4. MAGNITUDES OF CH$_4$-EXCHANGE BETWEEN FOREST ECOSYSTEMS AND THE ATMOSPHERE

In Table 1 published rates of CH$_4$ exchange between forest ecosystems and the atmosphere are summarized. Forest ecosystems which are for long periods of the year or permanently flooded are in all climate zones of the world strong sources for atmospheric CH$_4$. The magnitude of fluxes from such ecosystems is in the range of mg CH$_4$ mg m^{-2} h^{-1} and maximum observed CH$_4$ emission values exceed in all climate zones at least 20 mg CH$_4$ m^{-2} h^{-1} (e.g. Crill et al. 1988; Moore et al. 1990; Delmas et al. 1992). Uptake of atmospheric CH$_4$ by forest soils is a wide spread observation in all climate zones of the world if at least the uppermost soil layers are well aerated. In general uptake rates of atmospheric CH$_4$ by forest soils are mostly < 300 µg CH$_4$ m^{-2} h^{-1} and are therefore approx. one magnitude lower than rates of CH$_4$-emissions from flooded forest soils under comparable climate conditions. In high latitudes and also in mid-latitudes CH$_4$ uptake mainly occurs in the summer months, whereas in winter uptake rates even in well aerated soils are significantly lower or even zero (e.g. Crill 1991; Dong et al. 1998; Steinkamp et al. 2001; Alm et al. 1999) or soils may even turn to weak sources for atmospheric CH$_4$ (Fig. 2 and Yavitt et al. 1995). From all climate zones of the world, forest soils of the temperate zone are most likely the strongest sink for atmospheric CH$_4$, since due to the moderate climate conditions, time periods with unfavorable conditions for microbial atmospheric CH$_4$ uptake, like e.g. periods of droughts, high soil moisture or long lasting severe frost, are relatively rare to occur. Therefore, major concern has arisen, that increasing rates of atmospheric N-deposition especially to forest ecosystems of the temperate zones will most likely reduce the sink strength of these ecosystems for atmospheric CH$_4$ (Steudler et al. 1989; Sitaula et al. 1995; Butterbach-Bahl et al. 1998).

Table 1. Published CH$_4$ fluxes derived from field measurements for different forest ecosystems (negative values = deposition; positive values = emission)

Forest ecosystem	CH$_4$-flux [µg CH$_4$ m^{-2} h^{-1}]		Source
	Range	Mean	
High latitudes			
Spruce forest (Alaska)	-12 – 2800	192	Whalen and Reeburgh (1990)
Forests of floodplain sites (Alaska)	-40 – 350		Whalen et al. (1991)
Forests of upland sites (Alaska)	0 – -75		
Birch Forest (Alaska)	-19 – -40		Schimel et al. (1993)
Forested rich fen (Canada)	0 – 207000	380	Moore et al. (1990)
Conifer swamps (Canada)	-8 – 9800	80	Roulet et al. (1992)
Mixed hardwood swamps (Canada)	-242 – 1160	29	
Forested bog (Canada)	-4 – 4400	242	
Conifer swamps (Canada)	-23 – 27000	114	Roulet et al. (1992)
Mixed hardwood swamps (Canada)	-120 – 3200	11	
Forested bog (Canada)	-11 – 12000		
Spruce lichen woodland (Canada)	-65 – -11	-33	Adamsen and King (1993)
Forested fen (USA)	340 – 30000		Harriss et al. (1985),
Forested bog (USA)	1250 – 7900		Crill et al. (1988)
Scots pine forest (Finland)	-79 – -110	89	Saari et al. (1998)
Drained pine fen	-4		Alm et al. (1999)
Undrained pine fen winter fluxes (Finland)	400 – 650		
Mid latitudes			
Deciduous forest (Belgium)		-16	Boeckx et al. (1997)
Spruce forest (Denmark)	-29 – -0.8	-10	Ambus and
Beech forest (Denmark)	-13 – -4.1	-4	Christensen (1995)
Deciduous/ coniferous woodland (Scotland)	-140 – -7	-58	Dobbie et al. (1996)
Deciduous woodland (Denmark)	-44 – -11	-26	
Deciduous/ coniferous woodland (Poland)		-42	
Spruce forest (Denmark)	-60 – -25	-39	Priemé and Christensen (1997)
Different deciduous/coniferous forests on different soils (Germany)	-148 – -10	-75	Born et al. (1990)
Different forests (Germany)	-350 – 0		Dörr et al. (1993)

Forest ecosystem	CH$_4$-flux [µg CH$_4$ m^{-2} h^{-1}]		Source
	Range	Mean	
Spruce forests (Germany)	-11 – -3	-7	Borken and Brumme
Deciduous forests (Germany)	-16 – -2	-7	(1997)
Mixed beech and oak forest (Germany)	-113 – -22	-73	Dong et al. (1998)
Spruce forest (Germany)	-116 – -40	-82	Steinkamp et al. (1999)
Spruce forests (Germany)		-4.6	Brumme and Borken
Deciduous forests (Germany)		-13	(1999)
Scots pine (Norway)	-85 – 0	-54	Sitaula et al. (1995)
Coniferous forests(Sweden)			Klemedtsson and
Wet area	-15 – 5	-1	Klemedtsson (1997)
Dry area	-40 – 12	-16	
Sitka spruce (Scotland, UK)	-9.4 – 21	-15	MacDonald et al. (1997)
Hardwood forest (USA)	-90 – 50	-11	Keller et al. (1983)
Forested bogs (USA)	790 – 8600	4150	Harriss et al. (1985)
Forested fens (USA)	125 – 7100	3540	
Seasonal flooded forests (USA)	-125 – 11400	2460	Harriss et al. (1988)
Forested bogs (USA)	460 – 28900	3700	Crill et al. (1988)
Forested fens (USA)	2580 – 11000	5900	
Pine forest (USA)	-180 – -130	-147	Steudler et al. (1989)
Hardwood forest (USA)	-230 – -140	-173	
Mixed mesophytic forest (USA)	-271 – 1630	2.5	Yavitt et al. (1990a, b)
	-271 – 1350	-0.8	
Red pine (USA)	-50 – 17	0	
Red spruce (USA)			
Deciduous-conifer forest (USA)	-200 – 8	-68	Crill (1991)
Mixed coniferous forests (USA)	-100 – -26		Castro et al. (1993)
Mixed pine-hardwood forest (USA)		-110	Adamsen and King (1993)
Northern hardwoods (USA)	-42 – 330		Yavitt et al. (1995)
Oak forest, rural site (USA)	-290 – -110	-208	Goldman et al. (1995)
Red pine plantation (USA)	-220 – 0	-106	Castro et al. (1995)
Mixed hardwood forest (USA)	-200 – -10	-120	
Subtropics and Tropics			
Terra firme seasonal forest (Brazil)	-47 – 100	-11	Keller et al. (1983, 1986)

Forest ecosystem	CH$_4$-flux [μg CH$_4$ m^{-2} h^{-1}]		Source
	Range	Mean	
Terra firme forest (Brazil)	-76 – 10	-37	Goreau and de Mello (1988)
Flooded forests in Amazonian floodplain (Brazil)	0 – 125000	7000	Bartlett et al. (1988, 1990)
Flooded forests in Amazonian floodplain (Brazil): high water	40 – 22000	3100	Devol et al. (1988, 1990)
Low water		300	
Flooded forest in Amazonian Floodplain (Brazil)	410 – 8300		Wassmann et al. (1992)
Cerrado Aberto (Brazil)		-54	Anderson and Poth (1998)
Semi-deciduous humid forest (Cameroon)		-53	Macdonald et al. (1998)
Flooded forests in Congo River Basin (Congo)			Tathy et al. (1992)
Wet conditions	9.9 – 550	106	
Dry conditions	-33 – 190	-8	
Mountain forests (Congo)			Delmas et al. (1992)
Dry sites	-120 – 41	-47	
Wet sites	-114 – 24000	8000	
Lowland forests (Costa Rica)	-36 – -84	-51	Keller et al. (1993); Keller and Reiners (1994)
Secondary lowland forest (Costa Rica)	-85 – 8	-42	Weitz et al. (1998)
Primary forest (Ecuador)	-20 – 115	26	Keller et al. (1986)
Secondary forest (Ecuador)	-20 – 300	-34	Keller et al. (1986)
Subtropical deciduous forest (India)	-750 – -200	-450	Singh et al. (1998)
Broad-leafed Savanna (South Africa)	-30 – 51	0	Zepp et al. (1996)
Broad-leafed Savanna (South Africa)	-92 – -12		Seiler et al. (1984)
Cypress swamps (USA)	400 – 40000	11750	Harriss and Sebacher (1981)
Semideciduous forest (Venezuela)		-48	Scharffe et al. (1990)

REFERENCES

Aber JD, Nadelhoffer KJ, Steudler P & Melillo JM (1989) Nitrogen saturation in northern forest ecosystems. BioScience 39: 378-386

Aber JD (1992) Nitrogen cycling and nitrogen saturation in temperate forest ecosystems. Tree 7: 220-224.

Adamsen APS & King G (1993) Methane consumption in temperate and subarctic forest forest soils: Rates, vertical zonation and responses to water and nitrogen. Appl Environ Microbiol 59: 485-490

Alm J, Saarnio S, Nykänen H, Silvola J & Martikainen P (1999) Winter CO_2, CH_4, and N_2O fluxes on some natural and drained boreal peatlands. Biogeochemistry 44: 163-186

Ambus P & Christensen S (1995) Spatial and seasonal nitrous oxide and methane fluxes in Danish forest-, grassland-, and agroecosystems. J Environ Qual 24: 993-1001

Anderson IC & Poth MA ´(1998) Controls on fluxes of trace gases from Brazilian Cerrado soils. J Environ Qual 27: 1117-1124

Aselmann I & Crutzen J P (1988) Global distribution of natural freshwater wetlands and rice paddies, their net primary productivity, seasonality and possible methane emissions. J Atmos Chem 8: 307-358

Bartlett KB, Crill PM, Sebacher DI, Harriss RC, Wilson JO & Melack JM (1988) Methane flux from the central Amazonian floodplain. J Geophys Res 93: 1571-1582

Bartlett KB, Crill PM, Bonassi JA, Richey JE & Harriss RC (1990) Methane flux from the Amazon river floodplain: emissions during rising water. J Geophys Res 95: 16773-16788

Bender M & Conrad R (1994) Methane oxidation activity in various soils and freshwater sediments: Occurrence, characteristics, vertical profiles, and distribution on grain size fractions. J Geophys Res 99: 16531 - 16540

Boeckx P, Van Cleemput O &Villaralvo I (1997) Methane oxidation in soils with different textures and land use. Nutr Cycl Agroecosyst 49: 91-95

Borken W & Brumme R (1997) Liming practice in temperate forest ecosystems and the effects on CO_2, N_2O and CH_4 fluxes. Soil Use Manage 13: 251-257

Born M, Dörr H & Levin I (1990) Methane consumption in aerated soils of the temperate zone. Tellus 42: 2-8

Brumme R & Borken W (1999) Site variation in methane oxidation as affected by atmospheric deposition and type of temperate forest ecosystem. Glob Biogeochem Cycle 13: 493-501

Butterbach-Bahl K, Papen H & Rennenberg H (1997) Impact of gas transport through rice cultivars on methane emission from rice paddy fields. Plant Cell Environ 20: 1175-1183

Butterbach-Bahl K, Gasche R, Huber C, Kreutzer K & Papen H (1998) Impact of N-input by wet deposition on N-trace gas fluxes and CH_4-oxidation in spruce forest ecosystems of the temperate zone in Europe. Atmos Environ 32: 559-564

Castro MS , Steudler PA, Melillo JM, Aber JD & Millham S (1993) Exchange of N_2O and CH_4 between the atmosphere and soils in spruce-fir forests in the northeastern United States. Biogeochemistry 18: 119-135

Castro MS, Steudler PA, Melillo JM, Aber JD and Bowden RD (1995) Factors controlling atmospheric methane consumption by temperate forest soils. Glob Biogeochem Cycl 9: 1-10

Conrad R (1996) Soil microorganisms as controllers of atmospheric trace gases (H_2, CO, CH_4, OCS, N_2O and NO). Microbiol Rev 60: 609-640

Crawford RMM (1982) Physiological responses to flooding, In: Lange OL, Nobel PS, Osmond CB & Ziegler H (eds) Encyclopedia of Plant Physiology, Physiological Plant Ecology II, Water relations and carbon assimilation., pp 453-477. Springer Verlag, Berlin, Germany

Crill PM, Bartlett KB, Harriss RC, Gorham E, Verry ES, Sebacher DI, Madzar L & Sanner W (1988) Methane flux from Minnesota peatlands. Glob Biogeochem Cycle 2: 371-384

Crill PM (1991) Seasonal patterns of methane uptake and carbon dioxide release by a temperate woodland soil. Glob Biogeochem Cycle 5: 319-334

Czepiel PM, Crill PM & Harriss RC (1995) Environmental factors influencing the variability of methane oxidation in temperate forest soils. J Geophys Res 100: 9359-9364

Delmas RA, Servant J, Tathy JP, Cros B & Labat M (1992) Sources and sinks of methane and carbon dioxide exchanges in mountain forest in Equatorial Africa. J Geophys Res 97: 6169-6179

Devol AH, Richey JE, Clark WA, King SL and Martinelli LA (1988) Methane emissions to the troposphere from the Amazon floodplain. J Geophys Res 93: 1583-1592

Devol AH, Richey JE, Forsberg BR & Martinelli LA (1990) Seasonal dynamics in methane emissions from the Amazon floodplain. J Geophys Res 95: 16417-16426

Dobbie KE, Smith KA, Priemé A, Christensen S, Degorska A & Orlanski P (1996) Effect of land use on the rate of methane uptake by surface soils in Northern Europe. Atmos Environ 30: 1005-1011

Dörr H, Katruff L & Levin I (1993) Soil texture parameterization of the methane uptake in aerated soils. Chemosphere 26: 697-713

Dong Y, Scharffe D, Lobert JM, Crutzen JP & Sanhueza E (1998) Fluxes of CO_2, CH_4 and N_2O from a temperate forest soil: the effects of leaves and humus layers. Tellus 50B: 243-252

Dunfield P & Knowles R (1995) Kinetics of inhibition of methane oxidation by nitrate, nitrite, and ammonium in a humisol. Appl Environ Microbiol 61: 3129-3135

Galchenko VF, Lein A & Ivanov M (1989) Biological sinks of methane. In: Andreae MO & Schimel DS (eds) Exchange of trace gases between terrestrial ecosystems and the atmosphere, pp 39-58. John Wiley & Sons Ltd., Chichester, New York, U.S.A.

Gilbert B & Frenzel P (1995) Methanotrophic bacteria in the rhizosphere of rice microcosms and their effect on porewater methane concentration and methane emission. Biol Fertil Soil 20: 93-100

Gilbert B & Frenzel P (1998) Rice roots and CH_4 oxidation: the activity of bacteria, their distribution and the microenvironment. Soil Biol Biochem 30: 1903-1913

Gill CJ (1970) The flooding tolerance of woody species – a review. Forest Abstracts 31: 671-688

Goldman MB, Groffman PM, Pouyat RV, McDonnell MJ & Pickett ST (1995) CH_4 uptake and N availability in forest soils along an urban to rural gradient. Soil Biol Biochem 27: 281-286

Goreau JT & de Millo WZ (1988) Tropical deforestation: Some effects on atmospheric chemistry. Ambio 17: 275-281

Gulledge J & Schimel JP (1998) Moisture control over atmospheric CH_4 consumption and CO_2 production in diverse Alaskan soils. Soil Biol Biochem 8/9: 1127-1132

Harriss RC & Sebacher DI (1981) Methane flux in forested freshwater swamps of the southeastern United States. Geophys Res Lett 8: 1002-1004

Harriss RC, Gorham E, Sebacher DI, Bartlett KB & Flebbe PA (1985) Methane flux from northern peatlands. Nature 315: 652-654

Harriss RC, Sebacher DI, Bartlett KB, Bartlett DS & Crill PM (1988) Sources of atmospheric methane in the South Florida environment. Glob Biogeochem Cycle 2: 231-243

Heyer J (1990) Der Kreislauf des Methans, 250 pp., Akademie Verlag Berlin, Berlin

Keller M, Goreau TJ, Wofsy SC, Kaplan WA & McElroy MB (1983) Production of nitrous oxide and consumption of methane by forest soils. Geophys Res Lett 10: 1156–1159

Keller M, Kaplan WA & Wofsy SC 1986 Emission of N_2O, CH_4, and CO_2 from tropical forest soils. J Geophys Res 91: 11791-11802

Keller M, Veldkamp E, Weitz AM and Reiners WA (1993) Effect of pasture age on soil trace-gas emissions from a deforested area of Costa Rica. Nature 365: 244-246

Keller M & Reiners WA (1994) Soil-atmosphere exchange of nitrous oxide, nitric oxide, and methane under secondary succession of pasture to forest in the Atlantic lowlands of Costa Rica. Glob Biogeochem Cycle 8: 399-409

King GM & Schnell S (1994a) Enhanced ammonium inhibition of methane consumption in forest soils by increasing atmospheric methane. Nature 370: 282-284

King GM & Schnell S (1994b) Ammonium and nitrite inhibition of methane oxidation by *Methylobacter albus* BG8 and *Methylosinus trichosporium* OB3b at low methane concentrations. Appl Environ Microbiol 60: 3508-3513

King GM & Schnell S (1998) Effects of ammonium and non-ammonium salt additions on methane oxidation by *Methylosinus trichospirium* OB3b and Maine forest soils. Appl Environ Microbiol 64: 253-257

Kludze HK, Pezeshkl SR & Delaune RD (1994) Evaluation of root oxygenation and growth in baldcypress in response to short-term soil hypoxia. Can J For Res 24: 804-809

Klemedtsson ÅK & Klemedtsson L (1997) Methane uptake in Swedish forest soil in relation to liming and extra N-deposition. Biol Fertil Soil 25: 296-301

Knowles R (1993) Methane: processes of production and consumption. In: Harper LA, Mosier AR, Duxbury JM & Rolston DE (eds) Agricultural Ecosystem Effects of Trace Gases and Global Climate Change, pp 145-165. ASA Special Publication 55, American Society of Agronomy, Madison, U.S.A.

Macdonald JA, Skiba U, Sheppard L, Hargreaves KJ, Smith KA & Fowler D (1996) Soil environmental variables affecting the flux of methane from a range of forest, moorland and agricultural soils. Biogeochem 34: 113 - 132

MacDonald JA, Skiba U, Sheppard LJ, Ball B, Roberts JD, Smith KA & Fowler D (1997) The effect of nitrogen deposition and seasonal variability on methane oxidation and nitrous oxide emission rates in an upland spruce plantation and moorland. Atmos Environ 31: 3693-3706

MacDonald JA, Eggleton P, Bignell DE, Forzi F & Fowler D (1998) Methane emission by termites and oxidation by soils, across a forest disturbance gradient in the Mbalmayo forest reserve, Cameroon. Glob Change Biol 4: 409-418

Martikainen PJ, Nykänen H, Alm J & Silvola J (1995) Change in fluxes of carbon dioxide, methane and nitrous oxide due to forest drainage of mire sites of different trophy. Plant Soil 168-169: 571-577

Moore TR, Roulet N & Knowles R (1990) Spatial and temporal variations of methane flux from subarctic/ northern boreal fens. Glob Biogeochem Cycle 4: 29-46

Ojima DS, Valentine DW, Mosier AR, Parton WJ & Schimel DS (1993) Effect of land use change on methane oxidation in temperate forest and grassland soils. Chemosphere 26: 675-685

Papen H & Butterbach-Bahl K (1999) A 3-years continuous record of N-trace gas fluxes from untreated and limed soil of a N-saturated spruce and beech forest ecosystem in Germany: 1. N_2O-emissions. J Geophys Res 104: 18487-18503

Priemé A & Christensen S (1997) Seasonal and spatial variation of methane oxidation in a Danish spruce forest. Soil Biol Biochem 29: 1165-1172

Roulet NT, Ash R & Moore TR (1992) Low boreal wetlands as a source of atmospheric methane. J Geophys Res 97: 3739-3749

Rusch H & Rennenberg H (1998) Black alder (*Alnus Glutinosa* (L.) Gaertn.) trees mediate methane and nitrous oxide emission from the soil to the atmosphere. Plant Soil 201: 1-7

Saari A, Heiskanen J & Martikainen PJ (1998) Effect of the organic horizon on methane oxidation and uptake in soil of a boreal Scots pine forest. FEMS Microbiol Ecol 26: 245-255

Schimel JP, Holland EA & Valentine D (1993) Controls on methane flux from terrestrial ecosystems. In: Harper LA, Mosier AR, Duxbury JM & Rolston DE (eds) Agricultural Ecosystem Effects on Trace Gases and Global Climate Change, pp 167-182. ASA Special Publication 55, American Society of Agronomy, Madison, U.S.A.

Scholander PF, Van Dam L & Scholander SI (1955) Gas exchange in the roots of mangroves. Am J Bot 42: 92-98

Schröder P (1989) Characterisation of a thermo-osmotic gas transport mechanism in *Alnus glutinosa* (L.) Gaertn. Trees 3: 38-44

Schnell S & King GM (1996) Responses of methanotrophic activity in soils and cultures to water stress. Appl Environ Microbiol 62: 3203-3209

Schütz H, Schröder P & Rennenberg H (1991) Role of plants in regulating the methane flux to the atmosphere. In: Sharkey TD, Holland EA & Mooney HA (eds) Trace Gas Emissions by Plants, pp 29-63. Academic Press, San Diego, U.S.A.

Seiler W, Conrad R & Scharffe D (1984) Field studies of methane emission from termite nests into the atmosphere and measurement of methane uptake by tropical forest soils. J Atmos Chem 1: 171-186

Sexstone AJ & Mains CN ´(1990) Production of CH_4 and ethylene in organic horizons of spruce forest soils. Soil Biol Biochem 22: 135-139

Singh JS, Raghubanshi AS, Reddy VS, Dingh S & Kashyap AK (1998) Methane flux from irrigated paddy and dryland rice fields, and from seasonally dry tropical forest and savanna soils of India. Soil Biol Biochem 30: 135-139

Sitaula BK, Bakken LR & Abrahamsen G (1995) CH_4 uptake by temperate forest soil: effect of N input and soil acidification. Soil Biol Biochem 27: 871-880

Steudler PA, Bowden RD, Melillo J M & Aber JD (1989) Influence of nitrogen fertilization on methane uptake in temperate forest soils. Nature 341: 314-316

Steinkamp R, Butterbach-Bahl K & Papen H (2001) CH_4 oxidation by soils of a N limited and N fertilized spruce forest soils in the Black Forest, Germany. Soil Biol Biochem 33´: 145-153

Tathy JP, Delmas RA, Marenco A, Gros B, Labat M & Servant J (1992) Methane emission from flooded forest in Central Africa, J Geophys Res 97: 6159-6168

Topa MA & McLeod KW 1986 Aerenchyma and lenticel formation in pine seedlings: a possible avoidance mechanism to anaerobic growth conditions. Physiol Plant 68: 540-550

Wassmann R, Thein UG, Whiticar MJ, Rennenberg H, Seiler W & Junk WJ (1992) Methane emissions from the Amazon floodplain: characterization of production and transport. Glob Biogeochem Cycle 6: 3-13

Whalen SC & Reeburgh WS (1990) A methane flux transect along the trans-Alaska pipeline haul road. Tellus 42B: 237-249

Whalen SC, Reeburgh WS and Kizer KA (1991) Methane consumption and emission by taiga. Glob Biogeochem Cycle 5: 261-273

Whiting GJ & Chanton JP (1992) Plant-dependent CH_4 emission in a subarctic Canadian fen. Glob Biogeochem Cycl 6: 225-231

Weitz AM, Veldkamp E, Keller M, Neff J & Crill PM (1998) Nitrous oxide, nitric oxide, and methane fluxes from soils following clearing and burning of tropical secondary forest. J Geophys Res 103: 28047-28058

Yavitt JB, Lang GE & Sexstone AJ (1990a) Methane fluxes in wetland and forest soils,
    beaver ponds and low-order streams of a temperate forest ecosystem. J Geophys Res 95:
    22463-22474
Yavitt JB, Downey DM, Lang GE & Sexstone AJ (1990b) Methane consumption in two
    temperate forest soils. Biogeochem 9: 39-52
Yavitt JB, Fahey TJ & Simmons JA (1995) Methane and carbon dioxide dynamics in a
    northern hardwood ecosystem. Soil Sci Soc Am J 59: 796-804
Zeikus JG & Henning DL (1975) *Methanobacterium arbophilicum* sp. nov. An obligate
    anaerobe isolated from wetwood living trees. Antonie van Leeuwenhoek 41: 543-552
Zeikus JG & Ward JC (1974) Methane formation in living trees: A microbial origin. Science
    184: 1181-1183
Zepp RG, Miller WL, Burke RA, Parsons DAB & Scholes MC (1996) Effects of moisture and
    burning on soil-atmosphere exchange of trace gases in a southern African savanna. J
    Geophys Res 101: 23699-23706

EXCHANGE OF TRACE GASES AT THE TREE-ATMOSPHERE INTERFACE

Chapter 3.1

Ammonia exchange at the tree-atmosphere interface

Kent Høier Nielsen[1], Jan Kofod Schjørring[1], Jan Willem Erisman[2] and John Pearson[3]

[1]*Plant Nutrition Laboratory, Royal Veterinary and Agricultural University, Copenhagen, Denmark*
[2]*Energy Research Centre of the Netherlands (ECN), Petten, The Netherlands*
[3]*Department of Biology, University College London, Gower Street, London WC1E 6BT, United Kingdom*

1. INTRODUCTION

Atmospheric ammonia plays a central role as the dominant source of external nitrogen input to many natural and semi-natural ecosystems including forests. The input of ammonia affects tree growth and viability either directly by affecting primary metabolic processes or indirectly via affecting soil properties. Ammonia deposition may in the short term stimulate tree growth, but will over a longer time scale inevitably lead to soil acidification and nutritional imbalances which must be counteracted by suitable management practices if tree viability is to be maintained (see Chapter 1.3 for further description of nitrogen toxicity).

Ammonia exchange over forests has only been measured in a few cases, starting at the end of the 1980s (Wyers et al. 1992). The primary reason for this has been the lack of precise and easily operated measurement techniques. Recent developments in micrometeorological techniques and detector technology open new possibilities for more and better measurements and such information is now starting to appear (e.g. Erisman et al. 2001). In combination with new bio-indicators such as apoplastic pH and NH_4^+ for assessment of NH_3 compensation points based on leaf tissue

R. Gasche et al. (eds.), Trace Gas Exchange in Forest Ecosystems, 159–173.

analysis, this also allows for studying the links to physiological processes involved in the NH₃ exchange between plants and the atmosphere.

This chapter gives an overview of recorded NH₃ fluxes between trees and the atmosphere and their physiological and environmental basis. The first half of the chapter describes the processes controlling NH₃ fluxes between single tree leaves and the atmosphere, while NH₃ fluxes on canopy scale are treated in the second half.

2.   AMMONIA EXCHANGE AT THE LEAF-ATMOSPHERE INTERFACE

The NH₃ flux, J_{NH_3}, between a single plant leaf and the atmosphere can simplified be described as:

$$J_{NH_3} = g_{leaf}(n_i - n_a) = g_{NH_3}(\gamma - n_a)$$ [1]

where g_{leaf} is the conductance to diffusion of ammonia between the atmosphere and the interior of the leaf, n_a is the NH₃ concentration in the ambient atmosphere and n_i is the NH₃ concentration of the air in the sub-stomatal cavities and intercellular air spaces within the leaf. Whether a leaf will act as a sink for or a source of atmospheric NH₃ depends on the difference in internal and external NH₃ concentration. If n_a exceeds n_i, NH₃ will be absorbed, while in the opposite case emission will occur. When n_a equals n_i, no net NH₃ flux occurs between the leaf and the atmosphere. The internal NH₃ concentration, γ, at which J_{NH_3} is zero is called the stomatal compensation point for NH₃ (Eq. 1; Farquhar et al. 1980; Husted et al. 1996). As evident from Equation 1, the rate and direction of NH₃ fluxes between plant leaves and the atmosphere at a given atmospheric NH₃ concentration are controlled by the conductance to NH₃ transfer and the compensation point for NH₃. These two parameters are treated in further details in the following paragraphs.

2.1   Leaf conductance to NH₃ transfer

The leaf resistance (inverse of conductance) to NH₃ transfer includes a stomatal and cuticular resistance in parallel, and usually also includes a boundary layer resistance in series with the two other terms. The latter is a function of the aerodynamic properties of the leaf, wind speed, and canopy turbulence, and will generally be an order of magnitude smaller than the maximum stomatal resistance. Boundary layer and canopy resistances to NH_3 transport are discussed in further details later in the chapter.

The stomatal density is species specific, and is not only determined by the number of stomates per unit surface area, but also dependent on whether stomates are present on both the upper and lower side of the leaf. Most deciduous trees only have stomates on the lower side of the leaf, while conifers have stomates on both sides of the needles (Weyers and Meidner 1990).

In addition to the density of stomatal apertures, the stomatal conductance is affected by the depth and diameter of the stomatal aperture. Diameter is an important factor due to the dynamics of opening and closing of the stomates. The degree of stomatal opening is controlled by environmental parameters such as CO_2 concentration, air temperature, air humidity, light intensity and soil water availability. Decreasing internal CO_2 concentration causes stomates to open, while drought stress make them close. The stomatal response to light intensity correlates with photosynthetic activity, with increasing stomatal aperture at increasing light intensity and photosynthetic activity. The temperature response parallels the light response with increasing stomatal aperture at increasing temperature. However, if temperature increases above a critical level the stomates start to close due to increased transpiration and water deficit. The intimate dependence on environmental and physiological conditions makes the stomatal resistance the most dynamic resistance in the NH_3 diffusion pathway between the atmosphere and the leaf interior. For e.g. beech, this results in a diurnal rhythm in NH_3 deposition with increased deposition rates during daytime when stomates are open (Geßler et al. 2000).

The cuticular resistance (R_{cut}) is extremely high for NH_3, probably in the range of 2.000-40.000 s m^{-1} (Van Hove et al. 1989). This means that hardly any NH_3 will pass through the cuticle. However, NH_3 can readily be deposited on the cuticular surface due to the presence of a surface water film. Since NH_3 is highly soluble in water, moist leaf surfaces can act as a storage compartment for atmospheric NH_3 making the leaf surface a temporary sink for NH_3. To include deposition to the cuticle, Equation 1 can be expanded to:

$$J_{NH_3} = g_{leaf}(n_i - n_a) + g_w(n_w - n_a) \qquad [2]$$

where g_w is the conductance for NH_3 deposition in the cuticular water film and n_w is the NH_3 concentration in the atmospheric interface of the water film (the gaseous NH_3 concentration associated with the water film). By combining Henry's law and the Henderson-Hasselbalch equation, n_w can be calculated as:

$$n_w = \frac{K_d \times K_H \times \left([NH_4^+] + [NH_3]\right)}{K_d + (H^+)} \quad [3]$$

where $[NH_4^+]$, $[NH_3]$ and (H^+) are the concentrations (activities) in the water film. When the leaf surface subsequently dries out, the NH_3 will volatilize from the canopy, then acting as source of atmospheric NH_3 (Sutton et al. 1995; Wyers and Erisman 1998). Van Hove (1989) found large NH_3 deposition velocities even under dry conditions due to the presence of micro-scale water layers on leaf surfaces that otherwise appeared dry. Benner et al. (1992) showed water films with a thickness of 2–3 molecular monolayers at 50% RH up to more than 40 molecular monolayers at 95% RH. In the presence of salts deposited on the leaf surface, the thickness of the water film can be significantly higher, leading to enhanced NH_3 deposition (Van Hove et al. 1989). It has been proposed that the cuticular water film *via* the stomatal aperture can form a continuum with the apoplastic water film inside the leaf, thereby extending the surface of the apoplast to the leaf surface (Burkhardt and Eiden 1994). Such an apoplast-cuticular continuum would enhance the NH_3 exchange between the atmosphere and the leaf interior. This is supported by the fact that a significant foliar NH_3 uptake occurs during wet deposition as the result of cation exchange.

After entering through the stomatal apertures, the NH_3 molecule has to pass through the intercellular air space (IAS) before reaching the apoplastic water film on the extracellular surface of the plasma membrane. The IAS, which represents between 25 and 40% of the leaf volume, depending on plant species and leaf age, acts as an unstirred air layer across which substances must diffuse. This means that the path length and the concentration gradient of NH_3 between the stomatal aperture and the apoplastic water film control the resistance for ammonia movement in this compartment (R_{ias}).

The combined leaf resistance (R_l) to NH_3 transport includes the stomatal resistance (R_{sto}), the cuticular resistance (R_{cut}), the resistance associated with passing through the intercellular air space (R_{ias}) and the resistance to NH_3 deposition in the cuticular water film (R_w). Following the traditional rules for calculating electrical resistances, R_l can be expressed as:

$$R_l = \frac{(R_{ias} + R_{sto})(R_w R_{cut})}{R_w R_{ias} + R_w(R_{sto} + R_{ias}) + R_{cut}(R_{sto} + R_{ias})} \quad [4]$$

A mesophyll resistance does usually not exist for NH_3 due to its high water solubility, which implies that the exchange is controlled by reactions in the extracellular water of the leaf cells (the apoplastic water; see below).

The importance of the leaf temperature for NH_3 and CO_2 exchange between beech trees and the atmosphere is shown in Figure 1. Peaks in temperature coincide with peaks in gas fluxes, the latter following temperature-mediated fluctuations in stomatal resistance.

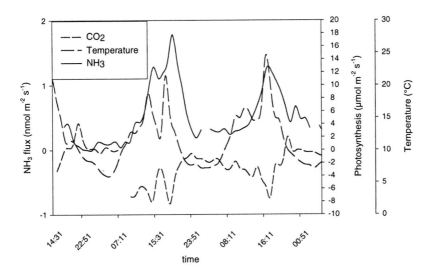

Figure 1. Gas flux measurements carried out early May 2000 in a Danish beech forest growing in an area with intensive agriculture. The nitrogen concentration in the dry matter of the foliage was around 2.5%. Flux measurements were carried out using a leaf cuvette mounted in the canopy and flushed with NH_3 free air (Nielsen and Schjoerring, unpublished)

2.2   The stomatal NH_3 compensation point

The stomatal compensation point is regulated by the NH_3 concentration in the apoplastic water film (Fig. 2) since, according to Henry's Law, the gaseous NH_3 concentration in the sub-stomatal air space is in equilibrium with the NH_3 concentration dissolved in the apoplastic water film. The pK_a value for NH_3/NH_4^+ in the apoplastic solution is approximately 9.32 (corrected for the ionic strength in the apoplastic solution; Husted and Schjoerring 1996). Since the pH in the apoplastic solution is between 5.5 and 6.5 it means, according to the Henderson-Hasselbalch equation, that the NH_3

concentration only represents a few ‰ of the total NH_3/NH_4^+ concentration found in the apoplastic water film. However, these few ‰ determine, together with the atmospheric concentration, whether trees will absorb or emit NH_3 through their stomates.

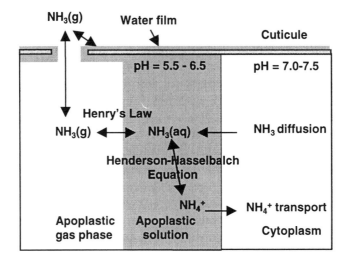

Figure 2. Schematic drawing of the physiological/chemical processes in the apoplast, which regulates the exchange of NH_3 between trees and the atmosphere

Ammonium is always present in plants because it is an essential intermediate in their nitrogen metabolism. The apoplastic NH_4^+ level is controlled by the balance between efflux of NH_3 from the cytoplasm and recapture of NH_4^+ by active transport systems (Fig. 3; Nielsen and Schjoerring 1998). The leaf apoplastic NH_4^+ concentration also depends on the xylem sap which contains NH_4^+ (Rennenberg et al. 1998). NH_3 is known to be able to freely diffuse across cell membranes (Kleiner 1981) while NH_4^+ influx into the leaf cells from the apoplast requires active transport. The direction for passive NH_3 transport is out of the cell due to the pH gradient between the cytoplasm (pH 7.0–7.5) and the apoplast (pH 5.5–6.5). In leaves of the herbaceous species oilseed rape apoplastic NH_4^+ is, depending on the concentration, re-absorbed by high-affinity NH_4^+ transporter or a low-affinity transporter with channel-like kinetic properties (Schjoerring et al. 1999; Nielsen and Schjoerring 1998). No information is presently available on the molecular basis for ammonium transport in tree leaves.

In addition to apoplastic NH_4^+ concentration, the NH_3 compensation point reflects apoplastic pH, which is regulated by the net proton flux across the plasma membrane (Fig. 3). The H^+ net flux is the balance between H^+ extrusion mediated by a proton pump (H^+-ATPase) and co-transporters transporting H^+ back into the cell together with anions, cations or neutral molecules. This leads to a fine-tuned balance between proton extrusion and proton uptake, which ensures a dynamic regulation of the apoplastic pH.

A vacuum infiltration-technique has been developed, making it possible to isolate leaf apoplastic solution with negligible (< 1.5%) contamination by cytoplasmic material as evaluated on the basis of marker enzyme activities and osmolality assessments (Husted and Schjoerring 1995). The NH_4^+ and H^+ concentrations in the apoplastic solution are used as a bio-assay, allowing the stomatal NH_3 compensation point to be calculated by Equation 3. In *Brassica napus*, NH_3 compensation points determined by this bioassay, were in good agreement with NH_3 compensation points determined on the basis of gas exchange measurements.

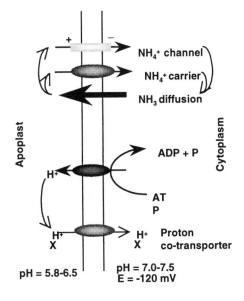

Figure 3. Processes regulating apoplastic pH and NH_4^+ concentration. Both pH and NH_4^+ concentration are highly dynamic and regulated by a well-tuned balance between efflux and uptake of respectively H^+ and NH_3/NH_4^+ (x denotes any ion co-transported with protons).

Measurements of NH_3 compensation points in tree leaves show values ranging from below 1 nmol NH_3 mol^{-1} air in canopies of pine, spruce and aspen growing at low N deposition (Langford and Fesenfeld 1992; Kesselmeier et al. 1993) to approx. 3.5 nmol NH_3 mol^{-1} air for spruce and fully developed green leaves of beech trees (Geβler et al 1998). Nielsen and Schjoerring (unpublished) studied the variation of the NH_3 compensation point throughout the growing season in a Danish beech forest growing in an area with high N deposition (Table 1). The stomatal compensation point for

NH$_3$ was high early in the season during foliation and again late in the season during senescence-induced defoliation. Both these growth stages are characterized by intensive turnover and transport of nitrogen compounds in leaves and storage organs.

Table 1. NH$_3$ compensation point in leaves of beech trees. All units are in nmol NH$_3$ mol^{-1} air. N.D. = not determined (Nielsen and Schjoerring, unpublished).

	NH$_3$ compensation based on cuvette measurements of NH$_3$ exchange fluxes (bottom of canopy)	*NH$_3$ compensation point calculated from apoplastic bioassay (bottom of canopy)*	*Ambient daytime NH$_3$ concentration*
Foliation	9.6 ± 1.0	7.0 ± 3.2	N.D.
Green-leaf-period	2.9 ± 1.3	3.7 ± 1.2	4.0 ± 0.2
Defoliation	7.3 ± 5.9	6.4 ± 2.8	1.9 ± 0.8

2.3   Co-deposition between NH$_3$ and other compounds

In addition to the presence of water layers on vegetation, the presence of SO$_2$ also considerably decreases the canopy resistance by decreasing the pH of the water film. Interactions between other acid gases such as HNO$_3$ and HCl and leaf surfaces may also be important. However, because the ammonium salts formed by reaction with these gases (NH$_4$NO$_3$ and NH$_4$Cl) have significant vapor pressures, they may possibly re-dissociate back to the precursor gases, limiting accumulation of salts on the leaf surface. This is in contrast to (NH$_4$)$_2$SO$_4$, which has negligible vapor pressure and, consequently, may accumulate.

The cuticular pH is also affected by the origin of the surface water, because pH in cloud and fog is lower than in pure water. Therefore, the origin of surface wetness through fog, cloud, rain, dew or guttation is important for deposition of NH$_3$ to a given surface.

3.   AMMONIA EXCHANGE OVER FORESTS

Measurements of NH$_3$ fluxes over forests show that such surfaces generally act as efficient NH$_3$ sinks (Erisman et al. 2001; Hovmand et al. 1998; Wyers and Erisman 1998; Duyzer et al. 1994; Andersen et al. 1993; Duyzer et al. 1992; Wyers et al. 1992). However, in agreement with the existence of a NH$_3$ compensation point in tree leaves (see above), episodes of NH$_3$ emission may occur as illustrated by micrometeorological measurements at the Speuld forest shown in Figure 4. Thus, Langford and Fehsenfeld (1992) demonstrated that under circumstances of high NH_3 concentrations a pine forest acted as a NH_3 sink, whereas at low NH_3 concentrations it acted as a NH_3 source. In agreement with the profound influence of temperature on NH_3 exchange, measurements over a mixed pine/spruce/fir/aspen forest showed that atmospheric NH_3 concentrations were temperature-dependent when the air had passed over the forest, but were independent of temperature when the air was sampled before it reached the forest (Langford and Fehsenfeld 1992).

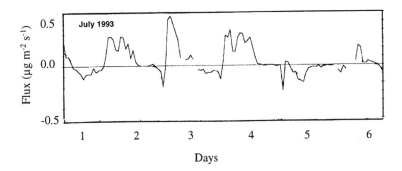

Figure 4. NH_3 fluxes at a high N status Douglas Fir forest in The Netherlands (Speuld) during a 6 days period in June 1993. Both emission (positive values) and deposition fluxes were recorded with a clear diurnal cycle resulting in emission during the day and deposition during the night. (Data from Wyers and Erisman 1998).

Deposition to forests may occur by wet, dry, or cloud/fog deposition. Wet deposition is the process by which atmospheric pollutants are delivered to a surface by rain, hail or snow. Cloud and fog deposition is the process where cloud and fog water droplets are directly intercepted by the canopy. This deposition process is often referred to as 'occult' deposition. By wet and occult deposition NH_4^+ is either deposited directly on the cuticle of the tree leaves or to the soil. Dry deposition is the process where gases and particles are deposited directly from the atmosphere onto vegetation, soil or other surfaces without a hydrological medium.

Dry deposition of NH_3 from the atmosphere to the leaf surface is regulated by two parameters, namely the transport through the boundary layer and the activity of the receptor (the trees). Generally, two resistances can be distinguished for transport of pollutants from the atmosphere to the receptor: the resistance in the fully turbulent layer, R_a, and the resistance in the quasi-laminar boundary layer, R_b (Fig. 5). Movement of gases in the fully turbulent air layer is regulated by movement of packets, or eddies, of air (Fig. 5). An eddy is a packet of air swirling around in the turbulent region above vegetation, which tends to increase in size with increasing height. Eddies carry all molecules that they contain more or less as a unit. They are continually breaking up, or coalescing with other eddies. This results in a high rate of movement of molecules, such as NH_3, and makes the movement of a gas approximately 10^4–10^5 times faster than movement by molecular diffusion (Nobel 1991). The level of atmospheric turbulence, generated by both wind shear and buoyancy, governs the transport of gases and particles

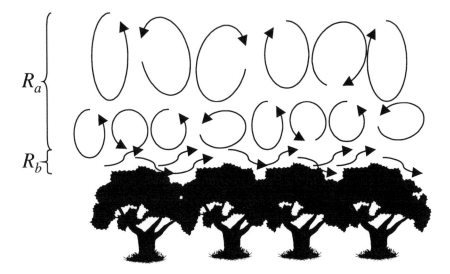

Figure 5. Graphical description of air movement above a canopy. The resistances for gas movement in the atmosphere are regulated by the turbulence in the air. In the boundary layer just above the canopy the direction of the wind is rectified by the surface of the canopy, resulting in the boundary layer resistance (R_b). Further away from the canopy, air movement is determined by eddies producing a well-mixed atmosphere with the resistance (R_a)

from the atmosphere to within a short distance from the receptor surface. The higher the level of atmospheric turbulence, the more efficiently gases and particles are transported to a given receptor surface. Given constant sink strength for a particular pollutant of interest, the concentration gradients above the receptor will vary according to the intensity of atmospheric turbulence. A well-mixed unstable boundary layer leads to relatively small concentration gradients and to low resistance to vertical transport. Quasi-laminar air layers are a result of a rectification of the wind direction above a surface and cause the quasi-laminar boundary layer resistance (R_b). The thickness of the quasi-laminar boundary layer is dependent on the architecture of the canopy with a decreasing thickness at increasing canopy roughness. Transport through the quasi-laminar layer is dominated by molecular diffusion and R_b is therefore regulated by the concentration gradient across the boundary layer and the thickness of the boundary layer. The total resistance (R_{tot}) for transport of NH_3 from the atmosphere to the apoplastic water film may now be considered as the sum of the 3 resistances described:

$$R_{tot} = R_a + R_b + R_l \qquad [5]$$

In analogy to Equation 1, the NH_3 flux over an intact forest canopy can be expressed as:

$$J_{NH_3} = g_{tot}(n_0 - n_a) = \frac{1}{R_{tot}}(n_0 - n_a) \qquad [6]$$

where n_0 is the NH_3 concentration at the sink (leaf cuticle or leaf apoplast) and n_a is the atmospheric NH_3 concentration at a given height above the canopy.

In many measurements of NH_3 deposition it has been assumed that the NH_3 concentration at the surface, i.e. the NH_3 compensation point, was zero and that the average NH_3 deposition flux could be expressed on the basis of a deposition velocity, V_d (cm s^{-1}) (see e.g. Duyzer et al. 1987, 1992):

$$R_{tot} = \frac{-n_a}{J_{NH_3}} = \frac{1}{V_d} \qquad [7]$$

V_d is normally larger over forest areas than over other terrestrial ecosystems (Table 2). This is due to a greater roughness of the forest canopy, which increases the turbulence in the boundary layer and thereby decrease R_a and R_b (Sutton et al. 1994; Ferm 1998). The assumption, that n_0 *is* equal to zero is, however, very rarely fulfilled for vegetation due to the NH_3 compensation point being larger than zero (see above). Deposition velocities should therefore be used with reservation and only taken as an indicator of the deposition potential.

Table 2. Deposition velocities (V_d) for forest and other semi-natural ecosystems. Both Speuld and Ulborg are spruce forests located in The Netherlands and Denmark, respectively.

Location	V_d (cm s^{-1})	Reference
Speuld	1.9-3.0	Duyzer et al. 1994
Speuld	3.2[1]	Wyers et al. 1992
Ulborg	-12.5-20.1 (mean 2.6)	Andersen et al. 1993
Dry Heathland	1.9	Duyzer et al. 1987
Neutral cut meadow	1.6	Sutton et al. 1993

[1]Presented as median

The distribution between dry and wet deposition is highly dependent on the distance between a given NH_3 source and the forest area (Table 3). After entering the atmosphere, NH_3 is readily converted to NH_4^+ or subjected to dry deposition, which means that high concentrations of atmospheric NH_3 and high NH_3 deposition occur relatively close to an emitting source. Long range transport mainly takes place as NH_4^+ in aerosols in the form of $(NH_4)_2SO_4$ and NH_4NO_3 (Harrison and McCartney 1980; Hedin et al. 1990). Due to a relatively long residence time of NH_4^+ in the atmosphere (4–15 days) transboundary transport of reduced nitrogen compounds can proceed over long distances.

Table 3. Annual deposition of N in different forest areas (kg N ha⁻¹ yr⁻¹). Niwot Ridge is located in the Rocky Mountains and is dominated by clean continental air. Oak Ridge is dominated by low agricultural but high industrial emission, while both Göttingen and Speuld are dominated by high agricultural and industrial emission

Location	Dry deposition		Wet deposition	Total deposition	References
	NH_3	NH_4^+	NH_4^+		
Niwot Ridge	0.4 (17 %)	0.06 (2.5 %)	0.5 (21 %)	2.4	Langford & Feshenfeld 1992
Oak Ridge	0.3 (6 %)	0.2 (3.7 %)	1.2 (22 %)	5.4	
Göttingen	3.6 (19 %)	0.9 (4.7 %)	1.1 (5.8 %)	19	
Speuld	17.9 (38 %)	4.7 (10 %)	11.3 (5.8 %)	47	Erisman et al. 1998 a, b

High deposition of NH_3 is therefore mainly occurring in areas with high agricultural emissions (Table 3). However, in industrialized countries, agricultural areas often surround forest areas. This means that the deposition of NH_3 can exceed the critical NH_3 load for these areas (Table 4). The consequence of high loads of nitrogen to forest ecosystems are further discussed in chapter 1.3.

Table 4. Critical nitrogen loads in different forest ecosystems

Forest type	Critical loads (kg N ha⁻¹ yr⁻¹)	References
Forest on silicate soil	3-14	Schulze et al. 1989
Forest on calcareous soil	1-48	Schulze et al. 1989
Acidic (managed) coniferous forest	15-20	Anon 1992
Acidic (managed) deciduous forest	<15-20	Anon 1992

ACKNOWLEDGMENTS

This work was supported by grants from the Danish Strategic Environmental Research Program II to J.K.S. The work was conducted as part of the SOROFLUX project under the Centre for Sustainable Land Use and Management of Contaminants, Carbon and Nitrogen.

REFERENCES

Andersen HV, Hovmand MF, Hummelshøj P & Jensen NO (1993) Measurements of ammonia flux to a spruce stand in Denmark. Atmos Environ 27: 189-202

Anon (1992) Critical loads for nitrogen. In: Grennfelt P & Thörnelöfe E (eds) Critical loads for nitrogen, a workshop report, pp 7-54. Nordic Council of Ministers, Lökeberg, Sweden

Benner W, Ogorevc B & Novakov T (1992) Oxidation of sulfur dioxides in thin water films containing ammonia. Atmos Environ 26: 1713-1723

Burkhardt J & Eiden R (1994) Thin water films on coniferous needles. Atmos Environ 28: 2001-2017

Duyzer JH, Verhagen HLM, Weststrate JH, Bosveld FC & Vermetten AWM (1994) Dry deposition of ammonia onto a douglas fir forest in The Netherlands. Atmos Environ 28: 1241-1253

Duyzer JH, Verhagen HLM & Weststrate JH (1992) Measurement of dry deposition flux of NH_3 on to coniferous forest. Environ Pollut 75: 3-13

Duyzer JH, Bouman AMH, Diederen HSMA & Van Aalst RM (1987) Measurement of dry deposition velocities of NH_3 and NH_4^+ over natural terrains. Report R 87/273. MT-TNO, Delft, The Netherlands

Erisman JW, Bleeker A & Van Jaarsveld H (1998a) Atmospheric deposition of ammonia to semi-natural vegetation in the Netherlands - methods for mapping and evaluations. Atmos Environ 32: 481-489

Erisman JW, Mennen MG, Fowler D, Flechard CR, Spindler G, Grüner A, Duyzer JH, Ruigrok W & en Wyers GP (1998b) Deposition monitoring in Europe. Environ Monit Assess 53: 279-295

Erisman JW, Otjes R, Hensen A, Jongejan P, van den Bulk, P, Khlystov A, Möls H & Slanina J (2001) Instrument development and application in studies and monitoring of ambient ammonia. Atmos Environ 35: 1913-1922

Farquhar G, Firth P, Wetselaar R & Weir B (1980) On the gaseous exchange of ammonia between leaves and the environment: determination of the ammonia compensation point. Plant Physiol 66: 710-714

Ferm M (1998) Atmospheric ammonia and ammonium transport in Europe and critical loads: a review. Nutr Cycl Agroecosys 51: 5-17

Geβler A & Rennenberg H (1998) Atmospheric ammonia: mechanisms of uptake and impacts on N metabolism of Plants. In: Kok LD & Stulen I (eds) Responses of Plant Metabolism to Air Pollution and Global Changes, pp 81-94. Backhuys Publishers, Leiden, The Netherlands

Geβler A, Rienks M & Rennenberg H (2000) NH_3 and NO_2 fluxes between beech trees and the atmosphere - correlation with climatic and physiological parameters. New Phytol 147: 539-560

Harrison R & McCartney H (1980) Ambient air quality at a coastal site in rural northwest England. Atmos Environ 14: 233-244

Hedin L, Granat L, Likens G & Rodhe H (1990) Strong similarities in seasonal concentration ratios of sulfate ion, nitrate ion and ammonium in precipitation between Sweden and the northeastern USA. Tellus 42: 454-462

Hovmand M, Andersen H, Løfstrøm P, Ahleson H & Jensen NO (1998) Measurement of the horizontal gradient of ammonia over a conifer forest in Denmark. Atmos Environ 32: 423-429

Husted S, Mattsson MA & Schjoerring JK (1996) Ammonia compensation points in two cultivars of Horedeum vulgare L. during vegetative and generative growth. Plant Cell Environ 19: 1299-1306

Husted S & Schjoerring JK (1996) Ammonia flux between oilseed rape plants and the atmosphere in response to changes in leaf temperature, light intensity and air humidity. Plant Physiol 112: 67-74

Husted S & Schjoerring JK (1995) Apoplastic pH and ammonium concentration in leaves of *Brassica napus* L. Plant Physiol 109: 1453-1460

Kesselmeier J, Meck L, Bliefernicht M & Helas G (1993). Trace gas exchange between terrestrial plants and atmosphere. Carbon dioxide, carbonyl sulfide and ammonia under the rule of the compensation point. In: Slanina J, Angeletti G & Bielke S (eds) General Assessment of Biogenic Emission and Deposition of Nitrogen Compounds, Sulfur Compounds and Oxidants in Europe. Air Pollution Research Report 47, pp 71-80. CEC, Brussels, Belgium

Kleiner D (1981) The transport of NH_3 and NH_4^+ across biological membranes. Biochim Biophys Acta 639: 41-52

Langford A & Fehsenfeld FC (1992) Natural vegetation as a source or sink for atmospheric ammonia - a case study. Science 255: 581-583

Nielsen KH & Schjoerring JK (1998) Regulation of apoplastic ammonium concentration in leaves of oilseed rape. Plant Physiol 118: 1361-1368

Nobel P (1991) Physiochemical and Environmental Plant Physiology. Academic Press, Inc, San Diego, U.S.A.

Rennenberg H, Kreutzer K, Papen H & Weber P (1998) Consequences of high loads of nitrogen for spruce (*Picea abies*) and beech (*Fagus sylvatica*) forests. New Phytol 139: 71-86

Schjoerring JK, Finnemann J, Husted S, Mattsson M, Nielsen KH & Pearson JN (1999) Regulation of ammonium distribution in plants. In: Gissel-Nielsen G & Jensen A (eds) Plant Nutrition - Molecular Biology and Genetics, pp 69-82. Kluwer Academic Publishers, Dordrecht, The Netherlands

Schulze E (1989) Air pollution and forest decline in a spruce (*Picea abies*) forest. Science 244: 776-783

Sutton MA, Fowler D, Burkhardt JK & Milford C (1995) Vegetation atmosphere exchange of ammonia: canopy cycling and the impact of elevated nitrogen input. Water Air Soil Pollut 85: 2057-2063

Sutton MA, Asman WAH & Schjoerring JK (1994) Dry deposition of reduced nitrogen. Tellus 46: 255-273

Sutton MA, Fowler D & Moncrieff JB (1993) The exchange of atmospheric ammonia with vegetated surfaces. 1. Unfertilized vegetations. Q J Roy Meteor Soc 119: 1023-1049

Van Hove L, Adema E, Vredenberg W & Pieters G (1989) A study of the adsorption of NH_3 and SO_2 on leaf surfaces. Atmos Environ 23: 1479-1486

Weyers J & Meidner H (1990) Methods in Stomatal Research. Longman Scientific & Technical, Essex, U.K.

Wyers GP & Erisman JW (1998) Ammonia exchange over a coniferous forest. Atmos Environ 32: 441-451

Wyers GP, Vermeulen AT & Slanina J (1992) Measurement of dry deposition of ammonia on a forest. Environ Pollut 75: 25-28

Chapter 3.2

Isoprene and other isoprenoids

Rainer Steinbrecher[1], Alex Geunther[2] and Guenther Seufert[3]

[1]*Fraunhofer Institut für Atmosphärische Umweltforschung, Kreuzeckbahnstr. 19, D-82467 Garmisch-Partenkirchen, Germany*
[2]*Atmospheric Chemistry Division, National Center for Atmospheric Research, Boulder CO 80307, United States of America*
[3]*Environment Institute, Joint Research Centre, T.P. 051, 21020 Ispra, Italy*

1.   INTRODUCTION

The biosphere is a major source of volatile organic substances (VOC) with about 1.15×10^{15} g carbon emitted annually and globally. Although this emission accounts for approximately 2% of global total C-exchange between biota and the atmosphere, it has not been considered in global C-cycling up to now. Isoprene dominates global trace gas emissions from vegetation and is about half of the global total. Most of the remainder is comprised of about twenty to thirty compounds. Since these compounds can have very different behaviour in the atmosphere, a relatively small emission could have an important role in global atmospheric chemistry. In addition, some compounds have a relatively small global emission but may dominate emissions in certain regions. Woodland landscapes cover 48% of all land surfaces and are estimated to contribute about three fourth of global isoprene and monoterpene emissions. Forests are the major source of isoprenoids in the biosphere with an estimated amount of 0.63×10^{15} g carbon emitted. These estimates show a considerable source of uncertainty with an uncertainty factor of up to 10. The initiation of collaborative efforts is leading to significant advances in emissions estimates and is mainly a result of improved measurement methods employing multi-scale and multi-disciplinary experiments.

R. Gasche et al. (eds.), Trace Gas Exchange in Forest Ecosystems, 175–191.

2.   ISOPRENOIDS

2.1   Definition of isoprenoids

The term 'isoprenoids' is pointing to the basic structural element of these compounds. The formally parent C_5 hydrocarbon 'isoprene' is combined in the isoprenoid biosynthetic pathway in manifold ways. Today most steps of this biosynthetic pathway are known (for review see e.g. Kreuzwieser et al. 1999). The group of monoterpenes are main constituents of essential oils in plants and are emitted in large quantities to the atmosphere especially from evergreen Mediterranean oaks (Seufert et al. 1997). The chemical structure

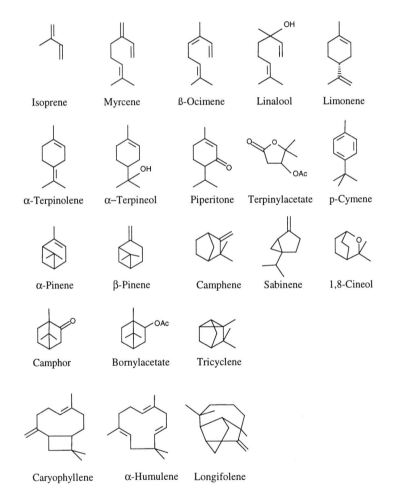

Figure 1. Chemical structures of selected plant isoprenoids

of selected isoprenoids is given in Figure 1. According to Ruzicka et al. (1953) who formulated the 'biogenetic isoprene rule' isoprenoids are synthesised by condensation of isoprene units (C_5H_8). Monoterpenes consist of 10 C-atoms in various arrangements: (1) linear molecules (e.g. myrcene), (2) mono-cyclic molecules (e.g. limonene), (3) bi-cyclic molecules (e.g. α-pinene, ß-pinene) and (4) tri-cyclic molecules (e.g. tricyclene). Apart from saturated and unsaturated hydrocarbon monoterpenes, oxygen containing derivates like aldehydes, keton esters and glycosides are formed.

2.2 Significance of isoprenoids for plants

The physiological function of volatile isoprenoids in plants is still under discussion. One of the most fascinating hypotheses is that monoterpenes like isoprene may protect leaves against high-temperature damage (Singsaas et al. 1997; Loreto et al. 1998) by enhancing membrane stability under heat stress. Also the formation of volatile, reduced and energy consuming compounds like isoprene and monoterpenes may be an efficient way of dispersing excess assimilatory power (ATP and NADPH) when stomata close under water limitation and heat stress (Steinbrecher et al. 1997).

2.3 Isoprenoids and the relevance for atmospheric chemistry

High photooxidant concentrations at ground level, e. g. ozone, have been shown to be detrimental to our environment. Ozone is not directly emitted by plants but rather produced photochemically in the presence of volatile organic compounds and nitrogen oxides (NO_x). Also reactions with the hydroxyl radical (HO) and the nitrate radical ($NO_3\bullet$) have to be considered. Such reactions lead to the formation of secondary substances (e.g., formaldehyde, peroxy radicals, carbonyl compounds etc.) that can enhance ozone as well as other photooxidant levels (e.g. peroxyacetylnitrate PAN) in the atmosphere. The importance of biogenic NMHC in tropospheric chemistry has been discussed extensively and the state of science is reviewed e.g. by Kley et al. (1999). Isoprenoids are of particular interest in atmospheric chemistry as they are highly reactive by usually containing one or more double bonds and they are emitted in large quantities from the biosphere in the tropics and during hot and sunny periods at northern and southern latitudes in relation to anthropogenic VOC (Guenther et al. 1995; Steinbrecher et al. 2000c; Derwent et al. 2001). Once reacted, isoprenoids can also form secondary organic particles that may impact the radiation balance of the earth with a feedback on the climate (Hoffmann et al. 1997; Leaitch et al. 1999).

2.4   Uncertainties with respect to biosphere/atmosphere exchange

The physiological basis of isoprenoid production in plants is well known (e.g.. Kreuzwieser et al. 1999). However, uncertainties still exist to what extent environmental factors (e.g. growth temperature, soil humidity, solar radiation intensity, nutrient levels etc.) modulate the isoprenoid emission potential. Currently, the temperature and light dependence of isoprene emission can be fairly well modelled and the current investigation of seasonal fluctuations in isoprene synthase activities and temperature as well as the energetic status of leaves help to improve modelling primary isoprene emission from vegetation (Guenther et al. 1997; Niinemets et al. 1999; Lehning et al. 2001; Petron et al. 2001). Once isoprenoids have escaped the plant they are transported, deposited as well as chemically degraded in the atmosphere. Therefore, only a fraction of primary emitted isoprenoids escape the forest canopy. Micrometeorological techniques can be used to quantify the above canopy flux. The uncertainty associated with isoprene and monoterpene flux measurements at a particular location and time can readily be assessed using error analysis and inter comparisons of independent techniques (e.g. eddy covariance). Generally, it is demonstrated that fluxes can be estimated within about 30% variation or better. A much greater challenge is the assessment of uncertainties associated with model predictions for locations and seasons that have not been characterized. There are few data available for developing a quantitative assessment of uncertainty. The situation is improving to some degree for certain compounds, seasons and locales but it remains extremely difficult to assign a valid quantitative uncertainty estimate for most biogenic emission estimates. Given the lack of quantitative uncertainty estimates, a qualitative assessment of the uncertainties associated with isoprene and monoterpene emissions is perhaps the best tool for describing the major uncertainties.

3.   DETERMINATION OF TRACE GAS EXCHANGE PROCESSES

For sound studies related to the exchange of trace gases between plant surfaces and the environment, sophisticated methods are required. The following chapter provides an up-to-date summary of different methods available for gas exchange studies on leaf as well as on synoptic scales.

3.1 Isoprenoid analysis in air samples

In air samples the mixing ratio of isoprenoids is generally in the range of some pmol mol^{-1} up to some nmol mol^{-1}. These low mixing ratios require a pre-concentration of the compounds prior to analysis in most cases, although sensitive detectors for isoprenoids became available making an online-analysis possible without pre-concentration. For isoprene an ozone chemiluminescence is available (Hills and Zimmerman 1990), for isoprene and the sum of monoterpenes a proton-transfer-mass-spectrometer can be used (Lindinger et al. 1998).

A detailed chemical speciation of the isoprenoids emitted into the atmosphere still requires the classical pre-concentration step followed by thermodesorption, separation by capillary gas chromatography and flame ionization and/or mass selective detection. Pre-concentration of the compounds is achieved by using a mixed-3-bed-adsorbent tube (e.g. CarbopackC, Carbopack, Carbopack X) in which the low volatile mono- and sesquiterpenes and the volatile isoprene are trapped. Separation of the trapped compounds after thermodesorption may be performed by a two dimensional gas chromatography. This methods allows the optimum choice of the separation columns for the low and high volatile organic compound fraction (e.g. column selection for compounds with less than six carbon atoms (<C6): aluminium oxide/potassium chloride film thickness 10 μm, inner diameter 0.53 mm, length 30 m; for compound with carbon number > C6: RTX-1701, film thickness 1 μm, inner diameter 0.25 mm, length 30 m). Further, this set up ensures that water present in every air sample does not affect the separation and detection performance of the complex compound mixture.

3.2 Enclosure techniques

For studying the exchange of gases between plants and the atmosphere several enclosure techniques are used. Depending on the scientific question to be solved ground, stem, branch and leaf enclosures have been developed.

The principle components are (1) air supply device, (2) the enclosure, (3) environmental sensor system, (4) gas analysis system (Fig. 2).

3.2.1   The air supply

For flushing the enclosure system ambient or conditioned air is used. The system may be operated in two ways: (1) by sucking air through the enclosures or (2) by pressing air through the system. In the following, the latter operation mode is described as it allows controlled impact studies of varying CO_2 concentration, air humidity or other gases on the isoprenoid emission. Pressurized pre-cleaned air needs additional water vapour and CO_2 for matching ambient air conditions provided by humidifying and CO_2-mixing devices. For investigating the influence of different air pollutants on the exchange behaviour of plant borne volatile organic compounds e.g. ozone may be added to the enclosure air.

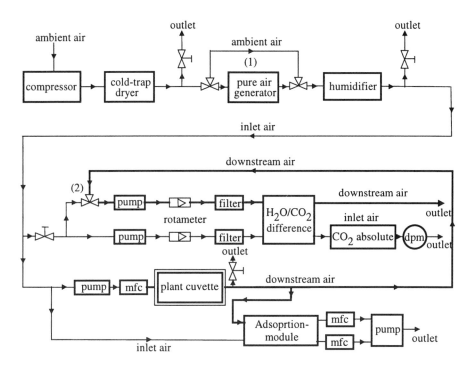

Figure 2. A enclosure system for studying trace gas exchange between plant parts and the atmosphere (dpm: dew point mirror; mfc: mass flow controller). (1) Solenoid valves for operating the system with ambient and purified air, respectively. (2) Solenoid valve for zero adjustment of the infrared gas analyzer.

As plants take up rapidly CO_2 and evaporate water during daytime especially under warm and sunny conditions they are changing the air gas composition in the enclosure dramatically if the air exchange rate of the enclosure is small. For avoiding feedback mechanisms to the plant metabolism resulting in changed isoprenoid exchange rates, the air flow through the enclosure is typically set to an value ensuring an exchange rate of less than one times per minute.

3.2.2   The enclosure

Different enclosure types are under operation. The size and shape of the enclosure is depending on the source type to be investigated. Common to all enclosures is that they are built of material which does not change the trace gas air composition in the enclosure as well as ensuring a natural light climate for the plant inside. For that reason e.g. Teflon or Tedlar membranes are used, if needed stabilized by an outer frame. The volume of the enclosure may vary from hundred of cubic-centimetres up to several cubic-meters. Typically, leaf enclosures have a size of up to 500 cm^3, whereas branch level and ground enclosures are sized up to 50 L. The biggest enclosures are used for whole plant studies with a volume of several cubic-meters. Fans inside the enclosures or high air flows ensure a mixing of the air volume avoiding unnatural high surface resistances which limit the gas exchange processes.

3.2.3   The environmental sensor and gas analysis system

Gas exchange between plants and the atmosphere is primarily controlled by light, temperature, and air concentrations of water vapour and CO_2. Therefore, it is essential to monitor these parameters in the enclosure during isoprenoid exchange studies. Special emphasis should be put on the temperature measurements. As plant surface temperature is not necessarily identical to the air temperature but plant temperature triggers substantially plant metabolism, plant surface temperature needs additionally to be recorded.

Differences in air mixing ratios before and after passing the enclosure are used to calculate the exchange rates of gases (e.g. Steinbrecher et al. 1997). The sensors for isoprenoids in air samples have been described above. Monitoring the CO_2 and water vapour exchange is needed for investigating relations between the emission of isoprenoids and photosynthesis. Photosynthetical parameters such as net photosynthesis, transpiration, conductances for water vapour and CO_2, and for CO_2 concentrations inside the leaf can be calculated according to Ball (1987). This information is also valuable for identifying any stress effects due to the transfer of plants into the enclosure and for avoiding artefact measurements of isoprenoid emission.

3.3   Micrometeorological methods

Micrometeorological flux measurement systems deployed on towers above a vegetation canopy are used to investigate the net trace gas flux (deposition and emission) from all components of a landscape. This is an important tool for evaluating biogenic emission estimates and for investigating diurnal and seasonal variations. Earlier investigations focus on the surface layer gradient. There has been a significant expansion in the types of tower flux measurement techniques in the past decade, including an improved surface layer gradient method and two additional methods: (1) relaxed eddy accumulation (REA) and (2) eddy covariance (Dabberdt et al. 1993; Oncley et al. 1993; Guenther et al. 1996; Steinbrecher et al. 2000a, b). The improved methods have been accompanied by more robust and less expensive equipment (e.g. sonic anemometers) which has greatly extended the application of these methods.

3.3.1   The flux gradient relationship

All micrometeorological techniques have certain requirements with respect to terrain and vegetation surface. The application of the flux gradient relationship requires a homogeneous surface such as grass- or crop-land. Only under these circumstances the constant flux assumption holds and the Muonin-Obukov theory can be applied resulting in reasonable reliable fluxes. If the terrain becomes more complex and the vegetation more rough the application of the flux gradient relationship results in large errors in estimated fluxes (Dabberdt et al. 1993). This method also requires that no degradation takes place when a compound is transported from one gradient level to the other. This assumption holds only for inert gases e.g. CO_2 or water vapour. Reactive gases like isoprenoids may be decomposed during the transport out of the canopy if the transport time is large compared to the life time of the compound in the atmosphere. Therefore, when using the flux gradient relationship for flux calculation of isoprenoids above forests, complex corrections have to be performed (Steinbrecher et al. 2000a; Rinne et al. 2000).

3.3.2   The eddy covariance technique

The most sophisticated method for determining turbulent surface fluxes is the eddy covariance technique (EC) where the eddy flux is directly determined by measuring the covariance of the chemical mixing ratio with the vertical wind speed. It has only been applied for a limited number of chemical species as the analytical system requires two constraints: (1) a fast response time (10 Hz) and (2) a specificity for the compound of interest. For the compound group of the isoprenoids, currently, only isoprene fluxes can be determined via EC technique using the ozone chemiluminescence detector (Hills and Zimmerman 1990).

3.3.3   Modifications of the eddy covariance technique

For investigating the turbulent isoprenoid exchange between the biosphere and the atmosphere, modifications of the EC technique are necessary. The eddy accumulation technique overcomes the need of fast response gas sensors. In this method, air is drawn from the vicinity of an anemometer measuring the vertical wind speed *w* and diverted into one of two accumulators on the basis of the sign of *w* at a pumping rate proportional to the magnitude of *w*. The mixing ratios of the specific compounds in the two accumulators can then be determined by slow response sensors. Businger and Oncley (1990) suggested that the demands of the eddy accumulation may be relaxed by sampling air at a constant rate for updrafts and downdrafts, rather than proportionally (relaxed eddy accumulation REA). Recent set up of REA-systems can be found e.g. in Steinbrecher et al. (2000b) and Schade and Goldstein (2001).

4.   FORESTS AND BIOSPHERE/ATMOSPHERE INTERACTION

Forests are the main source of isoprenoids at global as well as at regional scales (Guenther et al. 1995; Simpson et al. 1999; Schaab et al. 2001). The following section gives an overview of the potential isoprenoid emission in the major forest types around the world.

4.1   Boreal forests

Literature data of isoprenoid emission from Boreal plant species are rare. Isidorov (1994) reported mostly qualitative data. Janson (1993) reported some data on Scots pine (*Pinus sylvestris*) and Norway spruce (*Picea abies*). Hakola et al. (1998) surveyed the plant species *Salix phylicifolia*, *Betula pendula* and *Populus tremula*. A most recent investigation of Isidorov and Povarov (2000) on Russian forests gives only VOC totals and no speciated isoprenoid emission estimates. Speciated isoprenoid emission rate of typical boreal plant species are presented by Steinbrecher et al. (1999) which were used for a modelling exercise for Finnish forest emissions by Lindfors et al.

2000). Norway spruce is the most important isoprenoid emitter in the northern European boreal forests. In northern Finland a total of 516 kg isoprenoids per km^2 were calculated for a modelling period from April to October. In the southern part of the Finnish boreal zone isopreneoid fluxes are about a factor of two higher with α-pinene, β-pinene, Δ^3-carene and 1,8-cineole as most abundant emitted isoprenoids.

4.2   Temperate forests

Although isoprene-emitting trees comprise only about a third of all tree species in eastern North American mixed forests, isoprene is by far the dominant biogenic VOC in this region (Geron et al. 1994). Oak trees are responsible for about 75% of the total isoprene emission from these forests with *Liquidambar, Populus, Nyssa* and *Picea* species contributing most of the remainder. A large number of studies have been conducted in the mixed forest landscapes of the eastern United States (Geron et al. 2000; Guenther et al. 2000a). The total isoprene flux from North America amounts to 29.4 x 10^{12} g of carbon (Tg C) whereas the other isoprenoids contribute 21 Tg C to the total amount of 84 TgC non methane volatile organic compounds emitted. As the models have been parameterized in sufficient detail emission predictions are typically within a factor of two of field observations for isoprene.

Although significant monoterpene emissions occur, isoprenoid emissions from the pine forests of western North America are dominated by the oxygenated hemiterpene, 2-methyl-3-buten-2-ol (MBO). As a result, MBO is the dominant ambient VOC in some rural areas (Schade and Goldstein 2001; Baker et al. 2001). Harley et al. (1998) observed significant MBO emissions from about half of the 34 pine species that they investigated. All of the MBO emitters were North American species and were members of the subgenus *Pinus*. Of the species identified as MBO emitters, *P. ponderosa* and *P. contorta* are the most widespread. MBO has a midday lifetime of about 2 hours and should have an influence on OH and ozone that is similar to isoprene. An important difference is the large yield of acetone during the oxidation of MBO (Ferronoto et al. 1998).

European temperate forests are very heterogeneous in the type of the isoprenoid emission potential as beech (*Fagus sylvatica*), one of the dominant deciduous temperate forest trees in Europe, is a very weak isoprenoid emitter. Other forest forming trees are oaks (*Quercus robur*, and *Q. petraea*), Norway spruce (*Picea abies*), Scots pine (*Pinus sylvestris*), fir (*Abies alba*), and Douglas fir (*Pseudotsuga menziesii*). Chemically speciated isoprenoid emission factors for European tree and other plant species are summarized in Mannschreck et al. (2001).

For Germany a biogenic emission inventory was constructed for the year 1998 (Steinbrecher et al. 2000c) emphasising the importance of using plant species specific landuse data bases and emission factors. Most of the biogenic emissions occur in the middle and southern part of Germany with values of more than 400 kg C km^{-2} month^{-1} in the summer time. Compared to isoprenoid emission inventories which only consider general forest types (e.g. deciduous, coniferous, and mixed forest), the use of tree species distribution for Germany significantly reduced systematic errors associated with the assignment of emission factors to the source categories. Isoprene emissions using plant species data are in general 30% lower compared to emission estimates considering only forest types. Furthermore, the plant species specific emission inventory of Germany shows significantly different geographic distribution patterns shifting the focus from areas with high forest proportions to areas with large oak stands. The monthly isoprenoid amount estimated with leaf area index distributions actualized every ten days indicate significant lower amounts in springtime compared to calculations with literature-based biomass densities not considering seasonal effects. Further, forests dominated by evergreen species have a potentially longer period for emissions although there have been few investigations that have included measurements outside of the summertime season.

4.3   Mediterranean Forests

Emissions of isoprenoids from Mediterranean forests are characterized by some special features due to the specific type of vegetation growing in the Mediterranean region and in Mediterranean climate conditions, which are characterized by cool wet winters and hot and dry summers.

The isoprenoid emission factors for Mediterranean vegetation compiled by Seufert et al. (1997) were much different from the values of 1.2 and 16 µg C g^{-1} h^{-1} assigned to monoterpene and isoprene emissions, respectively, from Mediterranean ecosystems (Guenther et al. 1995). Evergreen oak species like *Q. ilex, Q. coccifera, Q. rotundifolia* emit monoterpenes in the range of 15–20 µg C g^{-1} h^{-1}. Isoprene emissions from deciduous oaks like *Q. pubescens, Q. frainetto, Q. pyrenaica, Q. canariensis* were in the range of 85 µg C g^{-1} h^{-1}, whereas the deciduous oak *Q. cerris* and the evergreen oak *Q. suber* did not show any measurable emissions during early summer (Steinbrecher et al. 1997; Csiky and Seufert 1999). The only monoterpene emitter among deciduous oaks was the East Mediterranean oak species *Q. ithaburiensis* (Csiky and Seufert 1999). Therefore, a first attempt to assigning more realistic emission rates to Mediterranean oakwoods could be about 70 µg C g^{-1} h^{-1} isoprene for deciduous oaks, and 15µg C g^{-1} h^{-1} monoterpenes for evergreen oak woods. It is obvious that oaks are the dominant source of isoprenoids in the Mediterranean area, considering that most parts of mountainous areas of Italy, France and Spain are covered by deciduous oak forests, and that unmanaged lands below 600 m are dominated by evergreen oakwoods and shrublands (e.g. Schaab et al. 2001).

The parameterizations of the temperature and light controls proposed by Guenther et al. (1993) have been shown to predict well the short term variation of isoprene emissions from Mediterranean species; interestingly, it performed also well in cases when monoterpene emissions originated from a short term photosynthesis related pool, e.g., *Q. ilex* (all monoterpenes, Staudt and Seufert 1995; Bertin et al. 1997) and *P. pinea* (t-β-ocimene, linalool, 1,8-cineole, Staudt et al. 2000). Weak points of the model appeared during the diurnal and seasonal onset and offset of emissions, suggesting that a diurnal and seasonal activity variation of enzymes is involved in emission control and a biochemical process based model would better predict the diurnal and seasonal variation of the isoprenoid emission (Lehning et al. 2001). The worst performance of the model became evident during Mediterranean summer conditions by a distinct non-linearity of isoprenoid emissions and of the environmental drivers, light and temperature. In the case of light driven monoterpene emissions, the temperature optimum has been identified at 35°C instead of 40°C in the original isoprene model of Guenther et al. (1993). Simultaneous exposure to drought and heat at reduced physiological activity is typical for Mediterranean plants (Steinbrecher et al. 1997) and is not treated by the presently used emission algorithm. Interestingly, the thermotolerance of *Q. ilex* and *Q. suber* leaves could be increased by fumigation with monoterpenes in controlled environments (Loreto et al. 1998).

The prediction of monoterpene emissions from plants containing monoterpenes in reservoirs suffers from a high variability of temperature dependencies. Instead of a β-value of 0.09 °C^{-1}, as proposed by Guenther et al. (1993) to be representative for all species, we observed a temperature sensitivity in the range between 0.06 and 0.43 °C^{-1}. The highest slopes were found for t-β-ocimene and linalool emissions from *P. pinea*, giving additional evidence that temperature alone is not sufficient to explain emissions of these compounds (Staudt et al. 2000). The calculation of emission factors on the basis of parameters taken from the literature caused significant over- and underestimates of emissions in comparison with the empirical emission rates observed near standard conditions.

4.4   Tropical Rainforests

Isoprene and monoterpene emissions from tropical rainforest landscapes have been characterized to a limited degree in Africa, South America and

Asia. These regions have some of the most dynamic, yet poorly understood biosphere-atmosphere interactions on earth. External factors, such as human induced biomass burning and the strong vertical mixing by convective systems, contribute to giving surface exchanges from these ecosystems an extremely important influence on global atmospheric chemistry. Jacob and Wofsy (1988) have shown that biogenic VOC emissions in the Amazon could have a strong impact on atmospheric chemistry within these regions. Global chemistry and transport models (Guenther et al. 2000b) suggest that biogenic VOC emissions in tropical regions can impact chemical composition at a global scale, especially in regions with biomass burning, which provides a source of oxides of nitrogen.

Guenther et al. (1995) estimated that tropical woodlands are responsible for almost half of total global biogenic VOC emissions. This is partly because biogenic VOC emissions are emitted from tropical systems throughout the year. Other reasons include the generally high foliar densities, temperatures and solar radiation fluxes. The recent study of Guenther et al. (2000b) in central Africa suggests that the annual isoprene emission in that region amounts to 35 Tg C which is only 14% less than that predicted by the earlier model.

5. CONCLUSIONS

It has been pointed out that we now have very sophisticated methods available to study the isoprenoid exchange between the biosphere and the atmosphere on different scales ranging from leaf, branch, tree, to canopy levels. These tools will further deepen our understanding in processes controlling the isoprenoid emission on different spatial scales leading to better estimates of the compound mix escaping the forests in different parts on earth on a hourly, daily, monthly or yearly basis. That is of particular interest as isoprenoids are reactive enough to disappear within the canopy in time-scales of transport from the place of emission to the above canopy layer where fluxes are measured (Ciccioli et al. 1999). The finding has important methodological implications with respect to isoprenoid sampling; and moreover, for calculating fluxes by the gradient method as the constant flux theory is not fullfilled (Steinbrecher et al. 2000a). Substantial efforts are required to understand better the within canopy transport, chemistry and deposition of reactive isoprenoids, and the formation of reaction products like carbonyls and organic particles.

As isoprenoids clearly are a source of atmospheric carbon they thus have a key role in the global carbon budget and cycling. Isoprenoid emission is triggered by environmental variables and feedback loops among global environmental change and hydrocarbon emissions are likely. Warming for example can increase the emission as isoprenoid production strongly depends on temperature. Also elevated CO_2 and nitrogen deposition may augment the isoprenoid emission through increased productivity in ecosystems as well as a shift in biodiversity. Perhaps the most important factor changing global isoprenoid emission is the ongoing land use change from forest to agriculture, especially in the tropical regions. In addition, some of the major species in agroforestry are high isoprenoid emitters (*Poplar*, *Liquidambar*, *Eucalyptus*) dramatically shifting the regional as well as the global budget of isoprenoid emission. Another aspect of isoprenoids, not very well studied so far, is the impact on the regional radiation balance. Two processes have to be considered: (1) direct absorption of thermal energy by e.g. α–pinene and (2) release of latent heat when condensation starts from secondary compounds formed by the chemical degradation of monoterpenes in a humid forest atmosphere. Both processes may lead to higher air temperatures above the forests initialising the same feed back loop described above. To address the processes described, integrated research efforts are required involving plant scientists, atmospheric chemists and physicists as well as foresters and geographers.

REFERENCES

Baker B, Guenther A, Greenberg J & Fall R (2001) Canopy level fluxes of 2-methyl-3-buten-2-ol, acetone, and methanol by a portable relaxed eddy accumulation system. Envion Sci Technol 35: 1701-1708

Ball TJ (1987) Calculations related to gas exchange. In: Zeiger E, Farquhar GD & Cowan JR (eds) Stomatal Function, pp 445-467. Stanford University Press, Stanford, U.K.

Bertin N, Staudt M, Hansen U, Seufert G, Ciccioli P, Foster P, Fugit J-L & Torres L (1997) Diurnal and seasonal course of monoterpene emissions by *Quercus ilex* L. under natural conditions - Application of light and temperature algorithms. Atmos Environ 31-S1: 135-144

Businger JA & Oncley SP (1990) Flux measurement with conditional sampling. J Atmos Ocean Tech 7: 349-352

Ciccioli P, Brancaleoni E, Frattoni M, Di Palo V, Valentini R, Tirone G, Seufert G, Bertin N, Hansen U, Csiky O, Lenz R & Sharma M (1999) Emission of reactive compounds from orange orchards and their removal by within-canopy processes. J Geophys Res 104: 8077-8094

Csiky O & Seufert G (1999) Terpenoid emissions of Mediterranean oaks and their relation to taxonomy. Ecol Appl 9: 1138-1146

Dabberdt WF, Lenschow DH, Horst TH, Zimmerman PR, Oncley SP & Delany AC (1993) Atmosphere-surface exchange measurements. Science 260: 1472-1482

Derwent RG, Jenkin EM, Saunders SM & Pilling MJ (2001) Characterization of the reactivties of volatile organic compounds using a master chemical mechanism. J Air Waste Manage 51: 699-707

Ferronato C, Orlando J & Tyndall GS (1998) The rate and mechanism of the reactions of OH and Cl with 2-methyl-3-buten-2-ol. J Geophys Res 103: 25579-25586

Geron C, Guenther A & Pierce T (1994) An improved model for estimating emissions of volatile organic compounds from forests in the eastern United States. J Geophys Res 99: 12772-12792

Geron C, Rasmussen R, Arnts RR & Guenther A (2000) A review and synthesis of monoterpene speciation from forests in the United States. Atmos Environ 34: 1761-1781

Guenther A (1997) Seasonal and spatial variations in the natural volatile organic compound emissions. Ecol Appl 7: 34-45

Guenther A, Geron C, Pierce T, Lamb B, Harley P & Fall R (2000a) Natural emissions of non-methane volatile organic compounds; carbon monoxide, and oxides of nitrogen from North America. Atmos Environ 34: 2205-2230

Guenther A, Greenberg J, Helmig D, Klinger L, Vierling L, Zimmerman P & Geron C (1996) Leaf, branch, stand and landscape scale measurements of volatile organic compound fluxes from U.S. woodlands. Tree Physiol 16: 17-24

Guenther AB, Baugh B, Brasseur G, Greenberg J, Harley P, Klinger L, Serca D & Vierling L (2000b) Isoprene emission estimates and uncertainties for the Central African EXPRESSO study domain. J Geophys Res 104: 30625-30639

Guenther AB, Hewitt CN, Erickson D, Fall R, Geron C, Graedel T, Harley P, Klinger L, Lerdau M, McKay WA, Pierce T, Scholes B, Steinbrecher R, Tallamraju R, Taylor J & Zimmerman P (1995) A global model of natural volatile organic compound emissions. J Geophys Res 100: 8873-892

Guenther AB, Zimmerman PR, Harley P, Monson RK & Fall R (1993) Isoprene and monoterpene emission rate variability: model evaluations and sensitivity analyses. J Geophys Res 93: 12609-12617

Hakola H, Rinne J & Laurila T (1998) The hydrocarbon emission rates of tea-leafed willow (*Salix phylicifolia*), silver birch (*Betula pendula*) and European aspen (*Populus tremula*). Atmos Environ 32: 1825-1833

Harley P, Fridd-Stroud V, Greenberg J, Geunther A & Vasconcellos P (1998) Emission of 2-methyl-3-buten-2-ol by pines: A potentially large natural source of reactive carbon to the atmosphere. J Geophys Res 103: 25479-25486

Hills AJ & Zimmerman PR (1990) Isoprene measurement by ozone-induced chemiluminescence. Anal Chem 62: 1055-1060.

Hoffmann T, Odum J R, Bowman F, Collins D, Klokow D, Flagan RC & Seinfeld JH (1997) Formation of organic aerosols from the oxidation of biogenic hydrocarbons. J Atmos Chem 26: 189-222

Isidorov VA (1994) Volatile emission of living plants: composition, emission rates, and ecological significance (in Russian). Alga Association, St. Petersburg: Russia, 188

Isidorov VA & Porvarov VG (2000) Phytogenic volatile organic compounds emission by Russian forests. Ecol Chem 9: 10-21.

Jacob DJ & Wofsy SC (1988) Photochemistry of biogenic emissions over the Amazon forest. J Geophys Res 93: 1477-1486

Janson RW (1993) Monoterpene emission from Scots Pine and Norwegian Spruce. J Geophys Res 98: 2839-2850

Kley D, Kleinmann M, Sandermann H & Krupa S (1999) Phtochemical oxidants: State of the science. Environ Poll 100: 19-42

Kreuzwieser J, Schnitzler J-P & Steinbrecher R (1999) Biosynthesis of organic compounds emitted by plants. Plant Biol 1: 149-159

Leaitch WR, Bottenheim JW, Biesenthal TA, Li SM, Liu PSK, Asalian K, Dryfhout-Clark H, Hopper F & Brechtel F (1999) A case study of gas-to-particle conversion in an eastern Canadian forest. J Geophys Res 104: 8095-8111

Lehning A, Zimmer W, Zimmer I, Schnitzler J-P (2001) Modelling of annual variations of oak (*Quercus robur* L.) isoprene synthase activity to predict isoprene emission rates. J Geophys Res 106: 3157-3166

Lindfors V, Laurila T, Hakola H, Steinbrecher R & Rinne J (2000) Modeling speciated terpenoid emissions from the European boreal forest. Atmos Environ 34: 4983-4996.

Lindinger W, Hansel A & Jordan A (1998) On-line monitoring of volatile organic compounds at ppt levels by means of Proton-Transfer-Reaction Mass Spectrometry (PTR-MS). Medical applications, food control, and environmental research. Int. Mass Spectrom. Ion Proc., 173: 191-241

Loreto F, Foster A, Durr M & Seufert G (1998) On the monoterpene emission under heat stress and on the increased thermotolerance on leaves of *Quercus ilex* L. fumigated with selected monoterpenes. Plant Cell Environ 21: 101-107

Mannschreck K, Bächmann K, Barnes I, Becker KH, Heil T, Kurtenbach R, Memmesheimer M, Mohnen V, Schmitz T, Steinbrecher R, Obermeier A, Poppe D, Volz-Thomas A & Zabel F (2002) A database for volatile organic compounds. J Atmos Chem 42: 281-286

Niinemets U, Tenhunen JD, Harley PC & Steinbrecher R (1999) A novel model of isoprene emission based on energetic requirements for isoprene synthesis and leaf photosynthetic properties for *Liquidambar* and *Quercus*. Plant Cell Environ 22: 1319-1336

Oncley SP, Delany AC, Horst TW & Tans PP (1993) Verification of flux measurement using relaxed eddy accumulation. Atmos Environ 27A : 2417-2426

Petron G, Harley P, Greenberg J & Geunther A (2001) Seasonal temperature variations influence isoprene emission. Geophys Res Lett 28: 1707-1710

Rinne J, Tuovinen J-P, Laurila T, Hakola H, Aurela M & Hypen H (2000) Measurements of hydrocarbon fluxes by the gradient method above a northern boreal forest. Agri For Meteorol 102: 25-37

Ruzicka L, Eschenmoser A & Heusser H (1953) The isoprene rule and the biosynthesis of terpenic compounds. Experientia 9: 357-396

Schaab G, Steinbrecher R, Lacaze B & Lenz R (2000) Assessment of long-term changes on potential isoprenoid emissions for a mediterranean typ ecosystem in France. J Geophys Res 105: 28863-28874

Schade WG & Goldstein AH (2001) Fluxes of oxygenated volatile organic compounds from a ponderosa pine plantation. J Geophys Res 106: 3111-3123

Schnitzler J-P, Lehning A & Steinbrecher R (1997) Seasonal pattern of isoprene sysnthase activity in *Quercus robur* leaves and its impact on modeling isoprene emission rates. Botanica Acta 110: 240-243

Seufert G, Bartzis JG, Bomboi-Mingarro R, Ciccioli P, Cieslik S, Dlugi R, Foster P, Hewitt N, Kesselmeier J, Kotzias D, Manes F, Perez-Pastor T, Steinbrecher R, Torres L, Valentini R & Versino B (1997) An overview of the Castelporziano experiments. Atmos Environ 31-S1 : 5-17

Simpson D, Winiwarter W, Börjesson G, Cinderby S, Ferreiro A, Guenther A, Hewitt N, Janson R, Khalil MAK, Owen S, Pierce T, Puxbaum H, Shearer M, Skiba U, Steinbrecher R, Tarrason L & Öquist MG (1999) Inventorying emissions from nature in Europe. J Geophys Res 104: 8113-8152

Singsaas EL, Lerdau M, Winter K & Sharkey TD (1997) Isoprene increases thermotolerance of isoprene-emitting species. Plant Physiol 115: 1412-1420

Staudt M, Bertin N, Frenzel B & Seufert G (2000) Seasonal variation in amount and composition of monoterpenes emittd by young *Pinus pinea* trees – Implications for emission modelling. J Atmos Chem 35: 77-99

Staudt M & Seufert G (1995) Light dependent emission of monoterpenes by holm oak (*Quercus ilex* L.) Naturwissenschaften 82: 89-92

Steinbrecher R, Hauff K, Rabong R & Steinbrecher J. (1997) Isoprenoid emission of oak species typical for the Mediterranean area: Source strenght and controlling variables. Atmos Environ 31-S1 : 79-88

Steinbrecher R, Hauff K, Hakola H & Rössler J (1999) A revised parametrisation for emission modelling of isoprenoids for boreal plants. In: Laurila T & Lindfors V (eds) Biogenic VOC Emission and Photochemistry in the Boreal Regions of Europe, pp 29-43. European Communities 1999

Steinbrecher R, Klauer M, Hauff K & Mayer H (2000b) Canopy fluxes of reactive volatile organic compounds over a Scots pine plantation in southern Germany. http://zdb-imk.physik.uni-karlsruhe.de/cgi-bin/fetch.pl/poster/lt2-c5.pdf

Steinbrecher R, Klauer M, Hauff K, Stockwell W, Jaeschke W, Dietrich T & Herbert F (2000a) Biogenic and anthropogenic fluxes of non-methane-hydrocarbons over an urban-impacted forest, Frankfurter Stadtwald, Germany. Atmos Environ 34: 3779-3788

Steinbrecher R, Smiatek G, Schoenemeyer T & Schaab G (2000c) Development of Biogenic VOC-Emission Inventories for Germany. http://zdb-imk.physik.uni-karlsruhe.de/cgi-bin/fetch.pl/poster/lt2-a3.pdf

Chapter 3.3

Aldehydes and organic acids

Jürgen Kreuzwieser
Institut für Forstbotanik und Baumphysiologie, Professur für Baumphysiologie, Georges-Köhler-Allee, Geb. 053/054, D-79110 Freiburg i. Br., Germany

1.   INTRODUCTION

Organic acids and aldehydes are ubiquitous chemical constituents in the atmosphere which are partially emitted from vegetation. Numerous publications indicate that formic and acetic acid as well as formaldehyde and acetaldehyde are of particular quantitative and qualitative importance for tropospheric chemistry. Other aldehydes and organic acids in the atmosphere which are possibly derived from biogenic sources include oxalic, propionic and pyruvic acids (Lunde et al. 1977; Norton et al. 1983; Kawamura et al. 1985; Andreae et al. 1987, 1988; Hofmann et al. 1997) as well as propanal, butanal and isobutanal (Isidorov et al. 1985; König et al. 1995). In remote areas, some semi-volatile aldehydes (hexanal, heptanal, octanal, nonanal, decanal, undecanal) were present ubiquitously in the atmosphere (Yokouchi et al. 1990). Acetaldehyde and formaldehyde concentrations measured in the atmosphere amounted to 0.3 to 5.0 ppbv at rural and forested sites, but increased up to 176 ppbv in urban areas (Table 1). The concentrations of acetic and formic acid are in a similar range between approximately 0.1 and 10.5 ppbv at urban and 0.1 and 7.5 ppbv at semi-rural and rural sites (Table 2). Less information is available for the concentrations of higher molecular weight aldehydes and carboxylic acids in ambient air. Values measured for these compounds varied between 0.01 and 0.26 ppbv (organic acids, Table 2) and 0.02 and 81 ppbv (aldehydes, Table 3). Due to their high reactivity (and therefore short half-life), the ambient mixing ratios of these compounds depend directly on the rates of production and destruction and

R. Gasche et al. (eds.), Trace Gas Exchange in Forest Ecosystems, 193–209.
© 2002 *Kluwer Academic Publishers. Printed in the Netherlands.*

only to a minor extent on long-range transport. In addition to anthropogenic sources, vegetation is considered a significant source, as well as a sink of these compounds. This chapter summarizes the actual knowledge on exchange processes of organic acids and aldehydes between trees and the atmosphere and discusses factors controlling these processes.

2.   EVIDENCE FOR BIOGENIC EMISSION OF ALDEHYDES AND ORGANIC ACIDS

The importance of emissions by vegetation has been assumed from (1) the observation of considerable amounts of formaldehyde and acetaldehyde, as well as formic and acetic acid, in the air of forests (Isidorov et al. 1985; Guenther et al. 1994), (2) the generally higher concentrations of acetaldehyde, formaldehyde and acetic, as well as formic acid during the growing season in summer as compared to the winter (Talbot et al. 1988; Mosello and Tartari 1992) and (3) the finding that the atmospheric concentrations of organic acids and aldehydes are often highest in the afternoon and lowest during the night (Yokouchi et al. 1990; Hahn et al. 1991; Khwaja 1995; Khare et al. 1997). More recent studies in which cuvette systems were used to investigate the emission of trace gases by various tree species clearly indicated that formic and acetic acid, as well as their aldehydes, are directly emitted by the leaves of plants (Talbot et al. 1990; Hahn et al. 1991; Kesselmeier et al. 1992, 1997, 1998; Kreuzwieser et al. 1999). These results are strongly supported by a recent study at a semi-rural site in Italy (Larsen et al. 1998). The isotope signatures (^{14}C-abundance) of different carbonyls (acetone, formaldehyde, acetaldehyde, propanal, butanal) in ambient air indicate that around 41–57% of these compounds must have been derived from biogenic sources and were not a product of the combustion of fossil resources.

Qualitative information on the emission of higher weight organic acids and aldehydes from trees is available from cuvettes studies. Isidorov et al. (1985), for example, reported the emission of propanal (sorb), i-butanal (Northern white cedar), crotonal (Chinese arbor vitae) and i-butenal (willow, aspen, Balsam poplar, European oak, Scots pine, European larch). Working mainly with crops, other authors identified emissions of the aldehydes hexanal, hexenal, nonanal and benzaldehyde (Buttery et al. 1982, 1985; Arey et al. 1991; König et al. 1995).

Table 1. Ambient concentrations (ppbv) of aldehydes at rural, semi-urban and urban sites.

Location	HCHO	CH₃CHO	Reference
Urban:			
Los Angeles, CA	2 – 40		Grosjean et al. (1983)
New York city	2		Cleveland et al. (1977)
Claremont, CA	16 – 19		Tuazon et al. (1981)
Claremont, CA	11.2		Grosjean (1988)
Warren, MI	1.3 – 6.5		Lipari et al. (1984)
Denver, CO	12 ± 6		Salas and Singh (1986)
Denver, CO	2.3 – 3.9		Anderson et al. (1996)
Sao Paulo, Brazil	10.7		Grosjean et al. (1990)
Mexico	5.9 – 9.7		Baez et al. (1995)
Copenhagen, DK	0.2 – 6.4	0.2 – 1.8	Granby et al. (1997)
Paris, France	4 – 32	2 – 9	Kalabokas et al. (1988)
Budapest, Hungary	7 – 176	0-35	Haszpra et al. (1991)
Rome, Italy	8.2 – 17	2.9 – 6.6	Possanzini et al. (1996)
Urawa. Japan	2.5 – 11	1.3 – 6.8	Satsumabayashi et al. (1995)
Atlanta, GE		0.4 – 8.4	Grosjean (1993)
Semi-urban:			
Lille Valby, DK	0.1 – 2.8	0.1 – 1.4	Granby et al. (1997
Vienna, Austria	6.4 – 13		Puxbaum et al. (1988)
Albany, NY, US	0.6 – 3.7		Khwaja (1995)
Rural:			
The Alps, FRG	0.5 – 3.2	0.3 – 1.2	Slemr (1992)
Jülich, FRG	0.1 – 6.5		Platt and Perner (1980)
Jülich, FRG	0.9 – 5		Schubert et al. (1984)
Eifel, FRG	0.03 – 5		Lowe et al. (1981)
Deuselbach, FRG	2.3 – 3.9		Schubert et al. (1984)
Deuselbach, FRG	0.4 – 5.0		Neitzert and Seiler (1981)
Black Forest, FRG	0.1 – 1.8		Slemr et al. (1996)
Sarnia, Ontario	1.5		Harris et al. (1989)
Pennsylvania	1.3 – 4.3		Martin et al. (1991)
Virginia, US	0.5 – 2.7		Hastie et al. (1993)
Gopalpura, Agra	0.98		Munger et al. (1995)
Dorset, Ontario	0.3 – 4.2		Khare et al. (1997)
Egbert	1.4 – 2.2	0.5 – 0.7	Shepson et al. (1991)
Southeastern US	0.9 – 3.2	0.6 – 0.7	Lee et al. (1995)
Northeastern Germany	1.7 – 1.9	1.3 – 1.6	Benning and Wahner (1998)

Table 2. Concentrations (ppbv) of organic acids in ambient air of rural, urban and semi-urban sites.

Location	Formic	Acetic	Propionic	Pyruvic	Reference
Urban:					
Claremont, CA	0.02 – 20	0.2 – 18			Tuazon et al. (1981)
Claremont, CA	0.4 – 15	0.2 – 15			Grosjean (1988)
Southwestern U.S.	3.0	4.0			Dawson and Farmer (1988)
Southern CA	1 – 13	2 – 16			Grosjean (1989)
Yokohama, Japan	7.3	3.8			Tokos et al. (1992)
Yokohama, Japan	0.2 – 4.8	0.1 – 3.2			Morikami et al. 1991)
Fukaya, Japan		1.0 – 7.6			Satsumabayashi et al. (1995)
Nagoya, Japan	0.8 – 2.9		0.04–0.26		Hoshika (1982)
Los Angeles	0.1 – 3.0	0.3 – 3.9			Kawamura et al. (1985)
Los Angeles	1.3 – 13	1.9 – 16			Grosjean (1989)
Boston	1.8 – 15	0.8 – 5.4			Lawrence and Koutrakis (1994)
Copenhagen, DK	0.3 – 1.5	0.4 – 2.8			Granby et al. (1997)
Brussels, Belgium	0.9 – 8.5	1.3 – 8.1			Granby et al. (1997)
Wilmington, US	2.5 – 3.4	1.0 – 2.5			Willey and Wilson (1993)
Semi-urban					
Schenectady, U.S.	0.8 – 2.5	0.6 – 3.4			Khwaja (1995)
Lille Valby, DK	0.1 – 2.3	0.2 – 2.1			Granby et al. (1997)
Frankfurt, FRG	0.1 – 2.8	0.1 – 2.2			Hartmann et al. (1989)
Vienna, Austria	0.3 – 3.8	0.4 – 0.8			Puxbaum et al. (1988)
Albany, NY, US	0.8 – 2.5	0.6 – 3.4			Khwaja (1995)
Eastern US	0.5 – 21	0.2 – 6.3		0.04–0.27	Talbot et al. (1995)
Mainz, FRG	1.1 – 2.1	1.1 – 2.2	0.07–0.21		Hofmann et al. (1997)
Rural:					
Bavaria, FRG	0.6 – 7.0	1.2 – 7.5			Enders et al. (1992)
Virginia, US	0.7 – 1.9	0.7 – 1.3			Talbot et al. (1988)

Location	Formic	Acetic	Propionic	Pyruvic	Reference
Southwestern US	0.7	0.6			Dawson and Farmer (1988)
Austria	0.9	0.5			Puxbaum et al. (1988)
Germany	0.2	0.7			Hartmann et al. (1989)
Amazon (wet)	0.1 – 1.0	0.1 – 1.8		0.025	Talbot et al. (1990)
Amazon (dry)	1.3 – 2.9	1.0 – 3.0			Andreae et al. (1988)
Amazon (dry)	1.2 – 2.9			0.08–0.4	Andreae et al. (1987)
Pennsylvania	0.1 – 10	0.1 – 6.5		0.01–0.1	Martin et al. (1991)
Congo	0.5	0.6			Helas et al. (1992)
Hawaii	0.5	0.4			Norton (1992)
Venezuela	0.8 – 1.7	0.5 – 1.4			Sanhueza et al. (1992)
Virginia	5.6	2.1			Talbot et al. (1995)
Gopalpura, Agra	1.7	1.6			Khare et al. (1997)

Table 3. Ambient concentrations (ppbv) of higher molecular weight aldehydes possibly emitted by vegetation. Data compiled from (1) Ciccioli et al. (1993), (2) Solberg et al. (1993) and (3) Yokouchi et al. (1990).

Compound	Reference	Concentration
Propanal	1	0.02 – 0.8 (1)
Pentanal	1	0.15 – 3.0
	2	0.04 – 0.2
Hexanal	3	0.1 – 3.0
	2	0.5 – 10.0
	1	0.07 – 0.4
Heptanal	3	0.2 – 3.0
	2	0.7 – 14.0
Octanal	3	0.4 – 6.0
	2	1 – 18.0
Nonanal	3	0.8 – 10.0
	2	1 – 81.0
Decanal	3	0.6 – 7.0
	2	1 – 31
Undecanal	3	0.05 – 1.0

3.   FACTORS INFLUENCING GAS EXCHANGE PROCESSES

The flux rate of gases between leaf cells and the atmosphere is controlled by physico-chemical and plant internal factors. The physico-chemical factors include (1) the resistances on the individual gas between the site of production and the atmosphere, (2) the concentration gradients between these sites and (3) and the solubility of the gas in the aqueous phase of the apoplastic space. Plant internal factors influencing gas exchange processes are the rate of biosynthesis of the specific compound and the developmental stages of plant organs and whole plants.

3.1   Internal resistances

Gas exchange between cells, tissues or whole plants and the atmosphere requires to overcome various plant internal resistances. Since physiological processes mainly take place either in the cytoplasm or in cell compartments, the bordering membranes (plasmalemma, tonoplast, plastid envelopes) have to be passed prior to emission. The diffusive flux through these membranes is determined by the molecular size and the lipophilic character of the specific compound. Since aldehydes and organic acids are polar molecules, they are not likely to be dissolved in the lipophilic membrane and, therefore, the diffusive flux should be slow. It would be interesting to know whether carrier proteins are involved in the membrane transport of these polar compounds from the cytoplasm into the apoplastic space as discussed by Gabriel et al. (1999) for the transport of organic acids.

The apoplastic space is the site where the emitted compounds are transferred from the liquid into the gaseous phase. Since the apoplastic space is slightly acidic with a pH of 5 to 6.5 (Kesselmeier et al. 1998), weak organic acids will be protonated during uptake. The volatilization into the internal gas phase of a leaf depends on (1) the chemical properties of the aqueous phase and the individual compound to be emitted, e.g. its solubility in the apoplastic solution (Gabriel et al. 1999), and (2) physical factors such as the ambient concentration of the individual gas, its vapour pressure and temperature. As discussed by Gabriel et al. (1999), a reduction of the apoplastic pH, which may occur during the light period, reduces the apoplastic resistance for organic acids since they will be protonated and therefore become more volatile. When present in the gaseous phase of the apoplast, aldehydes and organic acids may theoretically escape from the leaves either by passing through the cuticles or the stomata. Due to their polarity, the lipophilic cuticles have to be considered a strong barrier (Schönherr and Riederer 1989) and therefore the stomata are a superior escape path. For this reason, factors influencing the stomatal aperture should widely determine the magnitude of gas exchange of these compounds between plants and the atmosphere. Because light triggers the opening of stomata, considerable emission of aldehydes and organic acids can be expected during daytime. This assumption is supported by the finding of significantly higher emission rates of acetaldehyde during the day than at night (Hahn et al. 1991; Steinbrecher et al. 1997). A light dependent emission of acetaldehyde and formaldehyde was also observed in Mediterranean oak and pine (Kesselmeier et al. 1997). Consistent with these results, light also triggered the emission of formic and acetic acid in ash, beech, spruce Holm oak, birch and Mediterranean oak (Kesselmeier et al. 1997, 1998; Gabriel et al. 1999); especially formic acid emission rates could be related to transpiration and stomatal conductance. However, non-stomatal emission of formic acid in Holm oak (*Quercus ilex*) was higher than expected (Gabriel et al. 1999) and the possibility of a direct transport of formic acid through the plant cuticle or through cracks in the cuticle therefore has to be taken into account (Gabriel et al. 1999).

One of the major difficulties in studying trace gas exchange is posed by the measuring techniques applied. Conventional methods are often based on an enrichment of the specific compound of interest which usually causes long sampling times. The poor time resolutions (0.5 to several hours) that result from such approaches can prevent a proper analysis of correlation between the observed emission rates and the simultaneously determined plant physiological parameters such as stomatal conductance, rate of photosynthesis and transpiration. This problem can be overcome in the future through the application of new techniques with higher sensitivities. In a recent study Hofmann et al. (1997) described an efficient trapping technique for organic acids which resulted in improved time resolutions. In order to determine acetaldehyde emission from flooded poplar trees, Kreuzwieser et al. (2000) used the highly sensitive laser-based photoacoustic detection technique which has been applied earlier in post-harvest analysis of trace gases (mainly ethene). With time resolutions of 5-15 min, diurnal rhythms of acetaldehyde could be measured and it was found that acetaldehyde emission rates closely followed stomatal conductance (Kreuzwieser et al. 2000). In similar investigations with flooded *Quercus robur* trees, clear diurnal patterns of acetaldehyde emission were observed, with high emission rates during the day and almost no emission during the night when the stomata were closed (Fig. 1). Similar results were also obtained by Holzinger et al. (2000) using the highly sensitive PTR-MS (Proton Transfer Reaction Mass spectrometry) technique. For future studies, such techniques allow precise parameterization of the exchange of organic acids and aldehydes and the influence of environmental (light, temperature, relative humidity, ambient concentrations of the individual compound of interest) and plant physiological (transpiration, stomatal conductance) factors on gas exchange could be studied in more detail.

3.2   Concentration gradients

The direction of gas exchange via the stomata is determined by the concentration gradient between ambient air and the substomatal gaseous phase. Diffusive gas flux and net flux direction can therefore be described by Fick's law. When substomatal concentrations are identical to ambient concentrations net flux is zero. This concentration is the compensation point for the particular compound. When VOC concentrations outside the leaves are higher than those in the substomatal cavities, a net flux of VOC will take place from the atmosphere into the leaves; in principle trees may therefore act both as a source of a specific trace gas (if ambient concentrations are lower than substomatal concentrations) or as a sink of the same compound. A deposition of aldehydes and organic acids has been shown only in a few studies. Gabriel et al. (1999) observed a deposition of organic acids into the apoplast directly after stomatal opening in the morning due to (1) high ambient concentrations of the organic acids in the atmosphere and (2) high pH values of the apoplastic phase which favour solubility but reduce the emission of organic acids. Results of Kesselmeier et al. (1998) indicate that trees usually have to be considered as sources of organic acids, whereas crop plants generally act as sinks. Since ambient concentrations of organic acids are more or less the same for trees and crop plants, differences in plant internal processes (apoplastic pH, production and metabolization of organic acids) must be the reasons for the higher compensation points in crops.

Even less information is available on the deposition of aldehydes. Studying the detoxification of formaldehyde by Spider Plant (*Chlorophytum comosum*) Giese et al. (1994) observed a significant deposition into the leaves and a subsequent metabolization of formaldehyde. However, these results were obtained with formaldehyde concentrations of 7.1 ppm which are far from ambient concentrations.

3.3   Plant physiology

The transiently high acetaldehyde emission from oak trees in the morning hours (Fig. 1) suggests that storage pools of acetaldehyde are filled during the night and acetaldehyde is set free immediately when stomatal conductance increases. Thus, the strength of trace gas emission from leaves is not only controlled by factors determining stomatal aperture but also by the size of storage pools which are influenced by the plant's physiology. The

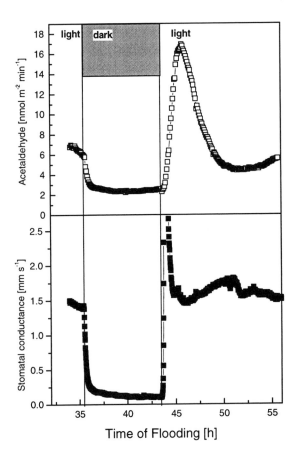

Figure 1. Diurnal rhythms of acetaldehyde emission (A) and stomatal conductance (B) of submerged oak trees. The shoot of an intact six month old *Quercus robur* tree was placed into a Teflon cuvette which was flushed with ambient charcoal filtered air. The tree's root system was submerged with tap water. Two days after flooding, acetaldehyde concentrations in the air of the plant cuvette and an empty control cuvette were measured simultaneously using a laser-based photoacoustic detection unit. CO_2 and H_2O concentrations were measured continuously with a BINOS (Heraeus-Leibold, Germany) in both cuvettes. Rates of gas exchange were calculated from the concentration differences in both cuvettes on a leaf area basis. Stomatal conductance was calculated from transpiration, temperature and relative humidity.

biosynthetic pathways so far known for specific organic acids and aldehydes have been described in chapter 1.5 of the present issue. These pathways, and as a consequence the emissions of the specific compounds, are influenced (1) by a variety of biotic and abiotic factors (e.g. stress factors) and (2) by the developmental stage of the plant. The potential of biotic and abiotic stress factors to influence trace gas emission has been demonstrated by Kimmerer and Kozlowski (1982). Wounding or induction of oxidative stress by fumigation with ozone or SO_2 caused a production of considerable amounts of ethene, ethane, ethanol and acetaldehyde. A release of ethanol and acetaldehyde from the leaves of numerous plant species, including trees, was also observed if the plants were stressed by anaerobiosis (Kimmerer and MacDonald 1987). A potential mechanism which causes trees to emit large quantities of acetaldehyde has been demonstrated by Kreuzwieser et al. (1999, 2000). Oxygen shortage, e.g. by root submergence, leads to the production of ethanol in the roots by alcoholic fermentation. The ethanol

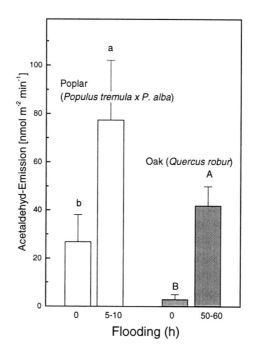

Figure 2. Acetaldehyde emission of poplar and oak trees in response to root submergence. The root systems of three month old poplar and six month old oak trees were flooded for the times indicated and acetaldehyde emission from the leaves of the plants was analysed by applying the 2,4-dinitrophenylhydrazine-coated silica gel cartridge method.

produced is transported to the leaves by the transpiration stream and oxidized in the leaves, thereby producing acetaldehyde. The results shown in Figure 2 indicate that this mechanism exists at least for oak and poplar trees; although the same mechanism seems to be utilized by both species, clear differences in the time response to flooding as well as in the rates of acetaldehyde emission were observed between the two species. Another factor which leads to the synthesis of considerable amounts of acetaldehyde in plants is postanoxic stress which is induced when plants are re-supplied with oxygen after anaerobic periods (e.g. by flooding). During this phase, considerable amounts of acetaldehyde are produced in the individual tissues (e.g. Monk et al. 1987; Pfister-Sieber and Brändle 1994; Zuckermann et al. 1997). However, no emission data have been determined so far for trees under postanoxic stress. This is also true for the emission of formaldehyde and formic acid. It is known that stress factors like anaerobiosis, drought, chilling, wounding and iron deficiency induce the enzymes for formaldehyde and formate detoxification (Colas des Francs-Small et al. 1993; Hourton-Cabasse et al. 1998; Suzuki et al. 1998; see also Chapter 1.5, this issue), which indicate formic acid and formaldehyde production under these conditions. Still emission data of these compounds under such conditions are not available.

3.4   Developmental stage

Physiological processes are not only influenced by environmental conditions but are also controlled by the developmental stage of individual plant organs. The activity of the enzymes responsible for the synthesis of hexenal and hexanal (see Chapter 1.5, this issue) in leaves, for example, was shown to change during the growing season, with the highest activity recorded during late summer (Hatanaka 1993). A dependency on the developmental stage was also observed for the emission of methanol from *Populus deltoides* leaves (Nemecek-Marshall et al. 1995); methanol emission declined with increasing leaf age after leaf expansion. Since formic acid and formaldehyde production are thought to be closely related to methanol synthesis, a similar emission pattern could be expected for these compounds. In contrast to this assumption, a large seasonal increase in acetic and formic acid emissions was detected in ash (Kesselmeier et al. 1998). This effect was probably due to microbial leaf decomposition.

3.5   Emission rates

There are only few published studies providing emission rates for organic acids and aldehydes by trees. Kesselmeier and Staudt (1999) recently compiled available data on VOC emissions in different plant species. In recent years Kesselmeier's group performed field and laboratory studies in order to quantify the exchange of C1 and C2 aldehydes and organic acids between trees and the atmosphere. Acetic and formic acid are released by the leaves of different tree species in the range of 0.01–0.25 µg g^{-1} LDW h^{-1} (Kesselmeier et al. 1997, 1998). Similar values were obtained for the emission of acetaldehyde by non-flooded poplars (c. 0.4 µg g^{-1} LDW h^{-1} Kreuzwieser et al. 1999a), *Pinus pinea* (0.4–1.5 µg g^{-1} LDW h^{-1}, Kesselmeier et al. 1997) and *Quercus ilex* (0.4–0.9 µg g^{-1} LDW h^{-1}, Kesselmeier et al. 1997). Emission rates of formaldehyde were in a similar range (*P. pinea* and *Q. ilex*: 0.1–0.9 µg g^{-1} LDW h^{-1}, Kesselmeier et al. 1997). In contrast, acetaldehyde emission from flooded poplars was significantly higher (up to c. 17 µg g^{-1} LDW h^{-1}) (Kreuzwieser et al. 1999, 2000). Quantitative emission data on higher molecular weight aldehydes and organic acids are scarce. Emission rates of c. 0.1 to 0.5 µg g^{-1} LDW h^{-1} have been reported for different woody and herbaceous plant species for hexanal, heptanal, pentanal, nonanal and decanal (Owen et al. 1997; Kirstine et al. 1998).

4.   CONCLUSION

Although isoprene and monoterpenes certainly are the main VOC released by vegetation, oxygenated compounds are also emitted in considerable amounts. Data on organic acid and aldehyde emission are scarce and more data are urgently required for a better understanding of oxidative processes in the atmosphere. Further studies are necessary to quantify the emission rates and to get more information on environmental conditions controlling the emission of organic acids and aldehydes by trees.

REFERENCES

Anderson LG, Lanning JA, Barrell R, Miyagishima J, Jones RH & Wolfe P (1996) Sources and sinks of formaldehyde and acetaldehyde: an analysis of Denver´s ambient concentration data. Atmos Environ 30: 2113-2123

Andreae MO, Talbot RW, Andreae TW & Harriss RC (1988) Formic and acetic acid over the Central Amazon region, Brazil. 1. Dry Season. J Geophys Res 93: 1616-1624

Andreae MO, Talbot RW & Li S-M (1987) Atmospheric measurements of pyruvic and formic acid. J Geophys Res 92: 6635-6641

Arey J, Winer AM, Atkinson R, Aschmann SM, Long WD & Morrison CL (1991) The emission of (Z)-3-hexen-1-ol, (Z)-3-hexenylacetate and other oxygenated hydrocarbons from agricultural plant species. Atmos Environ 25A, 1063-1075

Baez AP, Belmont R & Padilla H (1995) Measurements of formaldehyde and acetaldehyde in the atmosphere of Mexico City. Environ Pollut 89: 163-167

Benning L & Wahner A (1998) Measurements of atmospheric formaldehyde (HCHO) and acetaldehyde (CH3CHO) during POPCORN 1994 using 2.4-DNPH coated silica cartridges. J Atmos Chem 31:105-117

Buttery RG, Ling LC & Wellso SG (1982) Oat leaf volatiles: possible insect attractants. J Agr Food Chem 30: 791-792

Ciccioli P, Brancaleoni E, Frattoni M, Cecinato A & Brachetti A (1993) Ubiquitous occurrence of semi-volatile carbonyl compounds in tropospheric samples and their possible sources. Atmos Environ 27: 1891-1901

Cleveland WS, Graedel TE & Kleiner B (1977) Urban formaldehyde: observed correlation with source emissions and photochemistry. Atmos Environ 11: 357-360

Colas des Francs-Small C, Ambard-Bretteville F, Small ID & Remy R (1993) Identification of a major soluble protein in mitochondria from nonphotosynthetic tissues as NAD-dependent formate dehydrogenase. Plant Physiol 102: 1171-1177

Dawson GA & Farmer JC (1988) Soluble atmospheric trace gases in the southwestern United States 2. Organic Species HCHO, HCOOH, CH_3COOH. J Geophys Res 93: 5200-5206

Enders G, Dlugi R, Steinbrecher R, Clement B, Daiber R, Eijk J v., Gäb S, Haziza M, Helas G, Herrmann U, Kessel M, Kesselmeier J, Kotzias D, Kourtidis K, Kurth H-H, McMillan RT, Roider G, Schürmann W, Teichmann U & Torres L (1992) Biosphere/atmosphere interactions: integrated research in a European coniferous forest ecosystem. Atmos Environ 26: 171-189

Gabriel R, Schäfer L, Gerlach C, Rausch T & Kesselmeier J (1999) Factors controlling the emissions of volatile organic acids from leaves of *Quercus ilex* L. (Holm oak). Atmos Environ 33: 1347-1355

Giese M, Bauer-Doranth U, Langebartels C & Sandermann H (1994) Detoxification of formaldehyde by the spider plant (Chlorophytum comosum L.) and by soybean (*Glycine max* L.) cell-suspension cultures. Plant Physiol 104: 1301-1309

Granby K, Christensen CS & Lohse C (1997) Urban and semi-rural observations of carboxylic acids and carbonyls. Atmos Environ 31: 1403-1415

Grosjean D, Miguel AH & Tavares TM (1990) Urban air pollution in Brazil: Acetaldehyde and other carbonyls. Atmos Environ 24: 101-106

Grosjean D (1988) Aldehydes, carboxylic acids and inorganic nitrate during NSMCS. Atmos Environ 22: 1637-1648

Grosjean D (1989) Organic acids in southern California air: Ambient concentrations, mobile source emissions, in situ formation and removal processes. Environ Sci Technol 23: 1506-1514

Grosjean D, Swanson RD & Ellis EC (1983) Carbonyls in Los Angeles air: contribution of direct emissions and photochemistry. Sci Total Environ 28: 65-85

Guenther AP, Zimmerman P & Wildermuth M (1994) Natural volatile organic compound emission rate estimates for U.S. woodland landscapes. Atmos Environ 28: 1197-1210

Hahn JR, Steinbrecher R & Slemr J (1991) Study of the emission of low molecular-weight organic compounds by various plants. EUROTRAC Annual Report, Part 4, BIATEX. 230-235

Harris GW, Mackay GI, Iguchi T, Mayne LK & Schiff HI (1989) Measurements of formaldehyde in the troposphere by tunable diode laser absorption spectroscopy. J Atmos Chem 8: 119-137

Hartmann WR, Andreae MO & Helas G (1989) Measurements of organic acids over Central Germany. Atmos Environ 23: 1531-1534

Hastie DR, Shepson PB, Sharma S & Schiff HI (1993) The influence of nocturnal boundary layer on secondary trace species in the atmosphere. Atmos Environ 27: 533-541

Haszpra L, Szilagyi I, Demeter A, Turanyi T & Berces T (1991) Non-methane hydrocarbon and aldehyde measurements in Budapest, Hungary. Atmos Environ 25: 2103-2110

Hatanaka A (1993) The biogeneration of green odour by green leaves. Phytochemistry 34: 1201-1218

Helas G, Bingemer H & Andreae MO (1992) Organic acids over Equatorial Africa: Results from DECAFE 88. J Geophys Res 97: 6187-6193

Hofmann U, Weller D, Ammann Ch, Jork E & Kesselmeier J (1997) Cryogenic trapping of atmospheric organic acids under laboratory and field conditions. Atmos Environ 31: 1275-1284

Holzinger R, Sandoval-Soto L, Rottenberger S, Crutzen PJ & Kesselmeier J (2000) Emissions of volatile organic compounds from Quercus ilex L. measured by Proton Transfer Reaction Mass spectrometry (PTR-MS) under different environmental conditions. J Geophys Res 105: 20573-20579

Hoshika Y (1982) Gas chromatographic determination of lower fatty acids in air at part-per-trillion levels. Anal Chem 54: 2433-2437

Hourton-Cabassa C, Ambard-Bretteville F, Moreau F, Davy de Virville J, Rémy R & Colas des Francs C (1998) Stress induction of mitochondrial formate dehydrogenase in potato leaves. Plant Physiol 116: 627-635

Isidorov VA, Zenkevich IG & Ioffe BV (1985) Volatile organic compounds in the atmosphere of forests. Atmos Environ 19: 1-8

Kalabokas P, Carlier P, Fresnet P, Mouvier G & Toupance G (1988) Field studies of aldehyde chemistry in the Paris area. Atmos Environ 22: 149-155

Kawamura K, Ng LL & Kaplan IR (1985) Determination of organic acids (C1-C10) in the atmosphere, motor exhausts, and engine oils. Environ Sci Technol 19: 1082-1086

Kesselmeier J, Bode K, Gerlach C & Jork E-M (1998) Exchange of atmospheric formic and acetic acids with trees and crop plants under controlled chamber and purified air conditions. Atmos Environ 32: 1765-1775

Kesselmeier J, Bode K & Helas G (1992) Exchange of organic acids between trees and the atmosphere. EUROTRAC Annual Report 1992, BIATEX. 151-153

Kesselmeier J, Bode K, Hofmann U, Müller H, Schäfer L, Wolf A, Ciccioli P, Brancaleoni E, Cecinato A, Frattoni M, Foster P, Ferrari C, Jacob V, Fugit JL, Dutaur L, Simon V & Torres L (1997) The BEMA-Project: Emission of short chained organic acids, aldehydes and monoterpenes from *Quercus ilex* L. and *Pinus pinea* L. in relation to physiological activities, carbon budget and emission algorithms. Atmos Environ 31: 119-133

Kesselmeier J & Staudt M (1999) Biogenic volatile organic compounds (VOC): An overview on emission, physiology and ecology. J Atmos Chem 33: 23-88

Khare P, Satsangi GS, Kumar N, Maharaj Kumari K & Srivastava SS (1997) HCHO, HCOOH and CH$_3$COOH in air and rain water at a rural tropical site in North Central India. Atmos Environ 31: 3867-3875

Khwaja H (1995) Atmospheric concentrations of carboxylic acids and related compounds at a semiurban site. Atmos Environ 29: 127-129

Kimmerer TW & Kozlowski TT (1982) Ethylene, ethane, acetaldehyde, and ethanol production by plants under stress. Plant Physiol 69: 840-847

Kimmerer TW & MacDonald RC (1987) Acetaldehyde and ethanol biosynthesis in leaves of plants. Plant Physiol 84: 1204-1209

Kirstine W, Galbally I, Ye Y & Hooper M (1998) Emissions of volatile organic compounds (primarily oxygenated species) from pasture. J Geophys Res 103: 10605-10619

König G, Brunda M, Puxbaum H, Hewitt CN, Duckham SC & Rudolph J (1995) Relative contribution of oxygenated hydrocarbons to the total biogenic VOC emissions of selected mid-European agricultural and natural plant species. Atmos Environ 29: 861-874

Kreuzwieser J, Kühnemann F, Martis A, Rennenberg H & Urban W (2000) Diurnal pattern of acetaldehyde emission by flooded poplar trees. Acta Fac Rerum Phy 108: 79-86

Kreuzwieser J, Scheerer U & Rennenberg H (1999) Metabolic origin of acetaldehyde emitted by trees. J Exp Bot 50: 757-765

Larsen BR, Brussol C, Kotzias D, Veltkamp T, Zwaagstra O & Slanina J (1998) Sample preparation for radiocarbon (^{14}C) measurements of carbonyl compounds in the atmosphere: quantifying the biogenic contribution. Atmos Environ 32: 1485-1492

Lawrence JE & Koutrakis P (1994) Measurement of atmospheric formic and acetic acids: methods evaluation and results from field studies. Environ Sci Technol 28: 957-964

Lee YN, Zhou X & Hallock K (1995) Atmospheric carbonyl compounds at a rural southeastern United state site. J Geophys Res 100: 25933-25944

Lipari F, Dasch M & Scruggs WF (1984) 2,4-Dinitrophenylhydrazin-coated florisil sampling cartridges for the determination of formaldehyde in air. Environ Sci Technol 19: 70-74

Lowe DC, Schmidt U, Ehhalt DH, Frischlorn CGB & Nurnberg HW (1981) Determination of formaldehyde in clean air. Environ Sci Technol 15: 819-823

Lunde G, Gether J, Gjos N & Stobet-Lande MB (1977) Organic micro-pollutants in precipitation in Norway. Atmos Environ 11: 1007-1014

Martin RS, Westberg H, Allwine E, Asman L, Farmer JC & Lamb B (1991) Measurements of isoprene and its atmospheric oxidation products in central Pennsylvania deciduous forest. J Atmos Chem 13: 1-32

Monk LS, Brändle R & Crawford RMM (1987) Catalase activity and post-anoxic injury on monocotyledonous species. J Exp Bot 38: 233-246

Morikami T, Tanaka S, Hashimoto Y, Inomata T & Hanaoka Y (1991) An automatic system for the measurement of carboxylic acids (HCOOH, CH$_3$COOH) in the urban atmosphere by combination of an aqueous mist chamber and ion exclusion chromatograpy. Anal Sci 7: 1033-1036

Mosello R & Tartari GA (1992) Formate and acetate in wet deposition at Pallanza (NW Italy) in relation to major ion concentrations. Water Air Soil Poll 63: 397-409

Munger JW, Jacob DJ, Daube BC & Horowitz LW (1995) Formaldehyde, glyoxal and methylglyoxal in air and cloudwater at a rural mountain site in central Virginia. J Geophys Res 100: 9325-9333

Neitzert V & Seiler W (1981) Measurements of formaldehyde in clean air. J Geophys Res 8: 79-82

Nemecek-Marshall M, MacDonald RC, Franzen JJ, Wojciechowski CL & Fall R (1995) Methanol emission from leaves. Plant Physiol 108: 1359-1368

Norton RB (1992) Measurements of gas phase formic and acetic acids at the Mauna Loa, Observatory, Hawaii during the Mauna Loa observatory photochemsitry experiment 1988. J Geophys Res 97: 10389-10393

Norton RB, Roberts JM & Huebert BJ (1983) Tropospheric oxalate. Geophys Res Lett 10: 517-520

Owen S, Boissard C, Street RA, Duckham C, Csiky O & Hewitt CN (1997) Screening of 18 Mediterranean plant species for volatile organic compound emissions. Atmos Environ 31: 101-118

Pfister-Sieber M & Brändle R (1994) Aspects of plant behavior under anoxia and post-anoxia. P Roy Soc Edinb B 102: 313-324

Platt U & Perner D (1980) Direct measurements of atmospheric CH$_2$O, HNO$_2$, O$_3$, NO$_2$, and SO$_2$ by differential optical absorption in the near UV. J Geophys Res 85: 7453-7458

Possanzini M, Di Palo V, Petricca M, Fratarcangeli R & Brocco D (1996) Measurements of lower carbonyls in Rome ambient air. Atmos Environ 30: 3757-3764

Puxbaum H, Rosenberg C, Gregori M, Lanzerstorfer C, Ober E & Winiwarter W (1988) Atmospheric concentrations of formic and acetic acid and related compounds in eastern and northern Austria. Atmos Environ 22: 2841-2850

Salas LJ & Singh HB (1986) Measurements of formaldehyde and acetaldehyde in the urban ambient air. Atmos Environ 20: 1301-1304

Sanhueza E, Santana M & Hermoso M (1992) Gas and aqueous-phase formic and acetic acids at a tropical cloud forest site. Atmos Environ 26: 1421-1426

Satsumabayashi H, Kurita H, Chang Y-S, Carmichael GR & Ueda H (1995) Photochemical formations of lower aldehydes and lower fatty acids under long-range transport in Central Japan. Atmos Environ 29: 255-266

Schönherr J & Riederer M (1989) Foliar penetration and accumulation of organic chemicals in plant cuticles. Rev Environ Contam T 108: 1-70

Schubert B, Schmidt U & Ehhalt DH (1984) Sampling and analysis of acetaldehyde in tropospheric air. In: Versino B & Angeletti G (eds) Physica-chemical behaviour of atmospheric pollutants, Proceedings of Bur. Symposium, Varese, Italy, pp 44-52. Reidel Publishers, Dordrecht, The Netherlands

Shepson PB, Hastie DR, Polizzi M, Bottenheim JW, Anlauf K, Mackay GI & Karecki DR (1991) Atmospheric concentrations and temporal variations of C1-C3 carbonylic compounds at two rural sites in Central Ontario. Atmos Environ 25: 2001-2015

Slemr J (1992) Development of techniques for the determination of major carbonyl compounds in clean air. In: EUROTRAC Annual Report of 1991, Part 9, 110-113

Slemr J, Junkermann W & Volz-Thomas A (1996) Temporal variations in formaldehyde, acetaldehyde and acetone and budget of formaldehyde at a rural site in southern Germany. Atmos Environ 30: 3667-3676

Solberg SN, Schmidbauer N, Pedersen U & Schaug J (1993) VOC measurements August 1992-June 1993. EMEP/CCC-Report 6/93, NILU, Lillestrom.

Steinbrecher R, Hahn J, Stahl K, Eichstädter G, Lederle K, Rabong R, Schreiner AM & Slemr J (1997) Investigations on emissions of low molecular weight compounds (C2-C10) from vegetation. In: Slanina S (ed) Biosphere-atmosphere exchange of pollutants and trace substances, pp 342-351. Springer Verlag, Berlin, Heidelberg, New York, U.S.A.

Suzuki K, Itai R, Suzuki K, Nakashiani H, Nishizawa N-K, Yoshimura E & Mori S (1998) Formate dehydrogenase, an enzyme of anaerobic metabolism, is induced by iron deficiency in barley roots. Plant Physiol 116: 725-732

Talbot RW, Andreae MO, Berresheim H, Jacob DJ & Beecher KM (1990) Sources and sinks of formic, acetic and pyruvic acids over Central Amazonia. 2. Wet season. J Geophys Res 95: 16799-16811

Talbot RW, Beecher K, Harriss RC & Cofer WR (1988) Atmospheric geochemistry of formic and acetic acid at a mid-latidue temperate site. J Geophys Res 93: 1638-1652

Talbot RW, Mosher BW, Heikes BG, Jacob DJ, Munger JW, Daube BC, Keene WC, Maben JR & Artz RS (1995) Carboxylic acids in the rural continental atmosphere over the eastern United States during the Shenandoah cloud and photochemistry experiment. J Geophys Res 100: 9335-9343

Tokos JJS, Tanaka S, Morikami T, Shigetani H & Hoshimoto Y (1992) Gaseous formic and acetic acids in the atmopshere of Yokohama. Jpn J Atmos Chem 14: 85-94

Tuazon EC, Winer AM & Pitts JN (1981) Trace pollutant concentrations in a multiday smog episode in the California south coast air basin by long path length Transform Infrared süpectroscopy. Environ Sci Technol 15: 1232-1237

Willey JD & Wilson CA(1993) Formic and acetic acids in atmospheric condensate in Wilmington, North Carolina. J Atmos Chem 16: 123-133

Yokouchi Y, Mukai H, Nakajima K & Ambe Y (1990) Semi-volatile aldehydes as predominant organic gases in remote areas. Atmos Environ 24A: 439-442

Zuckerman H, Harren FJM, Reuss J & Parker DH (1997) Dynamics of acetaldehyde production during anoxia and post-anoxia in red bell pepper studied by photoacoustic techniques. Plant Physiol 113: 925-932

Chapter 3.4

Ozone

Gerhard Wieser
Forstliche Bundesversuchsanstalt, Abteilung Forstpflanzenphysiologie, Rennweg 1, A-6020 Innsbruck, Austria

1.   INTRODUCTION

Analysis of the impact of ozone (O_3) on tree physiology and ecology requires the knowledge of O_3 formation, transport and uptake by the foliage (i.e., adsorption onto external leaf/plant surfaces and absorption or flux through the stomata into the leaf mesophyll). Within the canopy the leaves are the primary sites of O_3 deposition, with the stomata representing the interface for the O_3 taken up from the atmosphere into the tree. Analysis of O_3 uptake by plants is important because only O_3 absorbed through the stomata into the leaf mesophyll will directly affect biochemical, physiological and growth processes.

Since the foliage is a dominant sink for O_3 the structure and the density of the canopy has a noticeable effect on the O_3 concentration individual leaves are exposed to (Runeckles 1992). Several data suggest that a dense vegetation may act as a filter for O_3 moving from the atmosphere through the forest canopy (Enders 1975; Enders and Teichmann 1986; Coe et al. 1995; Fontan et al. 1992; Munger et al. 1996). Skelly et al. (1996) showed that the average growing season O_3 concentration was 13% lower near the forest floor as compared to the canopy. Duyzer et al. (1995) and Munger et al. (1996) demonstrated that O_3 deposition in forests is related to stomatal conductance, indicating flux through the stomata as the dominant pathway of O_3 uptake. Other factors expected to influence canopy O_3 gradients are resistance to vertical mixing, as well as scavenging of O_3 by reactive gases

R. Gasche et al. (eds.), Trace Gas Exchange in Forest Ecosystems, 211–226.
© 2002 *Kluwer Academic Publishers. Printed in the Netherlands.*

such as volatile organic compounds and nitrogen oxides (Ollinger et al. 1997).

Beside O_3 concentration other features concerning the environment of leaves within the canopy are also modified in relation to those outside the canopy, e.g. reduced irradiance and wind velocity as well as increased temperature and humidity. These factors therefore also influence stomatal conductance and hence also the potential for O_3 uptake.

This review reports research on the control of O_3 uptake in field grown trees by the stomata, addressing the following questions: How is O_3 uptake influenced by environmental factors? Are there differences in O_3 uptake within the canopy? To what extent do altitudinal changes in environmental parameters cause differences in O_3 uptake? The review will mainly focus on conifers because only little information is available on O_3 uptake by deciduous trees and hardwoods.

2.   ESTIMATION OF OZONE UPTAKE

Total O_3 uptake by the foliage is the sum of adsorption onto external leaf and plant surfaces and absorption or flux through the stomata into the leaf mesophyll (Winner 1994; Wang et al. 1995). Under certain circumstances, particularly when O_3 concentrations are low, O_3 adsorption to cuticles and other surfaces can be relatively high compared to the stomatal pathway (Coe et al. 1995; Wang et al. 1995). Adsorption rates are affected by the structure of the canopy, the density of the foliage, as well as the size, shape and texture of the leaves. Leaf hairs, trichomes and other grandular structures on the surfaces of leaves increase leaf area and the thickness of the boundary layer and retain leaf surface moisture which can act as an O_3 sink (Granz et al. 1995; Pleijel et al. 1995). Rates of O_3 flux through the stomata by contrast, can be highly variable, reflecting those factors influencing stomatal conductance.

Analysis of total O_3 uptake by leaves can be performed in cuvettes or fumigation chambers using a mass-balance approach (Skärby et al. 1987; Havranek and Wieser 1989, 1994; Wang et al. 1995). Total uptake rates can be calculated as the difference between the O_3 concentration entering and leaving a cuvette multiplied by the flow rate. As this approach calculates both the O_3 uptakes of the leaf and the chamber; blank cuvette fumigations are necessary to define the cuvettes sink strength for O_3.

Absorption through the stomata into the leaf mesophyll follows a gradient from the external gas-phase into the intercellular air spaces according to:

$$FO_3 = (C_a - C_i) * gO_3, \tag{1}$$

where FO_3 is the flux or uptake rate of O_3 into the leaves; C_a and C_i are the concentrations of O_3 outside and inside the leaf, respectively, and gO_3 is the stomatal conductance for O_3. The latter can be calculated by multiplying the conductance of water vapour by 0.613, the ratio of the diffusivities of water vapour and O_3 (Nobel 1983). Uptake of O_3 through the cuticle can be ignored because the cuticle is highly impermeable to O_3 compared to the open stomata (Kerstiens and Lendzian 1989). O_3 concentration inside the leaves (C_i) is undetectably low (Laik et al. 1989; Moldau et al. 1990) and the mesophyll resistance to O_3 is normally small (cf. Runeckles 1992). Therefore, O_3 flux into the leaves is governed by stomatal conductance (gO_3) and ambient air O_3 concentration (C_a).

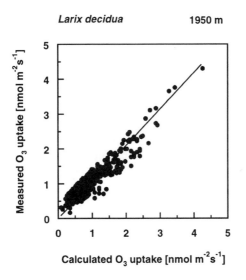

Figure 1. Correlation between calculated and measured ozone uptake rates in needles of European larch (*Larix decidua*). Each point represents a half-hour mean value taken from 13 diurnal courses of shoot gas exchange measured continuously during the growing season 1996. (G. Wieser unpubl.). The points were fitted by linear regression: $y = 0.85 * x + 0.24$, $r^2 = 0.86$.

Calculated O_3 uptake is in good agreement with uptake rates measured as the difference between the O_3 concentrations of the air entering and leaving the chamber as shown for *Larix decidua* in Figure 1. Similar results were obtained from experiments with *Picea abies* (Dobson et al. 1990; Wieser and Havranek 1993; Polle et al. 1995) and *Populus* (Wang et al. 1995). However, the results show that the measured data for O_3 uptake were higher than the calculated values by a constant of about 0.25 nmol m^{-2} s^{-1}. This can be attributed to an adsorption of O_3 on external surfaces of needles and twigs.

3.   OZONE UPTAKE AT THE LEAF LEVEL

Despite the many investigations on gas exchange in trees, continuous long lasting analysis (months, years) of foliar physiological processes in large trees combined with measurements of local O_3 concentrations within the forest environment are rare. So far, only a few studies have been carried out on 17– to 216-year old *Picea abies*, *Pinus cembra* and *Larix decidua* trees at several forest sites in Central Europe (Koch and Lautenschlager 1988; Häsler 1991; Koch 1993; Wieser and Havranek 1993, 1995; Götz 1996; Wieser et al. 2000).

3.1   Seasonal patterns and long term trends

Seasonal variations in O_3 uptake reflect variation in both ambient O_3 concentration and stomatal conductance. The atmospheric O_3 concentration generally exhibits a seasonal cycling with maximum values during the summer and minima during the winter (Fig. 2). The high summer values can be attributed to an accumulation of O_3 precursors during hot, dry and sunny periods (Stockwell et al. 1997; Vecci and Valli 1999), the latter favouring the photochemical production of O_3 in the lower troposphere (Chameides and Lodge 1992).

In evergreen conifers stomatal conductance and O_3 uptake reveal a pronounced seasonal variation (Fig. 2). During the summer maximum rates of O_3 uptake were recorded, which decrease in fall as a consequence of near freezing temperatures, lower irradiance and shorter days. During the winter gas exchange was almost completely suppressed. Recovery of O_3 uptake began in spring in response to the declining number and duration of frost events and an increase in temperature and irradiance.

In deciduous trees of the temperate and boreal zone such temperature induced reductions in O_3 uptake are of less importance. Rough estimations of uptake rates based on weekly to monthly measurements of stomatal conductance (Fredericksen et al. 1995, Samuelson and Kelly 1997) indicated that seasonal variations can mainly be attributed to alterations in stomatal conductance that are influenced by seasonal variations in leaf area and leaf age (Squire and Black 1981; Iacobelli and McCaughery 1993).

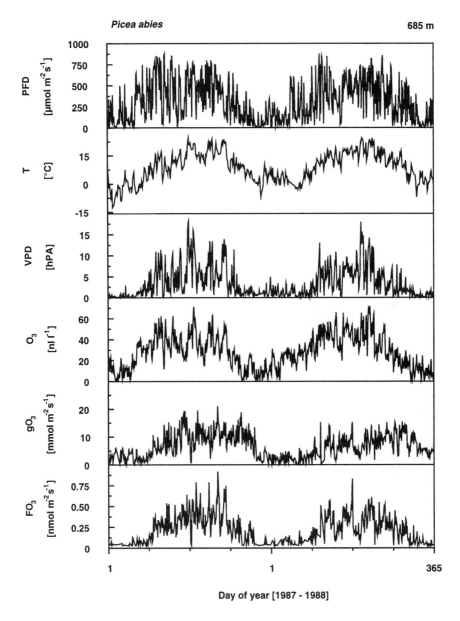

Figure 2. Annual courses of daily mean photon flux density (PFD), air temperature (T), vapour pressure deficit (VPD), ambient O_3 concentration (O_3), stomatal conductance for O_3 (gO_3) and O_3 uptake (FO_3) in the sun crown of a 127-year-old Norway spruce (*Picea abies*) tree at Lägeren, Switzerland during the period 1 January 1987 to 31 December 1988. (After Wieser et al. 2000).

Furthermore, forest trees of the temperate zone are often faced with conditions of increasing soil drought during the final part of the growing season that influence the water status of the tree. Soil moisture deficits causing mild drought stress significantly decreased O_3 uptake into the needles of *Larix decidua* and *Picea abies* (Havranek and Wieser 1993; Wieser and Havranek 1993, 1995). This is because the primary physiological response to water limitation is stomatal closure (Hinckley et al. 1978; Reich and Hinckley 1989).

Decreasing soil water availability might also be the determining long-term factor reducing O_3 uptake in Mediterranean-type climates, which are characterized by hot and dry summers connected with high ozone concentrations (Inclan et al. 1998). Recent investigations carried out on several Mediterranean tree species clearly indicated that during periods of prolonged drought, stomatal conductance significantly declined in parallel with decreasing predawn needle water potential (Epron and Dryer 1993; Damesin and Rambal 1995; Manes et al. 1997; Borghetti et al. 1998).

3.2   Short term responses

During the growing season, when stomatal aperture is not hindered by low temperatures and soil drought, O_3 uptake is strongly influenced by irradiance and leaf-air vapour pressure difference. Field data collected from the canopies of mature conifers (Wieser and Havranek 1993, 1995; Wieser et al. 2000) have emphasized leaf-air vapour pressure difference as the environmental factor most likely to control stomatal conductance and thus O_3 flux into the needles. This is because of the concomitant occurrence of high O_3 concentrations and high leaf-air vapour pressure difference (Grünhage and Jäger 1994), the latter causing stomatal closure (Fig. 3) and thus limiting O_3 flux into the needles at peak O_3 concentrations (Wieser and Havranek 1993, 1995; Wieser et al. 2000).

O_3 uptake also decreased as irradiance declined to values lower than 600 μmol m^{-2} s^{-1}, even when leaf-air vapour pressure difference remained at low values (Wieser and Havranek 1993, 1995; Wieser et al. 2000). Nocturnal O_3 uptake rates also varied considerably, and values ranged from 3 to 15% of maximum uptake rates obtained under full sunlight (Skärby et al. 1987; Wieser and Havranek 1993, 1995).

4.   SCALING FROM THE LEAF TO THE TREE AND STAND LEVEL

Canopies of trees and forests are not homogenous. Forest stands have complex vertical and horizontal environmental gradients that are created by differences in crown architecture and in the topography of the site, both influencing the microenvironment. Temperature, humidity, wind speed and O_3 concentrations all vary within the canopy and differ significantly from the conditions outside. As a result, it has to be expected that both stomatal conductance and O_3 uptake vary in relation to the position of the foliage within the canopy.

4.1   Variation within the canopy

Based on total leaf surface area, stomatal conductance varies between species and even within individual leaves of a tree. Conifers, except *Larix* have lower stomatal conductances than deciduous and hardwood trees (Körner et al. 1979; Reich et al. 1992) and therefore also lower O_3 uptake rates (Reich 1987). *Larix* displayed a higher maximum stomatal conductance as compared to *Picea abies* and *Pinus cembra* but stomatal narrowing in response to increasing leaf-air vapour pressure difference was also more pronounced (Matyssek et al. 1997) (Fig. 3).

Different microclimatic conditions in the sun and shade crown of a tree influence leaf morphology and the physiology of the stomata and hence also O_3 uptake. Stomatal conductance has been shown to decrease with increasing depth of the canopy (Leverenz et al. 1982; Iacobelli and McCaughery 1993; Fredericksen et al. 1995; Dang et al. 1997). Sun exposed needles of spruce and larch are known to posses higher stomatal conductances than shade needles (Fig. 3). Furthermore, in spruce and larch stomatal narrowing caused by increasing leaf-air vapour pressure difference (Fig. 3) limited O_3 uptake by the foliage earlier in the shade than in the sun crown (Wieser and Havranek 1993, 1995).

On a relative scale, however, all these responses are similar and an increase in leaf-air vapour pressure difference from 10 to 20 Pa kPa^{-1} results in a 30% reduction in stomatal conductance and hence also in O_3 uptake (Fig. 3). Similar leaf-air vapour pressure difference induced reductions in stomatal conductance were also reported for other tree species (Smith et al. 1984; Dang et al. 1997; Vygodskaya et al. 1997). Furthermore, these relationships will also hold under conditions of increasing soil water stress (Wieser and Kronfuß 1997).

**Larix decidua, Picea abies,
Pinus cembra**

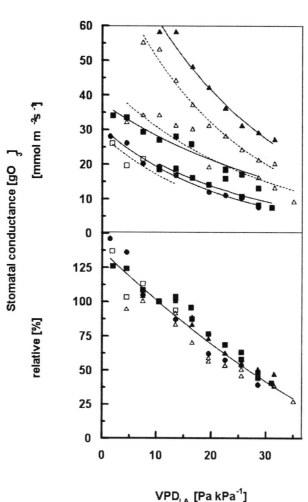

Figure 3. Absolute (top) and relative (bottom) values of stomatal conductance for O_3 (gO_3) in relation to leaf-air vapour pressure difference (VPDLA) in the sun (closed symbols) and shade (open symbols) crown of Norway spruce (*Picea abies*, (n)), cembran pine (*Pinus cembra*, (l)) and European larch (*Larix decidua*, (s)). Values are means of 10 to 45 half hour-intervals collected in situ from canopy access towers under conditions of light saturation (PFD ≥ 600 μmol m^{-2} s^{-1}). Relative values were obtained by setting stomatal conductance to 100% at a VPDLA of 10 Pa kPa^{-1} (After Wieser 1999). Values for relative stomatal conductance were fitted by polynomial regression: $y = 137.1 - 3.8 * x + 0.02 * x2$, $r^2 = 0.92$. Since at a given ambient O_3 concentration O_3 flux into the needles is exclusively dependent on stomatal aperture, these correlations will also hold for O_3 uptake.

4.2   Tree- and leaf-age related variations

Beside species related differences, stomatal conductance is also a function of biological factors such as tree (Kolb et al. 1997) and leaf age (Leverenz et al. 1982; Reich et al. 1992; Wieser et al. 2000).

In *Picea abies* grown under field conditions, including low altitude, montane, and subalpine forests, O_3 flux into the needles significantly increased with increasing tree age, both in the sun and in the shade crown (Fig. 4). Fredericksen et al. (1996) also observed higher rates of O_3 uptake in mature canopy trees as compared to seedlings and saplings of *Prunus serotina*. As in both studies O_3 concentrations did not differ significantly between the sun and shade environment for any tree size class, observed differences in O_3 uptake mainly can be attributed to the often observed decline in stomatal conductance with increasing tree age and tree size (for a review see Kolb et al. 1997).

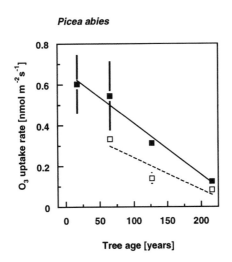

Figure 4. Correlation of tree age and ozone uptake rate during the growing season (May throughout October) in the sun (closed symbols) and shade crown (open symbols) of Norway spruce trees (*Picea abies*) in relation to tree age. Measurements were made during the period 1987 through 1992 at a mean ambient O_3 concentration of 36 ± 5 nL L^{-1}. n = 1 to $4 \pm$ SD. (Modified after Wieser et al. 1999). The points were fitted by linear regression: sun crown: y = -0.0025 * x + 0.665, r² = 0.98; shade crown: y = -0.0016 * x + 0.402, r² = 0.84.

In contrast, as a notable exception to this age and size related trend in O_3 uptake Samuelson and Kelly (1997) observed lower O_3 uptake rates as well as lower conductances in understory seedlings and saplings of *Prunus serotina, Acer rubrum* and *Quercus rubra* as compared to mature canopy trees. These greater O_3 fluxes in mature canopy trees might be due to O_3 depletion by the dense canopy and a lack of transport into the lower canopy layers (Fuentes et al. 1992).

4.3   Altitudinal trends

Altitudinal variations of climate in mountain regions also cause significant changes in the capacity for O_3 uptake. Long term measurements indicated that in evergreen conifers similar in age, average stomatal conductance and O_3 flux both increase with increasing altitude (Fig. 5). Restrictions of O_3 uptake by stomatal narrowing was also less pronounced in *Larix decidua* at the alpine timberline than at a low elevation site (Wieser and Havranek 1995). This is mainly because trees at the timberline are rarely forced to restrict their water loss (Tranquillini 1979), whereas in the valley stomatal conductance during midday and in the afternoon often declines because of increasing leaf-air vapour pressure difference, therefore discouraging O_3 uptake at peak O_3 concentrations (Wieser and Havranek 1993, 1996). Furthermore, higher elevation forests generally experience higher O_3 concentrations than low elevation forests (Miller et al. 1993; Smidt 1993) (Fig. 5).

These data also allow an assessment of the potential cumulative O_3 uptake throughout the vegetation period. However, one has to bear in mind that the vegetation period (i.e. the snow-free period, Havranek and Tranquillini 1995) decreases from approximately 250 days in the valley floor to about 180 days at the alpine timberline (Tranquillini 1979; Havranek and Tranquillini 1995). Cumulative O_3 uptake rates into needles of evergreen conifers obtained by this calculation were 11.4 ± 1.7 in the valley floor (600 m above sea level), and 14 mol m^{-2} of total needle surface area at the alpine timberline (1950 m), mean O_3 concentrations were 38 ± 5 and 46 nL L^{-1}, respectively (Wieser et al. 2000). Apparently the difference in cumulative O_3 uptake rate between the alpine timberline and the valley floor was about 20% for the whole vegetation period. On the other hand the detoxification capacity for O_3 also increases with increasing altitude (Polle et al. 1995; Tausz et al. 1996; Rennenberg et al. 1997). This might explain, why typical altitude-dependent effects on biochemical and physiological parameters could not be attributed to O_3 itself (Rennenberg et al. 1997; Tausz et al. 1999, 2001).

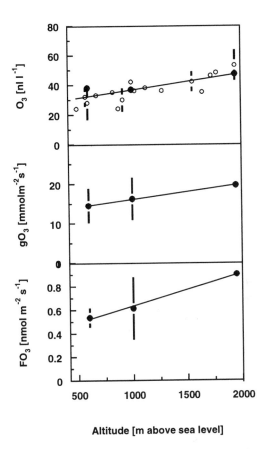

Figure 5. Mean ambient ozone concentration (O₃, top), average stomatal conductance for ozone (gO₃, middle) and average ozone uptake rate (FO₃) during the growing season (May throughout October) in 60- to 65-year-old evergreen conifers in relation to increasing altitude in the central European Alps. Measurements (solid symbols) were made during the period 1987 through 1996. n = 1 to 4 ± SD. (Modified after Wieser et al. 1999, 2000). O₃ data (open symbols) after Amt der Tiroler Landesregierung (1985 - 1998) and Smidt (1993). The points were fitted by linear regression: stomatal conductance for ozone (gO₃): y = 0.0037 * x + 12.69, r² = 0.89; ozone uptake rate (FO₃): y = 0.00027 * x + 0.38, r² = 0.99.

5.   CONCLUSIONS

Beside ambient air O_3 concentration the uptake or flux of O_3 into the leaf mesophyll is primarily influenced by stomatal conductance, which regulates O_3 flux from the ambient air into the leaf interior. Changes in O_3 uptake occur over time as a result of changes in external conditions. Seasonal changes in O_3 concentration in the lower troposphere as well as in stomatal conductance are strongly dependent on climatic conditions such as temperature, irradiance and humidity. Seasonal changes in phenology also need to be kept in mind, since they also affect conductance and hence the potential for O_3 uptake.

During the growing season however, periods of high O_3 often occur during dry weather conditions. Thus, when ambient O_3 concentrations are highest, O_3 flux into the leaves is severeiy reduced due to stomatal narrowing.

Beside climatic conditions O_3 flux is strongly influenced by species, the position of individual leaves within the canopy or even within a single tree, as well as leaf- and tree age. Deciduous trees generally have higher uptake rates than conifers. O_3 uptake rates decline with increasing depth of the canopy and also with increasing leave age. O_3 flux also declines with increasing tree age. These effects on O_3 uptake mainly can be attributed to lower stomatal conductance obtained in older and larger trees as compared to young and small individuals. In comparison to tree age, altitudinal effects on O_3 uptake are rather small, but evident in trees similar in age. The observed altitudinal increase of both average ambient O_3 concentration and stomatal conductance explains the higher O_3 uptake rates at higher habitats.

Although stomatal conductance provides the principal limiting factor for O_3 uptake additional field research on trees will be necessary in order to improve our understanding concerning the quantitative relationships between stomatal conductance, biological factors and site conditions, respectively . In addition, reassurance is needed that conductance values, usually measured on individual leaves or shoots, are representative for the canopy as a whole. Hence, gas exchange measurements on individual leaves or twigs that are linked to micrometeorological approaches should be intensified for estimating O_3 uptake rates on the tree and the stand level in order to determine the canopies sink strength for O_3.

REFERENCES

Amt der Tiroler Landesregierung (1985 - 1998) Zustand der Tiroler Wälder. Untersuchungen über den Waldzustand und die Immissionsbelastung. Berichte an den Tiroler Landtag 1985-1998, Innsbruck, Austria

Borghetti M, Cinnirella S, Magani F & Saracino A (1998) Impact of long-term drought on xylem embolism and growth in *Pinus halepensis*. Trees 12: 187-195

Chameides WL & Lodge JP (1992) Troposheric ozone. Formation and fate. In: Lefohn AS (ed) Surface level ozone exposures and their effects on vegetation, pp 5-30. Lewis, Chelsea, Mi., U.S.A.

Coe H, Gallagher MW, Choularton TW & Dore C (1995) Canopy scale measurements of stomatal and cuticular O_3 uptake by Sitka spruce. Atmos Environ 29: 1413-1423

Dang QL, Margolis HA, Coyea MR, Sy M & Collatz GJ (1997) Regulation of branch-level-gas exchange of boreal trees: roles of shoot water potential and vapour pressure difference. Tree Physiol 17: 521-535

Damesin C & Rambal S (1995) Field study on leaf photosynthetic performance by a Mediterranean deciduous oak (*Quercus pubescens*) during severe summer drought. New Phytol 131: 159-167

Dobson MC, Taylor G & Freer-Smith PH (1990) The control of ozone uptake by *Picea abies* (L.) Karst. and *P. sitchensis* (Bong.) Carr. during drought and interacting effects on shoot water relations. New Phytol 116: 465-474

Duyzer J, Westrate H & Walton S (1995) Exchange of ozone and nitrogen oxides between the atmosphere and coniferous forest. Water Air Soil Pollut 85: 2065-2070

Enders G (1975) Deposition of ozone to a mature spruce forest.: measurements and comparison to models. Environ Pollut 75: 61-67

Enders G & Teichmann U (1986) GASDEP - Gaseous deposition measurements of SO_2, NO_x and O_3 to a spruce stand: conception, instrumentation, and first results of an experimental project. In: Georgii HW (ed) Atmospheric pollutants in forest areas, pp 13-24. Reidel Publishers, Dordrecht, The Netherlands

Epron D & Dryer E (1993) Long term effects of drought on photosynthesis of adult oak trees [*Quercus petrea* (Mat.) Liebl. and *Quercus robur* L.] in a natural stand. New Phytol 125: 381-389

Fontan J, Minga A, Lopez A & Druilhet A (1992) Vertical ozone profiles in a pine forest. Atmos Environ 26a: 863-869

Fredericksen TS, Joyce BJ, Skelly JM, Steiner KC, Kolb TE, Kouterick KB, Savage JE & Snyder KR (1995) Physiology, morphology, and ozone uptake of seedlings, saplings, and canopy black cherry trees. Environ Pollut 89: 273-283

Fredericksen TS, Kolb TE, Skelly JM, Steiner KC, Joyce BJ & Savage JE (1996) Light environment alters ozone uptake per net photosynthetic rate in black cherry trees. Tree Physiol 16: 485-490

Fuentes JD, Gillespie TJ, Hartog G & Neumann HH (1992) Ozone deposition onto a deciduous forest during dry and wet conditions. Agr Forest Meteorol 62: 1-18

Götz B (1996) Ozon und Trockenstreß. Wirkungen auf den Gaswechsel von Fichte. Libri Botanici 16. IHW-Verlag, München, 149p

Granz DA, Zhang XJ, Massman WJ, den Hartog G, Neumann HH & Pederson JR (1995) Effects of stomatal conductance and surface wetness on ozone deposition in field grown grape. Amos Environ 29: 3189-3198

Grünhage L & Jäger HJ (1994) Influence of the atmospheric conductivity on ozone exposure of plants under ambient conditions: considerations for establishing ozone standards to protect vegetation. Environ Pollut 85: 125-129

Häsler R (1991) Vergleich der Gaswechselmessungen der drei Jahre (Juli 1986 - Juni 1989). In: Stark M (ed) Luftschadstoffe und Wald. Lufthaushalt, Luftverschmutzung und Waldschäden in der Schweiz, Vol. 5, pp 177-184. Verlag der Fachvereine, Zürich, Switzerland

Havranek WM & Tranquillini W (1995) Physiological processes during winter dormancy and their ecological significance. In: Smith WK & Hinckley TM (eds) Ecophysiology of coniferous forests, pp 95-124. Academic press, San Diego, U.S.A.

Havranek WM & Wieser G (1989) Research design to measure ozone uptake and its effects on gas-exchange of spruce in the field. In: Payer HD, Pfirrmann T & Mathy P (eds) Environmental research with plants in closed chambers, pp 148-152. Air Pollution Research Report 26, Commission of the European Communities, Brussels, Belgium

Havranek WM & Wieser G (1993) Zur Ozontoleranz der europäischen Lärche (*Larix decidua* Mill.). Forstwiss Centralbl 112: 56-64

Havranek WM & Wieser G (1994) Design and testing of twig chambers for ozone fumigation and gas exchange measurements in mature trees. P Roy Soc Edinb B 102: 541-546

Hinckley TM, Lassioe JP & Running SW (1978) Temporal and spatial variations in the water status of forest trees. Forest Sci, Monograph 20, 72p

Iacobelli A & McCaughery JH (1993) Stomatal conductance in a northern temperate deciduous forest: temporal and spatial patterns. Can J For Res 23: 245-252

Inclan R, Alonso R, Pujades M, Teres J & Gimeno BS (1998) Ozone and drought stress: interactive effects on gas exchange in Aleppo pine (*Pinus halepensis* Mill.) Chemosphere 36: 685-690

Kerstiens G & Lendzian KJ (1989) Interactions between ozone and plant cuticles. I. Ozone deposition and permeability. New Phytol 112: 13-19

Koch W (1993) Langjähriger Reinluft/Standortsluftvergleich des Gaswechsels von Fichten unter Freilandbedingungen. - Ein Beitrag zur Waldschadensforschung. Forstliche Forschungsberichte München 130, 94p.

Koch W & Lautenschlager K (1988) Photosynthesis and transpiration in the upper crown of a mature spruce in purified and ambient atmosphere in a natural stand. Trees 2: 213-222

Kolb TE, Fredericksen TS, Steiner KC & Skelly JM (1997) Issues in scaling tree size and age response to ozone: a review. Environ Pollut 98: 195-208

Körner C, Scheel JA & Bauer H (1979) Maximum leaf diffusive conductance in vascular plants. Photosynthetica 13: 45-82

Laisk A, Kull O & Moldau H (1989) Ozone concentration in leaf intercellular air spaces is close to zero. Plant Physiol 90: 1163-1167

Leverenz B, Deans JD, Ford ED, Jarvis PG, Milne R & Whitehead D (1982) Systematic spatial variation of stomatal conductance in a sitka spruce plantation. J Appl Ecol 19. 835-851

Manes F, Seufert G & Vitale M (1997) Ecophysiological studies of Mediterranean plant species at the Castelporziano estate. Atmos Environ 31: 51-60

Matyssek R, Havranek WM, Wieser G & Innes JL (1997) Ozone and the forests in Austria and Switzerland. In: Sandermann H, Wellburn AR & Heath RL (eds) Forest decline and ozone. A comparison of controlled chamber and field experiments. Ecological Studies 137, pp 95-134. Springer Verlag, Berlin, Heidelberg, New York, Tokyo

Miller EK, Friedland AJ, Arons EA, Mohnen VA, Battles JJ, Panek JA, Kadlecek J & Johnson AH (1993) Atmospheric deposition to forests along an elevational gradient at Whiteface Mountain, NY, USA. Atmos Environ 27A: 2121-2136

Moldau H, Sober J & Sober A (1990) Differential sensitivity of stomata and mesophyll to sudden exposure of bean shoots to ozone. Photosynthetica 24: 446-458

Munger JW, Wofsy SC, Bakwin PS, Fan S, Goulden ML, Daube BC, Goldstein AH, Moore K & Fitzjarrald D (1996) Atmospheric deposition of reactive nitrogen oxides and ozone in a temperate deciduous forest and a sub-arctic woodland. 1. Measurements and mechanisms. J Geophys Res 101: 12639-12657

Nobel PS (1983) Biochemical plant physiology and ecology. Freeman and Co., New York

Ollinger SV, Aber JD & Reich PB (1997) Simulating ozone effects on forest productivity: interactions among leaf-, canopy-, and stand-level processes. Ecol Appl 7: 1237-1251

Pleijel H, Kalsson GP, Danielsson H & Sellden G (1995) Surface wetness enhances ozone deposition to a pasture canopy. Atmos Environ 29: 3391-3393

Polle A, Wieser G & Havranek WM (1995) Quantification of ozone influx and apoplastic ascorbate content in needles of Norway spruce trees (*Picea abies* L., Karst) at high altitude. Plant Cell Environ 18: 681-688

Reich PB (1987) Quantifying plant response to ozone: A unifying theory. Tree Physiol 3: 63-91

Reich PB & Hinckley TM (1989). Influence of pre-dawn water potential and soil-to-leaf hydraulic conductance on the maximum daily leaf diffusive conductance in two oak species. Funct Ecol 3: 719-726

Reich PB, Walters MB & Ellworth DS (1992) Leaf life-span in relation to leaf, plant, and stand characteristics among diverse ecosystems. Ecol Monogr 62: 365-393

Rennenberg H, Polle A & Reuther M (1997) Role of ozone in forest decline on Wank Mountain (Alps). In: Sandermann H, Wellburn AR & Heath RL (eds) Forest decline and ozone. A comparison of controlled chamber and field experiments. Ecological Studies 137, pp 135-162. Springer, Berlin, Heidelberg, New York, Tokyo

Runeckles VC (1992) Uptake of ozone by vegetation. In: Lefohn AS (ed) Surface level ozone exposures and their effects on vegetation, pp 157-188. Lewis Publishers Inc., Chelsea, Mi., U.S.A.

Samuelson LJ & Kelly JM (1997) Ozone uptake in *Prunus serotina, Acer rubrum* and *Quercus rubra* forest trees of different size. New Phytol 136: 255-264

Skärby L Troeng E & Bostrom CA (1987) Ozone uptake and effects on transpiration, net photosynthesis, and dark respiration in Scots pine. Forest Sci 33: 801-808

Skelly JM, Fredricksen TS, Savage JE & Snyder KR (1996). Vertical gradients of ozone and carbon dioxide within a deciduous forest in central Pennsylvania. Environ Pollut 94: 235-240

Smith WK, Young DR, Catret GA, Hadley JL & McNaughton GM (1984) Autumn stomatal closure in six conifer species of the central Rocky Mountains. Oecologia 63: 237-242

Smidt S (1993) Die Ozonsituation in alpinen Tälern Österreichs. Centralbl Gesamte Forstwes 110: 205-220

Squire GR & Black (1981) Stomatal behaviour in the field, In: Jarvis PG & Mansfield TA (eds) Stomatal physiology, pp 223-245. Cambridge University Press, Cambridge, U.K.

Stockwell WR, Kramm G, Scheel H-E, Mohnen VA & Seiler W (1997) Ozone formation, destruction and exposure in Europe and in the United States. In: Sandermann H, Wellburn AR & Heath RL (eds) Forest decline and ozone. A comparison of controlled chamber and field experiments. Ecological Studies 137, pp 1-38. Springer Verlag, Berlin, Heidelberg, New York, Tokyo

Tausz M, Batic F & Grill D (1996) Bioindication at forest sites - Concepts, practice and outlook. Phyton 36: 7-14

Tausz M, Bytnerowicz A, Arbaugh MJ, Weidner W & Grill D (1999) Antioxidants and protective pigments of *Pinus ponderosa* needles at gradients of natural stresses and ozone in the San Bernadino Mountains in California. Free Radical Res 31: 113-120

Tausz M, Bytnerowicz A, Arbaugh MJ, Wonisch A & Grill D (2001) Biochemical response patterns in Pinus ponderosa trees at field plots in the San Bernadino Mountains (Southern California. Tree Physiol 21: 329-336

Tranquillini W (1979) Physiological ecology of the alpine timberline. Ecological Studies 31, Springer, Berlin, Heidelberg, New York, Tokyo, 137p

Vecci R & Valli G (1999) Ozone assessment in the southern part of the Alps. Atmos Environ 33: 97-109

Vygodskaya NN, Milyuokova I, Varlagin A, Tatrinov F, Sogachev A, Kobak KI, Desyatkin R, Bauer G, Hollinger DY, Kelliher FM & Schulze E-D (1997) Leaf conductance and CO_2 assimilation of *Larix gmelinii* growing in eastern Siberian boreal forest. Tree Physiol 17: 607-615

Wang D, Hinckley TM, Cumming AB & Braatne J (1995) A comparison of measured and models ozone uptake into plant leaves. Environ Pollut 89: 247-254

Wieser G (1999) Evaluation of the impact of ozone on conifers in the Alps: a case study on spruce, pine and larch in the Austrian Alps. Phyton 39: 241-252

Wieser G & Havranek WM (1993) Ozone uptake in the sun and shade crown of spruce: quantifying the physiological effects of ozone exposure. Trees 7: 227-232

Wieser G & Havranek WM (1995) Environmental control of ozone uptake in *Larix decidua* Mill.: a comparison between different altitudes. Tree Physiol 15: 253-258

Wieser G & Havranek WM (1996) Evaluation of ozone impact on mature spruce and larch in the field. J Plant Physiol 148: 189-194

Wieser G & Kronfuß G (1997) The influence of vapour pressure deficit and mild soil water stress on the gas exchange of Norway spruce (*Picea abies* (L.) Karst). seedlings. Centralbl Gesamte Forstwes 114: 173-182

Wieser G, Häsler R, Götz B, Koch W, & Havranek WM (1999) Seasonal ozone uptake of mature evergreen conifers at different altitudes. Phyton 39: 233-240

Wieser G, Häsler R, Götz B, Koch W & Havranek WM (2000) Role of clmate, crown position, tree age and altitude in calculated ozone flux into needles of Picea abiea and Pinus cembra: a synthesis. Environ Pollut 109: 415-422

Winner WE (1994) Mechanistic analysis of plant response to air pollution. Ecol. Appl 4: 651-661

FOREST CANOPIES AS SOURCES AND SINKS OF ATMOSPHERIC TRACE GASES

Chapter 4.1

Scaling up to the ecosystem level

Dennis D. Baldocchi[1] and Kell Wilson[2]

[1]*Department of Environmental Science, Policy and Management, 151 Hilgard Hall, University of California, Berkeley, Berkeley, CA 94720*
[2]*Atmospheric Turbulence and Diffusion Division, NOAA/ATDD, PO Box 2456, Oak Ridge, TN 37831*

1. INTRODUCTION

The exchanges of solar energy, carbon dioxide, water vapor and trace gases between a forest and the atmosphere are among the most fundamental processes to be quantified when studying the physiological and ecological functioning of a forest and the chemistry and climate of its overlying atmosphere. A forest must attain energy to sustain the work that is needed to assimilate carbon dioxide, for biosynthesis, to evaporate water, and to transport nutrients from the soil to the plant. Concurrently, these activities require flows of substrate material, which are obtained from the atmosphere and soil.

The major trace gases that are exchanged between forests and the atmosphere are associated with chemical elements that are the principle constituents of organic matter. The Redfield ratio identifies these major elements and their relative importance to one another; for every unit of phosphorus in living organic matter, there are 80 units of carbon and 15 units of nitrogen. Reduced and oxidized forms of these elements, and micronutrients such as sulfur, constitute the bulk of trace gases that are evolved or assimilated by forests. Carbon dioxide, oxygen, ammonia, nitric oxide, sulfur dioxide, ozone and carbon monoxide are among the most notable trace gas compounds that are taken up by forests. Biological processes that cause the biosphere to be a sink for these trace gases include

229

R. Gasche et al. (eds.), Trace Gas Exchange in Forest Ecosystems, 229–242.

carbon assimilation and respiration, ammonification, nitrification and denitrification and pollutant deposition. Ozone is included in this group of gases because its presence in the atmosphere is linked to the biogenic emission of nitric oxide and volatile organic hydrocarbons.

In converse, water vapor, carbon dioxide, oxygen, isoprene, and monoterpenes are among the most common gaseous compounds that are emitted by plants. The emission of these gases is linked, respectively, to transpiration and evaporation, respiration, assimilation, and volatilization.

The rates at which trace gases are transferred between forests and the atmosphere depend upon a complex interplay among physiological, ecological, biochemical, chemical and edaphic factors and meteorological conditions. Information on fluxes of trace gases between the biosphere and atmosphere is needed at a variety of time and space scales by models that predict ecosystem carbon water and nutrient balances, weather and climate and tropospheric chemistry. The time scale of processes that are associated with the transfer of trace gases between forests and the atmosphere can range from the hour and day to season, year and decade. The range of spatial information that is needed by a forest-scale model spans the dimension of needles and leaves to height and breadth of tree crowns and their placement across a landscape. How to model the processes that govern trace gas fluxes throughout the spectrum of biologically relevant time and space scales remains a challenge to forest ecologists, biometeorologists and biogeochemists (see Rastetter 1996; Baldocchi and Wilson 2001).

The goal of this chapter is to discuss the integration and scaling of information on trace gas fluxes from the leaf and soil to the canopy and landscape scales. Specific topics to be covered in this chapter include the theory of trace gas exchange, model design and complexity and temporal and spatial factors affecting model parameterization and implementation.

2.   THEORY AND CONCEPTS

Any model of trace gas fluxes in the natural environment starts with the same fundamental principle, the conservation of mass. This equation states that the time rate of change of a gas' molar density (moles per unit volume, ρ_c) equals the difference between the molar flux in and out of the volume plus the rate of chemical production/destruction plus the rate of biological consumption or production. The time averaged equation for the conservation of mass at a point in space exposed to turbulent flow is expressed, using tensor notation, as:

$$\frac{d\overline{\rho_c(x,y,z,t)}}{dt} = \frac{\partial\overline{\rho}_c}{\partial t} + \overline{u_i}\frac{\partial\overline{\rho}_c}{\partial x_i} + \overline{\rho}_c\frac{\partial\overline{u_i}}{\partial x_i} =$$

$$-\frac{\partial\overline{u_i'\rho_c'}}{\partial x_i} + S_B(t,x_i) + S_{ch}(t,x_i)$$

[1]

The space, x_i and velocity, u_i variables are incremented from 1 to 3. For the space dimensions, this corresponds to the longitudinal (x), lateral (y) and vertical (z) dimensions. For the velocity vectors, this incrementing corresponds with u, v and w, the longitudinal, lateral and vertical velocity vectors. The biological source/sink term is denoted as S_B and the production or destruction of a trace gas by chemical reactions is denoted by S_{ch}. The overbar represents time averaging.

2.1   Evaluating the Conservation Equation

2.1.1   Turbulence Closure Schemes for Computing Scalar Fields

The conservation budget equation for a scalar cannot be solved readily because it does not form a closed set of equations and unknowns. The equation defining the time rate of change in ρ_c contains an additional unknown, the covariance between vertical velocity (w) and scalar concentration fluctuations ($w'\rho_c'$). To solve Equation 1, micrometeorologists use closure schemes to obtain an equal set of equations and unknowns.

Zero order closure is the simplest scheme used. It does not treat the prognostic equation for ρ_c directly. Instead, this closure scheme specifies the scalar fields in time and space. This approach is often adopted by ecosystem and ecophysiological models, which assume that temperature and humidity are constant within and above vegetation.

First order closure, called 'K-theory', is the next level of complexity. This closure level represents the flux covariance as the product of the scalar concentration gradient and a turbulent diffusivity (K):

$$F_c(z) = \overline{w'\rho_c'} = -K\frac{\partial\overline{\rho}_c}{\partial z}$$

[2]

'K-theory' is an appropriate concept in the surface boundary layer. On the other hand, it often fails to represent turbulent transfer inside forest canopies and within the roughness sublayer, where turbulent transport is dominated by large scale and intermittent eddies and turbulent diffusion is dominated by the distinct properties of 'near field' diffusion (Raupach, 1988). Near vegetative sources and sinks turbulent diffusion is linearly related to the time period that fluid parcels have traveled (Raupach 1988). Only after a long travel distance is the time-independent, 'far-field' limit of turbulent diffusion reached, the process that K-theory represents.

Higher-order closure models have been proposed as a means of circumventing the inherent limitation of first order closure models (Meyers and Paw U 1987). Higher order closure models rely on budget equations for mean horizontal wind velocity (u) and higher order moments, such as the scalar-velocity covariance, tangential momentum stress ($\overline{w'u'}$) and the turbulent kinetic energy components ($\overline{u'u'}$, $\overline{v'v'}$, $\overline{w'w'}$). The appeal of a higher order closure model includes its mechanistic basis and its ability to simulate counter-gradient transport. Unfortunately, the budget equations for the second order moments include additional unknowns of the third order (e.g., $\overline{w'w'u'}$, $\overline{w'w'c'}$). Deriving additional budget equations for third order moments introduces more unknowns, consisting of the next order moment, and so on. Hence, an equal set of equations and unknowns can only be obtained through parameterizing the highest order moment with an 'effective' eddy exchange coefficient (Meyers and Paw U 1987).

The Lagrangian framework circumvents the closure problem ailing Eulerian models. The Lagrangian approach analyzes the conservation equation by following parcels of fluid as they move with the wind, much like the trajectory of a neutrally buoyant balloon. Thereby, Lagrangian models are able to explicitly differentiate between near and far field diffusion (Raupach 1988). Lagrangian models, however, suffer from their own unique closure problem. The probability density function for the diffusion of fluid parcels depends only on the properties of the turbulent wind field, which must be prescribed or computed with a higher order turbulence closure model.

2.1.2   Quantifying Trace Gas Source-Sink Strengths

Functional relationships that quantify trace gas sources and sinks rates generally depend upon numerous micrometeorological and eco-physiological variables. To assess these functions, we must introduce micrometeorological modules that compute leaf and soil energy exchange, turbulent diffusion, scalar concentration profiles and radiative transfer through the canopy. Environmental variables, computed with the micrometeorological module, in turn, can be used to drive physiological modules that may compute leaf photosynthesis, stomatal conductance, transpiration and leaf, bole and soil/root respiration, gaseous deposition and emission rates. Products from a micrometeorological module are also needed to drive algorithms that compute trace gas fluxes from the soil.

3. MODEL DESIGN AND PARAMETERIZATION ISSUES

Forests may be tall or short, closed or open, and consist of shrubs or trees. The spatial distribution of leaves can be random or clumped and their shape can be needle-shaped or broad and planar. How much complexity to incorporate into a system of equations that quantifies trace gas fluxes between a forest and the atmosphere is a key issue to be considered when designing a model to quantify source/sink strengths. The complexity of the stand's physiognomy will dictate, in part, how complex the structure of a trace gas model needs to be. A simplified version of Equation 1, for example, can be used for the situation of horizontally homogeneous forests on level terrain. Full expansion of Equation 1 is needed to evaluate fluxes over forests that consist of patches of isolated trees on hill slopes. In the following sub-sections we discuss algorithms for computing source/sink strengths and issues relating to the parameterization of these algorithms.

3.1 Model Design Attributes: Scaling or Integrating Trace Gas Fluxes from Leaf to Canopy Dimensions

A hierarchy of model algorithms exists for computing trace gas fluxes. The simplest models treat the canopy as a single layer, and are denoted 'big-leaf' models. This concept is followed, in order of increasing complexity, by dual-source models, two-layer, one-dimensional multi-layer models and three-dimensional cube, ellipsoid or shell models.

3.1.1 Big-Leaf Models

Three types of 'big-leaf' trace gas models can be identified. The simplest 'big-leaf' model employs a series of multiplicative functions to a base flux rate.

$$F_c = S_{base} \cdot f(a) \cdot f(b) \cdot f(c)..... \qquad [3]$$

where a, b, c represent governing variables such as light (I), temperature (T), humidity (q) and soil moisture (θ). The isoprene emission model of Lamb et al. (1993) is a prime example of this model type. Technically Equation 3 is a scaling model, rather than integrative model (Jarvis 1995). The appeal of Equation 3 is its dependence on a limited number of variables that have a linear dependence upon one another. A perceived weakness of a multiplicative, 'big-leaf' model revolves around its dependence upon parameters that do not relate to measurable physiological or physical quantities. Such models must be tuned against stand-level, eddy flux measurements. On the other hand, this method has practical appeal for gap filling data records and for constructing long term sums of trace gas fluxes, as driven by meteorological variables.

A second version of a 'big-leaf' model borrows its heritage from an electrical analog; current flow (mass or energy flux density) is equal to the ratio between a potential and the sum of the resistances to the flow.

$$F_c = \frac{C_a - C_0}{R_a + R_b + R_c} \qquad [4]$$

This approach is popular for computing gaseous deposition over forests (Meyers and Baldocchi 1988) and canopy photosynthesis and evaporation (Amthor 1994; dePury and Farquhar 1997). In this case, C_a is the concentration of scalar in the atmosphere over the vegetation and C_0 is an 'internal' concentration. The major resistances are attributed to aerodynamics of the atmosphere (R_a), diffusion through quasi-laminar boundary layers (R_b) and resistances imposed by the vegetation and soil (R_c). The canopy resistance (R_c) is a function of the canopy stomatal resistance (R_{stom}), the canopy cuticle resistance ($R_{cuticle}$), and the soil resistance (R_{soil}). In turn, these plant and soil resistances are affected by leaf area, stomatal physiology, soil pH, and the presence and chemistry of liquid drops and films. The stomatal, leaf surface (cuticle) and soil resistances act in parallel.

A third type of a 'big-leaf' model is an analytical one. It is derived by integrating environmentally dependent, leaf-level functions for trace gas fluxes with respect to leaf area (L):

$$F_c = \int_0^L f(T(l), I(l), q(l))dl \qquad [5]$$

The Simple Biosphere (SIB) model of Sellers et al. (1986) is an example of this model class. A disadvantage of this model class is that compromising assumptions on the behavior of T, I and q with L may be needed to assemble a system of equations that can be integrated analytically.

'Big-leaf' models are susceptible to criticisms from micrometeorologists on three principles. First, they rely on K-theory, which is invalid within canopies (Raupach and Finnigan 1988). Second, many 'big-leaf' models do not account for the impacts that environmental and physiological gradients have on the scaling of photosynthesis, stomatal conductance, transpiration and the leaf energy balance. And third, many parameters required by 'big-leaf' models cannot be defined by mean leaf properties (Leuning et al. 1995; de Pury and Farquhar 1997).

In practice, a 'big-leaf' model is most susceptible to failure when attempting to compute hour-by-hour fluxes of clumps of trees or isolated trees, as in savanna woodlands. Model performance is improved when 'dual source' or 'two-layer' models are applied to such complex circumstances. 'Dual source' models are able to account for differential fluxes associated with sunlit and shade leaves (Meyers and Baldocchi 1988; dePury and Farquhar 1997) or a mixture of herb and shrub (Huntingford et al. 1995). 'Two layer' models are able to account for strong differences in mass and energy exchange that occur the vegetation and highly exposed soil, as is experienced over sparse woodlands and savannas (Huntingford et al. 1995).

3.1.2   Multi-Layer Models

A multi-layered model is an ideal means for computing trace gas fluxes to or from vertically inhomogeneous forests. The multi-layer model scheme is derived from Equation 1 by assuming steady state conditions, horizontal homogeneity and no chemical reactions. These assumptions yields an equality between the change, with height, of the vertical turbulent flux and the diffusive source/sink strength, $S_B(c,z)$:

$$\frac{\partial F(c,z)}{\partial z} = S_B(c,z) \qquad [6]$$

In practice, the net forest-atmosphere flux is computed by integrating Equation 6 with respect to height. The diffusive source strength is typically expressed in the form of a resistance-analog relationship:

$$S_B(c,z) = -\rho_a\, a(z)\, \frac{(c(z)-c_i)}{r_{bc}(z)+r_{sc}(z)} \qquad [7]$$

where $a(z)$ is the leaf area density, $(c(z) - c_i)$ is the concentration difference between air outside the laminar boundary layer of leaves and the air within the stomatal cavity, r_{bc} is the boundary layer resistance to molecular diffusion, r_{sc} is the stomatal resistance and ρ_a is air density. Normally, Equation 7 is evaluated by treating and weighting the sunlit and shaded portions of the canopy layer separately (Norman 1979). This activity requires the application of a radiative transfer model.

Chemical reactions are important when the time scale of the reactions are shorter than the turbulence time scale that determines the residence time of a parcel of air (Gao et al. 1993). In this case Equation 7 is expanded to include chemical production and destruction (S_{ch}):

$$\frac{\partial F(c,z)}{\partial z} = S_B(c,z) + S_{ch}(c,z) \qquad [8]$$

In the simplest circumstance, S_{ch} is parameterized using chemical kinetics, where the rate of reaction is proportional to the local concentration. The introduction of chemistry into a canopy trace gas exchange model increases the need to compute scalar profiles accurately. This is because errors attributed to the parameterization of turbulence and scalar profiles will translate directly into errors in the evaluation of chemical kinetics. Another issue involves what suite of chemical compounds to consider. Photochemical models tend to involve hundreds of reactions, which can be reduced to a suite of 20 to 40 key reactions (Gao et al. 1993).

Three-dimensional 'cubed' or 'shell' models (Wang and Jarvis 1990) treats trace gas fluxes to and from heteorogeneous and open stands most realistically. In practice, the approach is very difficult to parameterize and implement with fidelity.

3.2   Model Parameterization Issues

In ecological sciences, the philosophy dictated by Ocam's razor is often invoked as a guiding principle for designing a model. In other words, the simplest of competing theories is preferred to explain a phenomenon. One obvious question that is often raised, when modeling trace gas fluxes, is: do we need to worry about the attributes of every species in a forest or can we parameterize the system as a functional unit? In many cases, as with CO_2, energy and water vapor exchange, functional attributes of the forest stand (e.g. leaf area index, canopy conductance) are more important than the unique attributes of each species and tree in a stand (Valentini et al. 1999; Baldocchi and Wilson 2001). On the other hand, if we desire to predict hydrocarbon emissions from a forest we must know its species composition and their spatial distribution (Baldocchi et al. 1999).

Some model simplification can be achieved by restricting the span of time and space scales considered. Based on hierarchy theory, one generally uses information from adjacent time and space scales to design and implement a model (O'Neill 1989). Typically, the mechanics and the dynamics of the operational-scale are described at the smallest and fastest scales. For the case of a forest, this corresponds to the scale of leaves and how they respond to second-by-second variations in light and wind. Information at the operational-scale, i.e. the forest, is obtained by integrating reductionist-scale information in both time and space. For our case, this would correspond to hourly averages of stand-scale fluxes. The state variables that drive the operational-scale are imposed from the higher or macro-scale. In this case, a canopy-scale trace gas flux model would use weather and leaf area information as external inputs, rather than predicting the weather and forest growth.

As a model is applied for longer time-periods, other information on the structural and nutrient status of the plant canopy will be needed. Many trace gas model parameters, for example, vary significantly over the course of the year, as leaves age and resources change (Baldocchi and Wilson 2001). Yet often thes temporal dynamics are ignored.

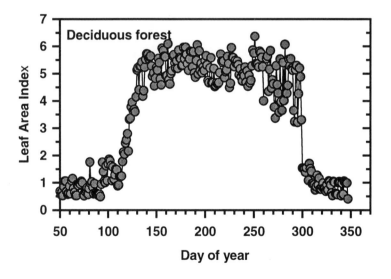

Figure 1. Seasonal variation of leaf area index of a temperate, broad-leaved deciduous forest growing near Oak Ridge, USA.

Leaf area index is a prominent characteristic that exhibits great variability over a year (Fig. 1). Deciduous broad-leaved forests, for instance, are leafless and dormant during the winter. In spring, they experience a rapid expansion of leaves and attain full-leaf within a month. The date of leaf initiation at a given site can vary by 20 days on a year-to-year basis and over a month along a north-south latitudinal gradient. Evergreen tropical and conifer forests exhibit seasonal variations in leaf area, too, though the changes are less dramatic. Over longer time scales, leaf area and leaf area profiles will be affected by disturbance history of the stand (Parker 1995; Hurtt et al. 1998). Physiological and structural characteristics that govern trace gas fluxes will depend on whether a forest is in the invading, aggrading or old growth stages.

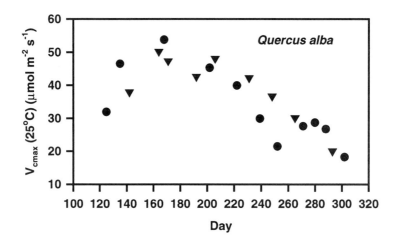

Figure 2. Seasonal variation of the maximum carboxylation velocity (V_{cmax}), a measure of photosynthetic capacity, of a temperate broad-leaved deciduous forest (Wilson et al. 2000).

Photosynthetic capacity is another model parameter that will vary over the course of the growing season (Fig. 2). Photosynthetic capacity increases in harmony with leaf expansion, during the spring. Exposure to frost, in the spring or fall will diminish it, as will subjection to drought and soil moisture deficits, during the growing season. Seasonal variation of photosynthetic capacity is attributed to changes in the amount and fraction of leaf nitrogen that is allocated to Rubisco. For instance, leaf nitrogen is re-mobilized into the stem of trees before senescence and leaf abscission. Maximum stomatal conductance exhibits a similar seasonal pattern as photosynthesis since these two variables are well-correlated with one another. Isoprene emission, on the other hand, is not initiated until the period when maximal is achieved.

The spatial variation of certain model parameters can be pronounced within a forest. Measurements on leaf nitrogen content and photosynthetic capacity, for example, vary by a factor of three between the top and bottom of a forest canopy (Ellsworth and Reich 1993; Harley and Baldocchi 1995). This adaptation of leaves to sun and shade causes the variation of photosynthetic capacity to be as great as what can be experienced across the globe, between tropical and boreal forest biomes (Schulze et al. 1994).

3.3   Validation and verification

Due to the multiplicity of time and space scales and processes that are associated with modeling trace gas fluxes, model testing is a necessary, but non-trivial, exercise. In practice, no will trace gas exchange model will pass the falsification criteria, which have been advocated by Popper (1959). For example, Rastetter (1996) shows that the Farquhar photosynthesis model, a key component of a coupled trace gas model, is capable of estimating photosynthesis responses to light and CO_2 correctly on hour-to-day time-scales. But the model fails to mimic seasonal and multi-year time-scales responses to CO_2, as plants acclimate or down-regulate. To correctly validate a canopy-scale trace gas model, the time and space scale of the model and validation data must match. It is inappropriate to test a model for conditions it was not intended for using (Rastetter 1996).

Data from a network of long term eddy flux measurement sites (FLUXNET, http://www-eosdis.ornl.gov/FLUXNET) are now available to test a hierarchy of trace gas flux models across a spectrum of forest types, on time scales from hours to years. In Figure 3, we show an example of a comparison between model calculations and measurements of carbon dioxide exchange over a broad-leaved deciduous forest for the duration of a year. Overall, the agreement between measurement and theory is good, as many of the data overlap. How well a model should agree with data is a matter of debate. A 1 μmol m^{-2} s^{-1} difference between calculated and measured carbon flux densities falls within expected measurement and modeling errors. Yet, a bias of this magnitude can cause annual sums of net carbon exchange to differ by 400 g C m^{-2}. There is also the issue relating to the accuracy of the test data, as eddy flux data suffer from bias errors at night and over complex terrain (Baldocchi and Meyers 1998).

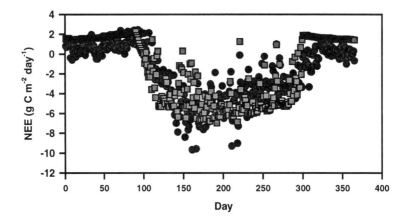

Figure 3. A comparison between measured and calculated fluxes of net ecosystem CO_2 exchange (NEE). The calculations were derived from the CANOAK model (Baldocchi and Wilson 2001). The measurements were derived from the eddy covariance method. The data are from a temperate broad-leaved deciduous forest growing near Oak Ridge, USA.

4.   CONCLUSIONS

We, as a community, possess a hierarchy of models for evaluating trace gas fluxes to and from forest canopies. The theory has matured enough that these models can simulate trace gas fluxes with reasonable fidelity, under ideal conditions. Into the future, we need to make use of networks of long-term flux measurement sites to validate a hierarchy of models over heterogeneous forests and over long time scales. To implement such models correctly, we will need better information on the seasonal and spatial variation of canopy model scaling parameters, i.e. leaf area index, photosynthetic capacity etc.

If we expect to model how trace gas fluxes of a forest will respond to future environmental perturbations, we will need to consider how a forest responds to disturbance (e.g. fire, insects) and how new genetic material is able to invade the stand and alter the genetic composition, and the functionality of the stand (see Hurtt et al. 1998). In the future, forest trace gas flux models should evolve toward a system that links micrometeorological, soil, ecophysiological, ecosystem, atmospheric chemistry and biogeochemical cycling models. A canopy micrometeorology model is needed to assess the light, temperature, humidity, CO_2, wind speed and the scalar trace gas environment within and above vegetation, which drives physiological functions described above. The soil model will be needed to compute information on soil temperature and moisture and on gas diffusion. The ecosystem and biogeochemical models will be needed to predict changes in leaf area index, canopy height, stand species composition and photosynthetic capacity of the stand. The chemistry modules are needed to compute rates of chemical reactions that are occurring within a forest and in its surface boundary layer. The most capable models will need to be able to simulate trace gas fluxes of open forests, growing on complex terrain that are subject to water deficits.

ACKNOWLEDGEMENTS

This research was supported by a grant from the U.S. Department of Energy's Terrestrial Carbon Program and the NASA GEWEX project and its Terrestrial Ecology Program. We thank Dr. Eva Falge for a review of this manuscript and discussion on this topic.

REFERENCES

Amthor JS (1994) Scaling CO_2-photosynthesis relationships from the leaf to the canopy. Photosynth Res 39: 321-350

Baldocchi DD & Meyers TP (1998) On using eco-physiological, micrometeorlogical and biogeochemical theory to evaluate carbon dioxide, water vapor and trace gas fluxes over vegetation: synthesis and application. Agr Forest Meteorol 90:1-25

Baldocchi DD & Wilson KB (2001) Modeling CO_2 and water vapor exchange of a temperate broadleaved forest across hourly to decadal time scales. Ecol Model 142: 155-184

Baldocchi DD, Fuentes JD, Bowling DR, Turnipseed AA & Monson RK (1999) Scaling isoprene fluxes from leaves to canopies: test cases over a boreal aspen and a mixed species temperate forest. J Appl Meteorol 38: 885-898

DePury D & Farquhar GD (1997) Simple scaling of photosynthesis from leaves to canopies without the errors of big leaf models. Plant Cell Environ 20: 537-557

Ellsworth DS & Reich PB (1993) Canopy structure and vertical patterns of photosynthesis and related leaf traits in a deciduous forest. Oecologia 96: 169-178

Gao W,Wesely ML & Doskey PV (1993) Numerical modeling of turbulent diffusion and chemistry of NO_X, O_3, isoprene and other reactive gases in and above a forest canopy. J Geophys Res 98: 18339-18353

Harley PC & Baldocchi DD (1995) Scaling carbon dioxide and water vapor exchange from leaf to canopy in a deciduous forest: parameterization. Plant Cell Environ 18: 1146-1156

Huntingford, C, Allen SJ & Harding RJ (1995) An intercomparison of single and dual-source vegetation-atmosphere transfer models applied to transpiration from Sahelian savanna. Bound Layer Meteorol 74: 397-418

Hurtt GC, Moorcroft PR, Pacala SW & Levin SA (1998) Terrestrial models and global change: challenges for the future. Glob Change Biol 4: 581-590

Jarvis PG (1995) Scaling processes and problems. Plant Cell Environ 18: 1079-1089

Lamb B, Guenther A, Gay D & Westberg H (1993) A biogenic hydrocarbon emission inventory for the U.S.A. using a simple forest canopy model. Atmos Environ 20: 1-8

Leuning R, Kelliher F, dePury D & Schulze ED (1995) Leaf nitrogen, photosynthesis, conductance and transpiration: scaling from leaves to canopies. Plant Cell Environ 18: 1183-1200

Meyers TP & Paw U KT (1987) Modeling the plant canopy micrometeorology with higher order closure principles. Agr Forest Meteorol 41: 143-163

Meyers TP & Baldocchi DD (1988) A comparison of models for deriving dry deposition fluxes of O_3 and SO_2 to a forest canopy. Tellus 40B: 270-284

Norman JM (1979) Modeling the complete crop canopy. In: Barfield BJ & Gerber JF (eds). Modification of the Aerial Environment of Crops, pp 249-280. American Society of Agricultural Engineers, St Joseph, MI, U.S.A.

O'Neill RV (1989) Perspectives in hierarchy and scale. In: May RM & Roughgarten J (eds) Ecological Theory, pp 140-156. Princeton University Press, Princeton, NJ. U.S.A.

Parker G (1995) Structure and microclimate of forest canopies. In: Lowman MD & Nadkarni N (eds) Forest Canopies, pp 73-106. Academic Press. San Diego, U.S.A.

Popper KR (1959) The logic of scientific discovery. p 46. Basic Books, New York, U.S.A.

Rastetter EB (1996) Validating models of ecosystem response to global change. Bioscience 46: 190-198

Raupach MR (1988) Canopy transport processes. In: Steffen WL & Denmead OT (eds) Flow and Transport in the Natural Environment: Advances and Applications, pp 95-127. Springer Verlag, Heidelberg, Germany

Raupach MR & Finnigan JJ (1988) Single layer models of evaporation from plant communities are incorrect, but useful, whereas multi-layer models are correct but useless: discussion. Aust J Plant Physiol 15: 705-716

Schulze ED, Kelliher F, Korner C, Lloyd J & Leuning R (1994) Relationships between maximum stomatal conductance, ecosystem surface conductance, carbon assimilation rate and plant nitrogen nutrition: a global exercise. Annu Rev Ecol Syst 25: 629-660

Sellers PJ, Mintz Y, Sud YC & Dalcher A (1986) A simple biosphere model (BIB) for use within general circulation models. J Atmos Sci 43: 505-531

Valentini R, Baldocchi D & Tenhunen J (1999) Ecological controls on land-surface atmospheric interactions. In: Tenhunen J & Kabat P (ed) Integrating Hydrology, Ecosystem Dynamics, and Biogeochemistry in Complex Landscapes, pp 117-145. John Wiley & Sons Ltd., Chichester, New Yoerk, U.S.A.

Wang YP & Jarvis PG (1990) Description and validation of an array model, MAESTRO. Agr Forest Meteorol 51: 257-280

Wilson KB, Baldocchi DD & Hanson PJ (2000) Spatial and seasonal variability of photosynthetic parameters and their relationship to leaf nitrogen in a deciduous forest. Tree Physiol 20: 565-578

ATMOSPHERIC CHEMISTRY OF TRACE GASES EXCHANGED IN FOREST ECOSYSTEMS

Chapter 5.1

Nitrogen oxides

J. Neil Cape

Centre for Ecology and Hydrology, Bush Estate, Penicuik, Midlothian, EH26 0QB, U.K.

1. INTRODUCTION

The nitrogen oxides, nitric oxide (NO) and nitrogen dioxide (NO_2), are collectively known as NO_X and play an important role in regulating atmospheric chemistry in the troposphere. On a global scale, natural sources (c. 20 Tg N year^{-1}) and anthropogenic sources (c. 30 Tg N year^{-1}) of NO_X contribute similar proportions of the total of ca. 40–50 Tg N year^{-1}, mostly at or close to the earth's surface (Lee et al. 1997; Olivier et al. 1998). Most primary emissions from combustion occur as NO, with conversion to NO_2 through reaction with ozone (O_3). A small proportion of NO_2 is also emitted directly from combustion sources, probably by reaction of NO with molecular oxygen (O_2). This reaction is only important at high NO concentrations, as the reaction rate is proportional to the square of the NO concentration, and plays no significant role except in highly polluted urban air with (NO) $\geq$ 1 ppm (parts in 10^6 by volume).

There is a small but significant natural source of NO_X from soils, arising from microbial activity. To some extent, this source may also be described as anthropogenic, as NO production is stimulated in soils to which N is added as fertilizer, or by N-pollution via atmospheric deposition. Although the flux of NO from soil is relatively small (c. 5 Tg N year^{-1}; Potter et al. 1996; Lee et al. 1997; Smith et al. 1997) compared with emissions from fossil fuel use in industrialized countries, it cannot be ignored in assessing the net flux and chemistry of oxidized nitrogen in natural ecosystems, such as forests, remote from major anthropogenic NO_X sources.

R. Gasche et al. (eds.), Trace Gas Exchange in Forest Ecosystems, 245–255.
© 2002 *Kluwer Academic Publishers. Printed in the Netherlands.*

2. THE NO$_X$/O$_3$ PHOTOSTATIONARY STATE

The importance of NO$_X$ to atmospheric chemistry arises because NO$_2$ is photolysed at wavelengths shorter than 400 nm, in the UV-B region of the spectrum. Photolysis of NO$_2$ produces NO, and an oxygen atom in the triplet state O(^{3}P). The O(^{3}P) reacts rapidly with an oxygen molecule to form ozone (O$_3$), which can react with NO to re-form NO$_2$.

This cycle of reactions (Reactions 1-3 below) is driven by sunlight which photolyses NO$_2$, at a rate represented by the j(NO$_2$). In clear skies at mid-latitudes in summer, j(NO$_2$) has a value around 10^{-2} s^{-1}, which means that 1% of NO$_2$ molecules are photolysed every second, or the characteristic lifetime of NO$_2$ (ϑ_{NO2}) is 100s.

$$NO_2 + h\nu \ (8 < 400nm) \rightarrow NO + O(^3P) \tag{1}$$

$$O(^3P) + O_2 \rightarrow O_3 \tag{2}$$

$$O_3 + NO \rightarrow NO_2 + O_2 \tag{3}$$

The reaction of the O atom with molecular O$_2$ to form O$_3$ in Reaction 2 is very rapid. Reaction 3 converts NO back to NO$_2$ at about 2% per second, for a typical tropospheric O$_3$ concentration of 50 ppb (parts in 10^9 by volume). NO, NO$_2$ and O$_3$ molecules are therefore inter-converted on timescales which are short relative to many atmospheric transport processes, and Reactions 1-3 are often described as a "photostationary steady state" or PSS.

Competing reactions which modify the PSS provide the eventual sinks for NO$_X$ in the atmosphere, and are also responsible for the photochemical production of ozone in polluted air.

3. NITROGEN OXIDES IN THE FREE TROPOSPHERE

The free troposphere extends upwards from the top of the boundary layer, the layer of air 1-2 km deep next to the earth's surface (see below), to the tropopause, where a marked temperature inversion denotes the boundary between the troposphere and stratosphere. The tropopause is a region with greatly reduced vertical mixing, restricting the exchange of gases and particles between the troposphere and the stratosphere. At its lower boundary the free troposphere is less restricted, and material can be transported upwards from the boundary layer by vertical mixing, e.g. in convective storms.

Nitrogen oxides in the free troposphere may originate at the earth's surface, from the combustion of fossil fuels or biomass burning, and be mixed upwards into the free troposphere. There is also direct emission of NO_X from aircraft, a source that is of increasing interest, given the projected increase in air traffic in the near future. Models of the contribution of aircraft to NO_X concentrations in the upper troposphere suggest that around 30% of the NO_X in the upper troposphere could come from aircraft emissions, even though the total emission strength (< 1 Tg N year^{-1}) is only a small proportion of global emissions (Brasseur et al. 1996; Lamarque et al. 1996; Kohler et al. 1997; Schumann 1997). Calculation of the amounts and associated chemistry of aircraft emissions relies on good estimates of natural sources of NO_X in the troposphere, of which the largest is from lightning (Penner et al. 1998). Estimates of the contribution of lightning to NO_X vary, but most are around 5 Tg N year^{-1} (Lamarque et al. 1996; Levy et al. 1996; Strand & Hov; 1996; Flatoy and Hov; 1997; Lee et al. 1997).

The chemistry of NO_X in the free troposphere is relatively simple compared with the boundary layer. In addition to Reactions 1-3 (above) the major sink for removing NO_X is the reaction of NO_2 with the hydroxyl radical (OH) to give nitric acid vapour (HNO_3):

$$NO_2 + OH \rightarrow HNO_3 \qquad\qquad [4]$$

The OH radical is produced by the photolysis of O_3 at wavelengths below 310 nm, and the subsequent reaction of the excited singlet oxygen atom $O(^1D)$ with water vapour:

$$O_3 + h\nu \ (\lambda < 310 \text{ nm}) \rightarrow O_2 + O(^1D)$$
$$O(^1D) + H_2O \rightarrow 2 \text{ OH} \qquad\qquad [5]$$

Reaction 5 occurs only in daylight, and most of the $O(^1D)$ produced from the photolysis of ozone is returned to the ground state $O(^3P)$ by collision with O_2 and N_2 molecules, and re-forms the original O_3 molecule through Reaction 2. The yield of OH radicals is proportional to the water vapour concentration, and so varies markedly with altitude and latitude.

In mid-latitudes in summer at noon the maximum removal rate for NO_2 by Reaction 4 is less than 10% per hour. In winter, and throughout most of the day, the removal rate is much slower. The removal rate is faster in the tropics, because greater sunlight intensity enhances the initial photolysis of O_3, and because the high temperatures lead to higher water vapour concentrations in the troposphere, and a greater efficiency in producing OH in Reaction 5.

At night-time, NO_2 reacts with O_3 to produce the nitrate radical NO_3:

$$NO_2 + O_3 \rightarrow NO_3 + O_2 \tag{6}$$

Although this reaction also occurs in daylight, the NO_3 radical is rapidly photolyzed, with a typical lifetime of a few seconds, to re-form NO_2 and O_3. The rate of Equation 6 is relatively slow, giving a removal rate for NO_2 of 14% per hour for an O_3 concentration of 50 ppb (Wayne et al. 1995).

The NO_3 radical will react with many organic trace gases to form products with an oxidized nitrogen group. NO_3 will also react further with NO_2 to give dinitrogen pentoxide N_2O_5 in thermal equilibrium:

$$NO_3 + NO_2 \equiv N_2O_5 \tag{7}$$

Reactions 6 and 7 combine, in the absence of other reactions, to give a night-time steady state concentration of NO_3 which depends on temperature and O_3 concentration, but is independent of NO_2 concentration (Wayne et al. 1995). The steady state concentrations are between 1 and 2 ppt (parts in 10^{12} by volume) for temperatures between 0 and 25°C. The importance of Reaction 7, though, is as another route to HNO_3. N_2O_5 will react rapidly with water vapour on particle or droplet surfaces, but not significantly in the gas-phase, to produce particle or droplet-bound acidic nitrate.

The relative simplicity of NO_X chemistry in the free troposphere shows the importance of the PSS in daytime, the role of NO_3 at night, and the gradual transformation of gaseous NO_X to gaseous or particle/droplet-bound nitric acid. The nitrate eventually falls to the earth's surface as precipitation after incorporation into cloud, or by sedimentation of large particles.

4.   NITROGEN OXIDES IN THE BOUNDARY LAYER

The NO_X chemistry of the boundary layer in daylight is also dominated by the PSS (Reactions 1–3). However, additional reactions begin to play an important role. Most important of these are the reactions of volatile organic compounds (VOCs). VOCs, and molecules such as CO, react with the OH radical to produce peroxy radicals, such as the hydroperoxy radical (HO_2) or organic peroxy radicals (RO_2, where R is an organic moiety, e.g. an alkyl group). Peroxy radicals, both HO_2 and RO_2, react with NO to produce NO_2:

$$RO_2 (HO_2) + NO \rightarrow NO_2 + RO (OH) \qquad [8]$$

This reaction continues the OH-driven oxidation of VOCs, and the coupling of NO_X chemistry with OH chemistry in this way leads to mutual catalysis of both reactive cycles. Because a peroxy radical oxidizes NO to NO_2 (Reaction 8) without using O_3 (Reaction 3) the net effect is that one molecule of O_3 is produced for every NO molecule oxidized in Reaction 8. This set of reactions is responsible for generating large O_3 concentrations in polluted air, in the presence of NO_X, VOCs and sunlight (Fig. 1).

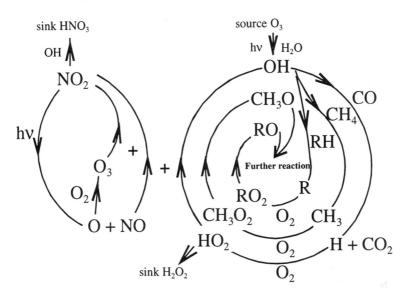

Figure 1. Schematic diagram of coupled NO_X and organic peroxy radical reactions

Another consequence of the coupling of VOC reactions with the NO_X/O_3 cycle is that OH concentrations may be enhanced significantly above that from primary production (Reaction 5), so that NO_2 removal rates, to give nitric acid (Reaction 4), may be similarly enhanced. Moreover, in air containing VOCs, other sink reactions for NO_X became possible through the direct reaction of RO and RO_2 radicals with NO and/or NO_2 to form stable organic nitrates (Lightfoot et al. 1993). Perhaps the best known of these is peroxyacetyl nitrate (PAN) formed by the reaction of NO_2 with the peroxyacyl radical ($CH_3CO \cdot O_2$):

$$CH_3CO.O_2 + NO_2 \equiv CH_3CO \cdot O_2 \, NO_2 \, (PAN) \qquad [9]$$

Reaction 9 is thermally reversible, so that PAN is more stable at low temperatures. Transport of PAN from the (warm) boundary layer to the (cool) free troposphere has been proposed as a mechanism for transporting reactive NO_X over long distances (Penkett and Brice 1986). In general, however, organic nitrates constitute a relatively small sink for NO_X, even in the boundary layer, with the major sink being HNO_3, which may be dry deposited at the ground, or may adsorb on particles or dissolve in droplets. Dry deposition of HNO_3 is very efficient, occurring as rapidly as atmospheric turbulence can transfer the gas to the surface.

At night-time, NO_X chemistry follows the processes discussed above for the free troposphere (Reactions 6 and 7), but there are many more possibilities for reaction of the NO_3 radical. In forested areas, a significant sink of NO_3 may be reaction with biogenic hydrocarbons emitted by the forest, such as isoprene or terpenes (e.g. Starn et al. 1998). The NO_3 radical reacts rapidly with these molecules by addition to unsaturated centres, to give complex organic nitrates.

The presence of such biogenic reactive hydrocarbons effectively determines the lifetime of the NO_3 radical, giving lifetimes of a few minutes or less. These night-time reactions of NO_3 with VOCs, if they occur by H atom abstraction rather than addition to double bonds, produce peroxy radicals which can form OH at night by reaction with NO (Equation 8). The presence of OH opens the pathway to the same linked VOC/NO_X chemical cycles that occur in daylight, but at much reduced rates (Wayne et al. 1995).

5. NO_X CHEMISTRY ABOVE AND WITHIN A FOREST CANOPY

The air above a forest canopy contains a mixture of NO_X, O_3 and anthropogenic VOCs that may have been transported long distances from their sources before being brought to the forest canopy surface by turbulence. To these gases may be added a different mixture of biogenic VOCs, from the forest canopy, and NO_X originating from the forest floor. NO emissions from forest soils vary greatly, and respond to deposition of pollutant nitrogen. Summer emission fluxes of NO up to 20 ng N m^{-2} s^{-1} have been measured in Denmark (Pilegaard et al. 1999), while monthly mean emission fluxes up to twice this rate have been measured for a spruce forest in Germany (Butterbach-Bahl et al. 1998a). These NO emission fluxes were linked to the deposition of ammonium and nitrate in throughfall (Butterbach-Bahl et al. 1998b). Fluxes in winter, and in less polluted areas, are much smaller, but the larger emission rates may be typical throughout much of central Europe, where N deposition exceeds the critical load (Kuylenstierna et al. 1998).

As well as additional sources of reactive gases, the canopy acts as a sink for both NO and NO_2, as well as O_3 (Chapter 3, this volume). The proximity of additional sources and sinks of the molecules involved in the PSS, and the rates of change in concentration caused by turbulent mixing close to the canopy, act to perturb the PSS. The time scales for transport and chemistry (several minutes) are similar, and the chemical processes can no longer be treated in isolation from the physical mixing processes. Consequently, appreciable vertical gradients in NO and NO_2 may be generated both within and above the forest canopy. These vertical gradients are not simply interpretable in terms of fluxes of the gases at the surface, and the interaction of deposition, transport and chemical processes has to be modelled to understand what is happening.

Such models of coupled physical and chemical processes have been constructed in order to interpret flux measurements of NO_2 and other gases over forest canopies. Models constructed to interpret field measurements may be relatively simple, including only the Reactions 1–3 (e.g. Duyzer et al. 1995; Joss and Graber 1996; Walton et al. 1997), or may be much more complex, with a large number of radicals and VOCs included (e.g. Gao et al. 1993; Gao and Wesley 1994). In general, however, there are insufficient measurements to allow such complex models to be rigorously tested, so the importance of the additional chemical reactions fuelled by OH and the presence of VOCs is still largely unknown. In the absence of significant loss reactions with OH (Equation 4), the total NO_X flux should be conserved, although $NO:NO_2$ ratios may change systematically above the forest canopy. In clean air, the net flux of NO_X from the forest may be detectable above the canopy, being the flux of NO from the soil, less any NO_2 absorbed by the canopy (e.g. Duyzer et al. 1995; Walton et al. 1997).

Below the canopy, the most obvious differences from the air above are the reduced light levels and the reduced turbulence. Both lead to longer lifetimes for NO_2. For example, a reduction in $j(NO_2)$ by a factor of ten would increase the photolysis lifetime of NO_2 to more than an hour. As O_3 is rarely totally depleted in the canopy space, the reaction of O_3 with NO essentially goes to completion. In rural forests, O_3 concentrations below the canopy are likely to greatly exceed NO_X concentrations, so the further reaction of NO_2 with O_3 to form the NO_3 radical (Equation 6) is likely. Moreover, the photolysis of NO_3 will also be greatly reduced, leading to the possibility of significant amounts of NO_3 reaction with VOCs, even in daytime, below a dense forest canopy. For example, the reaction of NO_3 with 5 ppb isoprene would give a removal rate for NO_3 of 10% s^{-1}; reaction with 1 ppb of α-pinene would remove NO_3 at approximately the same rate.

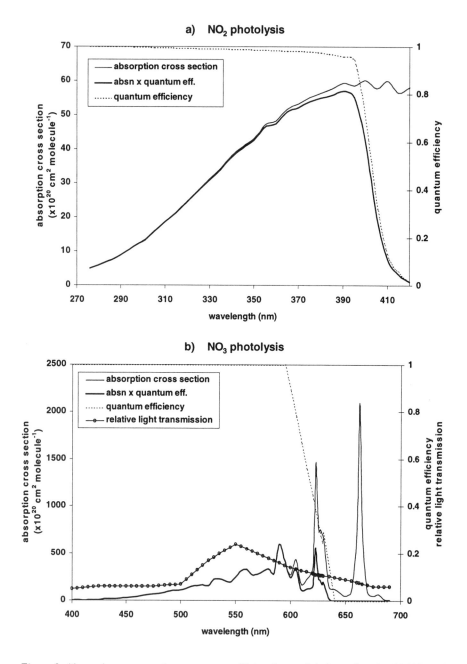

Figure 2. Absorption cross sections, quantum efficiencies, and their product for (a) NO_2 and (b) NO_3, showing the dependence of photolysis on wavelength (a: DeMore et al. 1997; b: Wayne et al. 1995). The relative transmission of light through plant canopies as a function of wavelength (McCree 1972) is also shown in (b).

By comparison, photolysis of NO_3 at one-tenth of the clear-sky sunlight rate would remove NO_3 at $< 2\%$ s^{-1}.

The photolysis of NO_3 will not necessarily be reduced by the same factor as for the photolysis of NO_2, because the forest canopy intercepts and scatters different frequencies of light with different efficiencies. The wavelengths of light which photolyse NO_3 are longer (450-630 nm) than those which photolyse NO_2 (< 410 nm; Fig. 2), so that the relative rates of photolysis depend on the change in spectral composition below the canopy as well as on changes in light intensity. In practice, the greater transmission of green light (c. 550 nm, McCree 1972; Fig. 2b) below the canopy means that the rate of photolysis of NO_3 below canopies will be relatively greater than that for NO_2. This difference is quantitatively important when modelling photolysis rates of NO_3 in canopies; a simple calculation for a 90% reduction of total photosynthetically active radiation (i.e. between 400 and 700 nm) suggests that the rate of photolysis of NO_2 below the canopy would be $< 5\%$ of the photolysis rate above the canopy, whereas the rate of NO_3 photolysis would be about 15% of the above-canopy photolysis rate. In this particular case, the assumption that all wavelengths of light are transmitted equally below the canopy (e.g. Gao et al. 1993) may not be appropriate.

6.   CONCLUSIONS

The chemistry of NO_X in forests cannot be separated from the deposition and emission fluxes of NO and NO_2 in the canopy or at the forest floor. Moreover, some of the chemical reactions that are unimportant in the sunlit air above the forest become of greater significance in the dim light below the canopy. The proximity to sources of biogenic hydrocarbons, particularly unsaturated molecules that react rapidly with NO_3, means that the forest atmosphere is a likely source of complex organic nitrates. Although some of these are relatively stable with respect to photolysis (e.g. PAN, peroxyacetyl nitrate), the reduced light levels may permit local transport and deposition of other organic nitrates which would otherwise be further fragmented by direct photolysis or reaction with OH. If these intermediate reaction products are phytotoxic, the possibility of local production, transport and deposition within the forest canopy may enhance the risk of damage. The extent of such reactions, however, is determined by the balance between the amount of NO_2 advected into the forest from outside, as a precursor of NO_3, and soil emissions of NO which will convert NO_3 back to NO_2. All of the potential chemical reactions are subject to the spatial heterogeneity of sources, sinks and light levels found in forests, and the large temporal variation in air concentrations caused by incursions of air through the canopy.

REFERENCES

Brasseur GP, Muller JF & Granier C (1996) Atmospheric impact of NO_X emissions by subsonic aircraft: A three-dimensional model study. J Geophys Res 101: 1423-1428

Butterbach-Bahl K, Gasche R, Breuer L & Papen H (1998a) Fluxes of NO and N_2O from temperate forest soils: impact of forest type, N deposition and of liming on the NO and N_2O emissions. Nutrient Cycl Agroecosyst 48: 79-90

Butterbach-Bahl K, Gasche R, Huber CH, Kreutzer K & Papen H (1998b) Impact of N-input by wet deposition on N-trace gas fluxes and CH_4 oxidation in spruce forest ecosystems of the temperate zone in Europe. Atmos Environ 32: 559-564

De More WB, Sander SP, Golden DM, Hampson RF, Kurylo MJ, Howard CJ, Ravishankara AR, Kolb CE & Molina MJ (1997) Chemical Kinetics and Photochemical Data for Use in Stratospheric Modeling, Evaluation Number 12, JPL Publication 97-4. National Aeronautics and Space Administration, Jet Propulsion Laboratory, Pasadena, California

Duyzer J, Weststrate H & Walton S (1995) Exchange of ozone and nitrogen oxides between the atmosphere and coniferous forest. Water Air Soil Poll 85: 2065-2070

Flatoy F & Hov Ø (1997) NO_X from lightning and the calculated chemical composition of the free troposphere. J Geophys Res 102: 21373-21381

Gao W & Wesely ML (1994) Numerical modeling of the turbulent fluxes of chemically reactive trace gases in the atmospheric boundary-layer. J Appl Meteorol 33: 835-847

Gao W, Wesely ML & Doskey PV (1993) Numerical modeling of the turbulent-diffusion and chemistry of NO_X, O_3, isoprene, and other reactive trace gases in and above a forest canopy. J Geophys Res 98: 18339-18353

Joss U & Graber WK (1996) Profiles and simulated exchange of H_2O, O_3, NO_2 between the atmosphere and the HartX Scots pine plantation. Theor Appl Climate 53: 157-172

Kohler I, Sausen R & Reinberger R (1997) Contributions of aircraft emissions to the atmospheric NO_X content. Atmos Environ 31: 1801-1818

Kuylenstierna JCI, Hicks WK, Cinderby S & Cambridge H (1998) Critical loads for nitrogen deposition and their exceedance at European scale. Environ Pollut 102: 591-598

Lamarque JF, Brasseur GP & Hess PG (1996) Three-dimensional study of the relative contributions of the different nitrogen sources in the troposphere. J Geophys Res 101: 22955-22968

Lee DS, Kohler I, Grobler E, Rohrer F, Sausen R, GallardoKlenner L, Olivier JGJ, Dentener FJ & Bouwman AF (1997) Estimations of global NO_X emissions and their uncertainties. Atmos Environ 31: 1735-1749

Levy H, Moxim WJ & Kasibhatla PS (1996) A global three-dimensional time-dependent lightning source of tropospheric NO_X. J Geophys Res 101: 22911-22922

McCree KJ (1972) The action spectrum, absorptance and quantum yield of photosynthesis in crop plants. Agr Meteorol 9: 191-216

Olivier JGJ, Bouwman AF, VanderHoek KW & Berdowski JJM (1998) Global air emission inventories for anthropogenic sources of NO_X, NH_3 and N_2O in 1990. Environ Pollut 102: 135-148

Penkett SA & Brice KA (1985) The spring maximum in photo-oxidants in the Northern Hemisphere troposphere. Nature 319: 655-657

Penner JE, Bergmann DJ, Walton JJ, Kinnison D, Prather MJ, Rotman D, Price C, Pickering KE & Baughcum SL (1998) An evaluation of upper troposphere NO_X with two models. J Geophys Res 103: 22097-22113

Pilegaard K, Hummelshoj P & Jensen NO (1999) Nitric oxide emission from a Norway spruce forest floor. J Geophys Res 104: 3433-3445

Potter CS, Matson PA, Vitousek PM & Davidson EA (1996) Process modeling of controls on nitrogen trace gas emissions from soils worldwide. J Geophys Res 101: 1361-1377

Schumann U (1997) The impact of nitrogen oxides emissions from aircraft upon the atmosphere at flight altitudes - Results from the AERONOX project. Atmos Environ 31: 1723-1733

Smith KA, McTaggart IP & Tsuruta H (1997) Emissions of N_2O and NO associated with nitrogen fertilization in intensive agricultural, and the potential for mitigation. Soil Use Manage 13: 296-304

Starn TK, Shepson PB, Bertman SB, Riemer DD, Zika RG & Olszyna K (1998) Nighttime isoprene chemistry at an urban-impacted forest site. J Geophys Res 103: 22437-22447

Strand A & Hov Ø (1996) The impact of man-made and natural NO_X emissions on upper tropospheric ozone: A two-dimensional model study. Atmos Environ 30: 1291-1303

Walton S, Gallagher MW & Duyzer JH (1997) Use of a detailed model to study the exchange of NO_X and O_3 above and below a deciduous canopy. Atmos Environ 31: 2915-2931

Wayne RP, Barnes I, Biggs P, Burrows JP, Canosa-Mas CE, Hjorth J, LeBras G, Moortgat GK, Perner D, Poulet G, Restelli G & Sidebottom H (1991) The nitrate radical: physics, chemistry and the atmosphere. Atmos Environ 25A: 1-203

Chapter 5.2

Ozone and volatile organic compounds: isoprene, terpenes, aldehydes, and organic acids

William R. Stockwell[1] and Renate Forkel[2]
[1]*Desert Research Institute (DRI), Reno, Nevada, United States*
[2]*Fraunhofer Institute for Atmospheric Environmental Research (IFU), Garmisch-Partenkirchen, Germany*

1.   INTRODUCTION

Emissions of volatile organic compounds from plants in the presence of nitrogen oxides can result in plant damage through the production of ozone and other air pollutants. It has long been known that photochemical air pollution can damage vegetation (Middleton et al. 1950). The first laboratory experiments by Haagen-Smit et al. (1952) showed that photolyzed mixtures of nitrogen dioxide (NO_2) and alkenes damage plants. Haagen-Smit and co-workers also showed that ozone was produced in these mixtures (Haagen-Smit 1952; Haagen-Smit et al. 1953; Haagen-Smit and Fox 1956).

More recent studies confirm that the emissions of volatile organic compounds (VOC) from plants contribute significantly to the formation of ozone and other photochemical air pollutants in forest and rural regions and even in urban areas (Chameides et al. 1988; Harley and Cass 1995). Ozone is produced by photochemical reactions involving volatile organic compounds and nitrogen oxides (Lin et al. 1988). Ozone formation may be limited by either the available volatile organic compounds or nitrogen oxides (Sillman 1995). The emissions of volatile organic compounds from plants are very reactive and may contribute a disproportionate share to ozone formation (Cardelino and Chameides 1995).

Ozone concentrations have at least doubled in the Northern Hemisphere over the last century and this increase is probably due to increased nitrogen

R. Gasche et al. (eds.), Trace Gas Exchange in Forest Ecosystems, 257–276.
© 2002 *Kluwer Academic Publishers. Printed in the Netherlands.*

oxide ($NO_X = NO + NO_2$) emissions (Fishman et al. 1979; Logan 1985; Hough and Derwent 1990; Isaksen and Hov 1987). Corrected ozone measurements by the Schönbein method show that around the year 1900 the boundary layer ozone concentrations typically were between 5–15 ppb (Volz and Kley 1988; Anfossi et al. 1991; Marenco et al. 1994). More recently increased ozone mixing ratios have found over large areas of forested land in the United States (EPA 1991; McKee 1994). During clear summer days ozone concentrations typically range between 50 to 70 ppb but the concentrations maybe as high as 80 to 110 ppb over forests in the eastern United States. Even remote Canadian sites may have O_3 concentrations as high as 45 to 63 ppb (Fuentes and Dann 1994). The dependence of ozone concentrations on NO_X emissions suggests that there are sufficient emissions of volatile organic compounds to make ozone production NO_X limited on the global scale.

Most biogenically emitted compounds such as isoprene and terpenes are highly reactive alkenes. Highly reactive alkenes react rapidly with HO• radicals, O_3 and nitrate radicals (NO_3) (Atkinson, 1997). Most of these reactions produce peroxy radicals that react with nitric oxide (NO) to convert it to nitrogen dioxide (NO_2). The photolysis of nitrogen dioxide in the troposphere produces ozone as discussed below.

2.   OVERVIEW OF THE CHEMISTRY OF TROPOSPHERIC OZONE FORMATION

The photolysis of NO_2 produces nitric oxide, NO, and ground state oxygen atoms, $O(^3P)$. The ground state oxygen atoms react with oxygen molecules to produce ozone.

$$NO_2 + h\nu \ (\lambda \le 425 \text{ nm}) \rightarrow NO + O(^3P) \tag{1}$$

$$O(^3P) + O_2 + M \rightarrow O_3 + M \tag{2}$$

In Reaction [1] hν represents photons of ultraviolet radiation and the symbol M in Reaction [2] is a second molecule of nitrogen or oxygen that carries away the excess energy of collision between $O(^3P)$ and O_2. The ozone formation reaction is in a pseudo-equilibrium with the reaction of ozone with NO, Reaction [3].

$$O_3 + NO \rightarrow NO_2 + O_2 \tag{3}$$

Reaction [3] destroys ozone almost as fast as it is produced by Reactions [1] and [2] and consequently reactions involving carbon monoxide or organic compounds are required to increase ozone concentrations. Organic

compounds react with hydroxyl radicals (HO•) to form hydroperoxy radicals
(HO_2•) and organic peroxy radicals (RO_2) that react with NO to convert it to
NO_2. This increases ozone concentrations by increasing the rate of the
production of ozone by Reactions [1] and [2] and by decreasing its rate of
destruction by Reaction [3]. Hydroxyl radicals are produced by an ozone
photolysis reaction that yields an excited oxygen atom, $O(^1D)$.

$$O_3 + hv \ (\lambda \leq 340 \ nm) \rightarrow O(^1D) + O_2 \qquad [4]$$

A few percent of the $O(^1D)$ react with water vapor to produce HO•
radicals.

$$O(^1D) + H_2O \rightarrow 2 \ HO• \qquad [5]$$

A sample reaction mechanism for ethane is given below.

$$CH_3CH_3 + HO• \rightarrow CH_3CH_2• + H_2O \qquad [6]$$

$$CH_3CH_2• + O_2 + M \rightarrow CH_3CH_2O_2• + M \qquad [7]$$

$$CH_3CH_2O_2• + NO \rightarrow CH_3CH_2O• + NO_2 \qquad [8]$$

$$CH_3CH_2O• + O_2 \rightarrow HO_2• + CH_3CHO \qquad [9]$$

$$HO_2• + NO \rightarrow HO• + NO_2 \qquad [10]$$

The net sum of Reactions [6] through [10] plus twice Reactions [1] and
[2] is:

$$CH_3CH_3 + 4 \ O_2 + 2 \ hv \rightarrow CH_3CHO + H_2O + 2 \ O_3 \qquad [11]$$

The analysis presented above underestimates the O_3 produced from
ethane because the acetaldehyde formed may also react with HO• or is
photolyzed to produce additional peroxy radicals (Finlayson-Pitts and Pitts,
1999). The HO• also consumes nitrogen oxides through the production of
nitric acid by the reaction of HO• with NO_2.

$$HO• + NO_2 \rightarrow HNO_3 \qquad [12]$$

Nitric acid is a sink for nitrogen oxides from the photochemical
production of O_3 in the lower troposphere because HNO_3 reacts slowly and it
is rapidly removed through deposition.

The amount of ozone formed from an organic compound is related to the
number of NO to NO_2 conversions affected by the compound and its
products when it degrades in the atmosphere (Leone and Seinfeld 1985).
Each organic compound has a different degradation mechanism and
therefore each has a different ozone formation potential. Ozone formation

potential has been quantified through the concept of incremental reactivity (IR), Equation [13],

$$IR = \Delta[O_3]/\Delta[VOC]$$ [13]

where $\Delta[O_3]$ is the change in the maximum O_3 concentration that results from an small change in the emissions of a volatile organic compound, $\Delta[VOC]$ (Carter 1984). Isoprene, terpenes and other biogenically emitted compounds have relatively high incremental reactivity values because they react rapidly with HO• to produce the peroxy radicals that convert NO to NO_2 which in turn undergoes photolysis to generate more ozone.

3. REACTIVITY AND EMISSIONS OF ISOPRENE, TERPENES AND OTHER BIOGENIC COMPOUNDS

Several thousand organic compounds have been identified as emitted from plants (Finlayson-Pitts and Pitts 1999). The structures of isoprene and a few typical biogenic compounds are shown in Figure 1. Many biogenically emitted organic compounds are alkenes but they may also include carbonyl, aromatic and alcohol functional groups.

Isoprene may be emitted from both deciduous and coniferous trees. Isoprene emissions (4.4 Mt per year) are about 15% of anthropogenic emissions (27 Mt per year) (Guenther et al. 1994). Coniferous trees also emit terpenes in significant quantities. Although isoprene is usually associated with rural and forested regions it is important in many urban areas. Figure 2 shows 3-hour average isoprene concentrations measured at Pico, California. Pico is located centrally within the Los Angeles urban region. Although there is some isoprene emitted from motor vehicles, the isoprene concentrations at Pico follow a diurnal cycle indicating that the isoprene was emitted from biogenic sources.

Staudt et al. (2000) used environmental chambers to measure the emission rates of terpenes from the Mediterranean pine, *Pinus pinea*, during July (Fig. 3). The emission rates roughly followed the diurnal cycles of temperature and photosynthetic photon flux. In general the emission rates tended to increase at a lower rate in the early part of the day and they tended to decrease more rapidly during the afternoon than either the temperature or the photosynthetic photon flux. For these trees myrcene, t-ß-ocimene and *d*-limonene were emitted at the greatest rates.

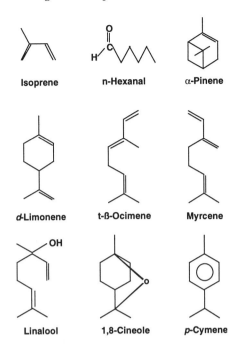

Figure 1. The chemical structure of isoprene and typical biogenic organic compounds (Finlayson-Pitts and Pitts 1999).

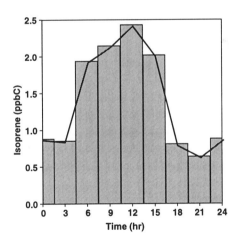

Figure 2. Average isoprene concentrations (ppbC) for summer 1997, as measured at Pico, California, are shown as cross hatched bars. The samples were taken over three hour intervals. The solid line is the time dependent interpolated average isoprene concentration.

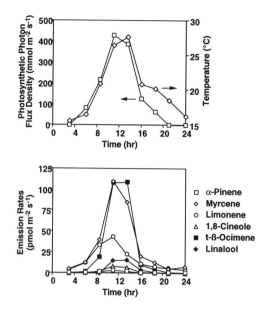

Figure 3. Emission rates of terpenes from the Mediterranean pine *Pinus pinea* for
measurements made in environmental chambers during July. Data from Staudt et al. (2000).

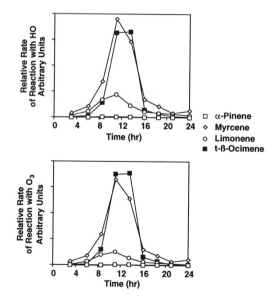

Figure 4. Relative reaction rates of terpenes from the Mediterranean pine *Pinus pinea* with
HO• and ozone. The rate constants for 1,8-cineole and linalool were not available. Data from
Staudt et al. (2000) and Atkinson (1997).

For the VOCs measured by Staudt et al. (2000) the product of their emission rates and their HO• and O_3 reaction rate constants (Atkinson 1997) was calculated to determine their relative reaction rates with HO• and ozone (Fig. 4). The rate constants for 1,8-cineole and linalool were not available. The relative importance of the reaction rates for myrcene and t-ß-ocimen were increased relative to the other terpenes.

Data from the **P**hoto-**O**xidant Formation by **P**lant Emitted **C**ompounds and **OH** **R**adicals in North-Eastern Germany (POPCORN) field experiment also illustrate the relative reactivity of biogenically emitted compounds (Plass-Dülmer et al. 1998). The POPCORN field measurements were made during August 1994 in rural Germany. The measured compounds included nitrogen oxides, organic compounds and HO• radicals. Figure 5 shows mean mixing ratios for several hydrocarbons (Koppmann et al. 1998). Although isoprene and the other alkenes are a relatively small fraction of these emissions they represent a relatively large sink for HO• radicals and ozone. About 30% of the HO• radicals and ozone that react with the measured hydrocarbons react with isoprene (Fig. 5). This is consistent with the

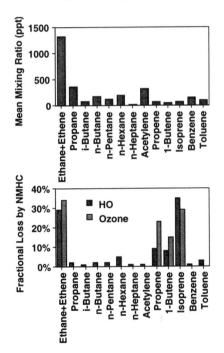

Figure 5. The upper plot gives mean mixing ratios of non-methane hydrocarbons and HO• radicals. Data from the Photo-Oxidant Formation by Plant Emitted Compounds and OH Radicals in North-Eastern Germany (POPCORN) field experiment (Koppmann et al. 1998). The lower plot gives the relative fraction of HO• and ozone reacting with these hydrocarbons.

relatively short lifetime of isoprene due to its reaction with HO• radicals. Typical lifetimes for each of the POPCORN non-methane hydrocarbons calculated on the basis of rate constants from Atkinson (1997) and an assumed HO• concentration of 5 x 10^6 cm^{-3} are shown in Figure 6.

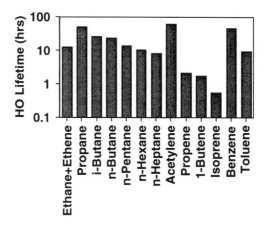

Figure 6. Typical lifetime for each of the POPCORN non-methane hydrocarbons calculated on the basis of rate constants from Atkinson (1997) and an assumed HO• concentration of 5 x 10^6 cm^{-3}.

Terpenes react even faster with HO• radicals than isoprene; the lifetime of *d*-limonene with respect to its reaction with HO• is about 4 hours. Terpenes also react rapidly with ozone. α-pinene and *d*-limonene have lifetimes with respect to their reactions with ozone of 4 hrs and 35 minutes, respectively (Finlayson-Pitts and Pitts 1999). The reaction mechanisms of isoprene and terpenes are relatively complicated and not completely understood. These reaction mechanisms are discussed below.

4.   REACTIONS OF ISOPRENE AND TERPENES

Many organic products are produced through the photochemical oxidation of biogenic compounds. The important primary reactions of isoprene are its reactions with HO•, NO_3, O_3 and O^3P. Products include formaldehyde, methyl vinyl ketone, methacrolein, 3-methylfuran and organic aerosols (Gu et al. 1985; Paulson and Seinfeld 1992; Carter and Atkinson 1996).

Isoprene reacts with ozone to directly produce excited Criegee intermediates and carbonyl species. A large fraction of the excited Criegee intermediates decompose and a small but non-negligible amount of HO• is produced. The known products of the reaction of ozone with isoprene include formaldehyde, methacrolein, methyl vinyl ketone, formaldehyde, HO$_x$• radicals and acyl radicals. Ozonolysis of methacrolein produces glyoxals, aldehydes, organic acids and radicals (Grosjean et al. 1993a, b, c). Methacrolein reacts with nitrate radical to produce organic nitrates and peroxyacetylnitrate analogs (PAN; R-(CO)-OO-NO$_2$).

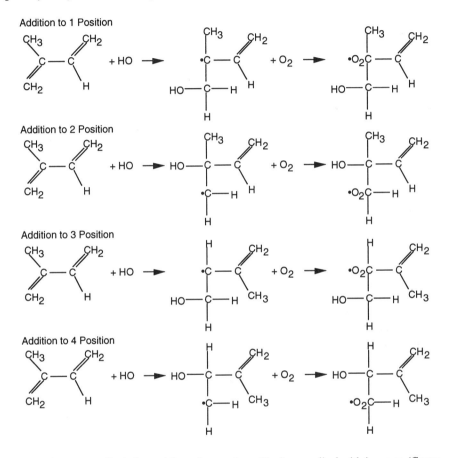

Figure 7. Peroxy radicals formed from the reaction of hydroxy radical with isoprene (Carter and Atkinson 1996).

HO• radical reacts with isoprene by addition to four possible sites (Fig. 7). About 66% of the HO• radicals add to the 1 or 2 position and about 34% add to the 3 or 4 position (Carter and Atkinson 1996). Following the addition of an HO• radical, the resulting alkyl radicals react with molecular oxygen to produce peroxy radicals. Although the detailed reaction mechanism is very complex, methyl vinyl ketone and formaldehyde are the major products if the HO• radical adds to the 1 or 2 position and methacrolein and formaldehyde are the major products if the HO• radical adds to the 3 or 4 position (Fig. 8). Figure 8 represents a highly simplified scheme for the secondary reactions of isoprene peroxy radicals.

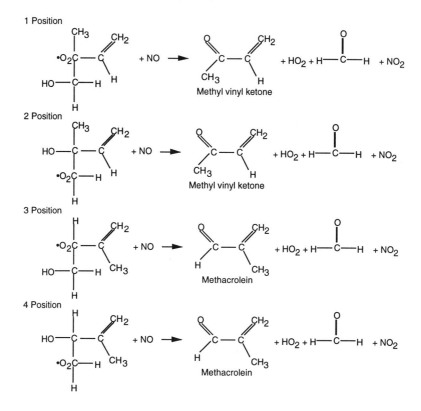

Figure 8. A highly simplified reaction scheme for the secondary reactions of isoprene peroxy radicals.

The unsaturated products such as methacrolein also react with HO•, O_3 and NO_3. For example, methacrolein reacts with HO• either by abstraction of its aldehydic hydrogen or by addition of the HO• radical to its double bond. Approximately half of the HO• radicals abstract methacrolein's aldehydic hydrogen atom and about half add to its double bond (Atkinson 1994). The products include unsaturated acyl peroxy radicals, hydroxy ketones and dicarbonyl compounds. The acyl peroxy radical reacts with NO_2 to produce methacrylic peroxynitrate, $CH_2=C(CH_3)C(O)OONO_2$, (MPAN), which is a PAN analog (Carter and Atkinson 1996).

The chemistry of methacrylic peroxynitrate and other PAN analogs produced from isoprene has been investigated by Grosjean et al. (1993a). Methacrylic peroxynitrate has two possible addition sites to its double bond for the HO• radical and the major products are hydroxyacetone, HCHO, HO$_2$• and CH$_3$C(O)C(O)OONO$_2$. Methacrylic peroxynitrate reacts with ozone and nitrate radical. Ozonolysis of methacrylic peroxynitrate produces formaldehyde but no other products have been identified although HO$_X$• radicals, acyl radicals CO$_2$ and NO$_2$ are likely products (Grosjean et al. 1993a). The products of the reaction of NO$_3$ with methacrylic peroxynitrate are unknown.

Nitrate radical reacts with isoprene with a mechanism that is similar to the hydroxyl radical reaction mechanism. It is known that about 78% of the NO$_3$ radicals add to the double bond in the 1 or 2 position and in that case 3-methyl-4-nitroxy-2-butenal is the major product (Skov et al. 1992). The NO$_3$-isoprene adducts may react either through the addition of oxygen molecules to form peroxy radicals or through decomposition (Finlayson-Pitts and Pitts 1999).

The reaction mechanisms for α-pinene and *d*-limonene are the most firmly established of the terpenes but product yields are uncertain. Most atmospheric chemistry mechanisms for terpenes are based on the experimental results for the degradation of lower alkenes (Stockwell et al. 1997). For the reaction of α-pinene with hydroxyl radical Arey et al. (1990) found that pinonaldehyde was a major product. The ozonolysis of α-pinene is assumed to produce some HO• and this complicated the analysis of early product studies. A more recent study of the ozonolysis of α-pinene (Hakola et al. 1994) used hydroxyl radical scavengers. The only product that they found was pinonaldehyde.

5. THE BEHAVIOR OF BIOGENIC ORGANIC COMPOUNDS IN CANOPIES

Biogenic organic compounds and their chemical reactions within and directly above the forest canopy affect the net flux of biogenic VOCs into the atmosphere. Deposition, i.e the loss of any atmospheric pollutant to a surface, may also have an effect on the fluxes of biogenic organic compounds and their reaction products from the canopy into the boundary layer. For example, the vegetation and the soil act as a sink for ozone, which is an important reactant for biogenic organic compounds. Therefore, emission rates for biogenic VOC that are directly measured from the trees on a leaf or branch basis, or emission rates derived from parameterizations of these measurements (e.g. Guenther et al. 1993; Steinbrecher et al. 1997) may not reflect the true emission into the boundary layer. Furthermore, it may be speculated if VOC reactions with the NO$_3$ radical may also occur during daylight hours within the lower parts of forest canopies.

Numerous measurements of leaf and branch emissions of biogenic VOC used enclosure techniques (e.g. König et al. 1995; Steinbrecher et al. 1997) as well as measurements of canopy fluxes with the relaxed eddy accumulation technique (e.g. Baker et al. 1999; Darmais et al. 2000). Canopy fluxes of BVOC can also be derived from measured vertical concentration gradients (e.g. Fuentes et al. 1996; Darmais et al. 2000). The dependence of the emissions of isoprene and monoterpenes on temperature and light can be described by comparatively simple parameterizations (e.g. Guenther et al. 1993; Steinbrecher et al. 1997, 1999). Comparison of canopy fluxes measured with the relaxed eddy accumulation technique and canopy emissions derived from branch enclosure techniques indicate an in-canopy removal of reactive compounds such as sequiterpenes (e.g. Ciccioli et al. 1999)

Chemical reactions within the canopy may affect the effective fluxes from forest canopies into the atmospheric boundary layer. In-canopy processing of biogenically emitted VOC is neglected in regional air quality models because the models' coarse vertical resolution does not allow explicit treatment. In order to investigate the effect of in-canopy processes on VOC and photooxidant concentrations several multi-layer canopy models with atmospheric chemistry have been developed (Gao et al. 1993; Walton et al. 1997; Makar et al. 1999; Forkel et al. 1999). The primary applications of these models are investigations of the effect of chemical reactions within the canopy on the net emission of VOC from forests and on the role of these emissions in ozone production over forested and rural areas.

The model by Gao et al. (1993) was the first coupled model that included turbulent vertical transport, deposition, isoprene emissions, radiation transfer in the canopy, and chemical reactions. They used a standard approach for deposition developed by Wesely (1989). A gradient approach, similar to most other canopy-chemistry models, was employed to calculate turbulent transport. The use of a second order closure scheme (e.g. Meyers and Paw U 1987) would be more physically realistic but it is too computationally expensive for most applications. For noontime simulations Gao et al. (1993) found large differences between profiles and fluxes of NO and NO_2 calculated with and without atmospheric chemical reactions. Chemistry had a weaker effect on gases that deposit rapidly such as O_3, HNO_3 and H_2O_2. The shapes of the vertical concentration gradients of O_3, HNO_3 and H_2O_2 were similar for the reactive and the non-reactive cases although their concentrations were different.

Isoprene emissions and chemical processing within and above a deciduous forest in Canada were investigated with a model using a Lagrangian approach to improve the description of vertical mixing but without treatment of deposition (Makar et al. 1999). A "Lagrangian approach" means that the model's reference frame moves with an air parcel. Makar et al. showed that chemical reactions within the canopy were required for the model to reproduce measured isoprene concentrations above the canopy. Methacrolein mixing ratios up to 0.2 ppb were predicted as a consequence of isoprene oxidation within the canopy. If chemical reactions within the canopy were ignored then the emission factors in the model had to be reduced by 40% to fit the measurements.

The model, Canopy Atmospheric Chemistry Emission Model (CACHE), by Forkel et al. (1999) is a prognostic 1-dimensional meteorological model with a soil module and an atmospheric chemistry module capable of calculating trace chemical concentrations within and above a forest canopy. Treatment of moisture within the canopy is included in the model along with the plant species, solar radiation and temperature to calculate emission behavior. Leaf surface temperatures are modeled by solving the leaf energy balance. A reduction in water supply results in an increase of the leaf surface temperature. An increase in the leaf surface temperature of only 2 degrees results in a 30% increase of isoprene emission. This increase may produce considerable ozone. The CACHE chemistry module is based on the Regional Atmospheric Chemistry Mechanism (RACM; Stockwell et al. 1997), a recent atmospheric chemistry mechanism with oxidation schemes for isoprene, α-pinene and *d*-limonene. RACM treats all biogenically emitted VOC by aggregating them together with these three compounds. RACM is used in many regional air quality models permitting CACHE results to be directly compared with regional air quality studies. The deposition of all chemical compounds is calculated according to the scheme of Wesely (1989) that includes the effects of leaf area, stomata response to atmospheric conditions and the solubility and the reactivity of chemical compounds.

The CACHE model has been applied to simulations for several boreal and Mediterranean forests in Europe. These results have shown that in-canopy processes may reduce the fluxes of isoprene, α-pinene and Δ^3-carene at the canopy top by 10 to 20% during daylight hours below the potential emission fluxes that would be estimated on the basis of the potential emissions with in-canopy processes neglected. Due to the high reactivity of *d*-limonene, in-canopy chemical reactions reduce the flux of this species by up to 30% as compared with the branch level potential flux. Figure 9 displays diurnal time series for the potential monoterpene flux, fluxes calculated for the canopy top and for 29 m. Relaxed eddy accumulation measurements made at 30 m over a 14.5 m high pine forest in southern Germany are plotted also. The data are highly scattered but the comparison suggests the model is producing reasonable results.

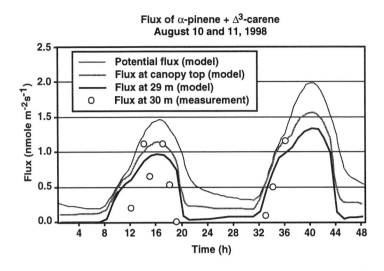

Figure 9. Diurnal variation in modeled monoterpene fluxes at different heights in comparison with relaxed eddy accumulation flux measurements above a 14.5 high pine forest in southern Germany.

These CACHE simulations along with all other canopy modeling studies show that the formation of ozone in forest areas depends strongly on the presence of anthropogenic NO_X. The formation of ozone has a large effect on the biogenically emitted VOC and the CACHE simulations show that ozone reactions are the most important sink for monoterpenes (Figure 10). Therefore it is necessary to simulate the ozone concentration profile accurately to simulate the profiles and fluxes of monoterpenes and *d*-limonene. Ozone deposition must be included in the model to account for the rapid loss of ozone to soil and plant surfaces. Otherwise the simulated ozone concentrations will be overestimated within the canopy and the effect of in-canopy chemistry on the concentrations and fluxes of biogenically emitted organic compounds will be overestimated, especially during the evening and nighttime hours.

Nitrate radical concentrations are negligible during the day because NO_3 is destroyed rapidly by photolysis. The attenuation of sunlight by dense forests can result reduce photolysis rate constants by up to 90% within the canopy. As a result of the decrease in the photolysis rate parameters significant NO_3 concentrations can be produced during daylight hours. Within the lower part of a forest canopy the reactions of NO_3 account for between 10 and 30% of oxidation of isoprene and monoterpenes during the day. The contribution of ozone, HO• and the NO_3 radical to the oxidation of α-pinene and other monoterpenes with one double bond are shown in Figure 10.

The potential importance of biogenically emitted organic compounds toward the deposition and concentrations of ozone can be found in deposition measurements made in the Bavarian forest in June 1987. The measurements were made at the Schachtenau site in the southern part of the "Grosse Ohe" watershed (Gietl and Rall 1986; Enders et al. 1989, 1992).

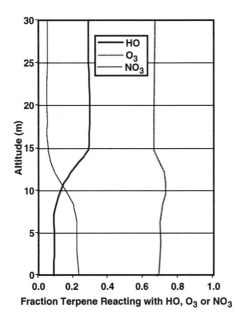

Figure 10. Contribution of ozone, HO• and the nitrate radical to the reaction of α-pinene and other monoterpines with a one double bond at noon.

The ozone concentrations at the measured levels (31 m, 41 m, 51 m) showed strong diurnal cycles with a maximum in the late afternoon and a minimum during the night. Usually the ozone concentrations decreased near the forest canopy during the night. However during the day the situation reversed and ozone concentrations were observed to increase with altitude (Enders et al. 1989). On 80% of the days of the field experiment the ozone concentrations increased with altitude over the canopy. The duration of such events lasted from only a few minutes to more than 10 hours. This complex behavior is probably due to the chemical reactions of biogenic organic compounds that may either produce or destroy ozone (Stockwell et al. 1996, 1997).

6.   EFFECTS OF BIOGENICALLY EMITTED ORGANIC COMPOUNDS ON LIQUID WATER AEROSOLS

There has been much new research on the effect of biogenically emitted organic compounds on liquid water aerosols. Aumont et al. (2000) showed that significant amounts of water soluble organic compounds could be produced from biogenically emitted organic compounds. This may have important implications for micrometeorology because the liquid water content of a parcel of air is a strong function of the relative humidity, temperature and the composition of the particles (Kim et al. 1993; Saxena et al. 1995). Saxena and Hildemann (1996) have shown that the partitioning of water soluble organic compounds is strongly affected by existing aqueous aerosols. The products of biogenically emitted organic compounds include compounds with the different functional groups: nitrate (NO_3^-), hydroperoxide (OOH), organic peroxides (OOR), acids (COOH), peroxyacylnitrate ($CO(OONO_2)$) and peroxy-acids (CO(OOH)). These compounds are soluble and therefore they are among the most important that influence the secondary organic composition of aerosols.

7.   CONCLUSIONS

There are several thousand organic compounds that are emitted into the air by plants (Finlayson-Pitts and Pitts 1999). Biogenic compounds, such as isoprene, are important in rural and urban regions. Isoprene is emitted both by deciduous trees and coniferous trees while terpenes are emitted mainly from coniferous trees. Emission rates for biogenic VOC estimated on a leaf or branch basis (e.g. Guenther et al. 1993; Steinbrecher et al. 1997) may not reflect the true emission into the boundary layer due to interactions between the emissions, vertical transport and chemical reactions. Most of the biogenically emitted organic compounds are alkenes but some contain carbonyl, aromatic and alcohol functional groups. These compounds react rapidly with HO• radicals, O_3 and nitrate radicals (NO_3) (Atkinson 1997). Depending upon the availability of nitrogen oxides these reactions may produce ozone. Most terpenes react even faster with HO• radicals than isoprene and the lifetime of terpenes can be as low as a few minutes. The known products of the biogenically emitted compounds include formaldehyde, methacrolein, methyl vinyl ketone, glyoxals, organic acids, methacrylic peroxynitrate, HO_X• radicals and acyl radicals. Simulations made with canopy models show that chemistry within a forest canopy may reduce the fluxes of isoprene at the canopy top by 10 to 20% during daylight hours below the potential emission fluxes that would be estimated on the basis of the branch level emissions alone. The fluxes of more reactive species at the canopy top, such as *d*-limonene, may be reduced by a much greater factor. For dense forest canopies reactions of nitrate radical may become important within the lower part of the canopy. The products of biogenically emitted organic compounds include many highly soluble compounds that may affect the meteorology of forest canopies by modifying the properties of liquid water aerosols (Aumont et al. 2000).

REFERENCES

Anfossi D, Sandroni S & Viarengo S (1991) Tropospheric ozone in the nineteenth century: the Moncalieri series. J Geophys Res 96: 17349-17352

Arey J, Atkinson R & Aschmann SM (1990) Product study of the gas-phase reactions of monoterpenes with the OH radical in the presence of NO_X. J Geophys Res 95: 18539-18546

Atkinson R (1997) Gas-phase tropospheric chemistry of volatile organic compounds: 1. alkanes and alkenes. J Phys Chem Ref Data 26: 215-290

Atkinson R (1994) Gas-phase tropospheric chemistry of organic compounds: a review. J Phys Chem Ref Data Monograph 2: 1-216

Aumont B, Madronich S, Bey I & Tyndall GS (2000) Contribution of secondary VOC to the composition of aqueous atmospheric particles: a modeling approach. J Atmos Chem 35: 59-75

Baker B, Geunther A, Greenberg J, Goldstein A & Fall R (1999) Canopy fluxes of 2-methyl-3-buten-2-ol over ponderosa pine forest by relaxed eddy accumulation: field data and model comparison. J Geophys Res 104: 26107-26114

Carter WPL (1994) Development of ozone reactivity scales for volatile organic compounds. J Air Waste Manage Assoc 44: 881-899

Carter WPL & Atkinson R (1996) Development and evaluation of a detailed mechanism for the atmospheric reactions of isoprene and NO_X. Int J Chem Kinet 28: 497-530

Cardelino CA & Chameides WL (1995) An observation-based model for analyzing ozone precursor relationships in the urban atmosphere. J Air Waste Manage Assoc 45: 161-180

Chameides WL, Kindsay RW, Richardson J & Kiang CS (1988) The role of biogenic hydrocarbons in urban photochemical smog: Atlanta as a case study. Science 241: 1473-1475

Ciccioli P, Brancaleoni E, Frattoni M, DiPalo V, Valentini R, Tirone G, Seufert G, Bertin N, Hansen U, Csiky O, Lenz R & Sharma M (1999) Emission of reactive compounds from orange orchards and their removal by within-canopy processes. J Geophys Res 104: 8077-8094

Darmais S, Dutaur L, Larsen B, Cieslik S, Luchetta L, Simon V & Torres L (2000) Emission fluxes of VOC by orange trees determined by both relaxed eddy accumulation and vertical gradient approaches. Chemosphere Glob Change Sci 2: 47-56

Enders G, Teichmann U & Kramm G (1989) Profiles of ozone and surface-layer parameters over a mature spruce stand. In: Georgii H-W (ed) Mechanisms and Effects of Pollutant-Transfer into Forests, pp 21-35. Kluwer Academic Publishers, Dordrecht, The Netherlands

Enders G, Dlugi R, Steinbrecher R, Clement B, Daiber R, v. Eijk J, Gäb S, Haziza M, Helas
    G, Hermann U, Kessel M, Kesselmeier J, Kotzias D, Kourtidis K, Kurth H-H, McMillen
    RT, Roider G, Schürmann W, Teichmann U & Tores L (1992) Biosphere/atmosphere
    interactions: integrated research in a European coniferous forest ecosystem. Atmos
    Environ 26A: 171-189
EPA (1991) National Air Quality and Emissions Trends Report, 1989. Research Triangle
    Park, North Carolina: U.S. Environmental Protection Agency
Finlayson-Pitts BJ & Pitts JN (1999) Chemistry of the Upper and Lower Atmosphere, Theory,
    Experiments and Applications. Academic Press, New York, U.S.A.
Fishman J, Solomon S & Crutzen PJ (1979) Observational and theoretical evidence in support
    of a significant in-situ photochemical source of tropospheric ozone. Tellus 31: 432-446
Forkel R, Stockwell WR & Steinbrecher R (1999) Multilayer canopy/chemistry model to
    simulate the effect of in-canopy processes on the emission rates of biogenic VOCs. In:
    Borell PM & Borell P (eds) Proceedings of the EUROTRAC Symposium '98, Volume 2,
    pp 45-49. WITPRESS, Southampton, U.K.
Fuentes JD & Dann TF (1994) Ground-level ozone in eastern Canada: seasonal variations,
    trends, and occurrences of high concentrations. J Air Waste Manage Assoc 44: 1019-1026
Fuentes JD, Wang D, Neumann HH, Gillespie TJ, den Hartog G & Dann TF (1996) Ambient
    biogenic hydrocarbons and isoprene emissions from a mixed deciduous forest. J Atmos
    Chem 25: 67-95
Gao W, Weseley ML & Doskey PV (1993) Numerical modeling of the turbulent diffusion and
    chemistry of NO_X, O_3, Isoprene, and other reactive trace gases above a forest canopy. J
    Geophys Res 98: 18339-18353
Gietl G & Rall A (1986) Bulk deposition into the catchment "Grosse Ohe". Results of
    neighbouring sites in the open and under spruce at different altitudes. In: Georgii H-W (ed)
    Atmospheric Pollutants in Forest Areas, pp 79-88. Reidel Publishers, Dordrecht, The
    Netherlands
Grosjean D, Grosjean E & Williams II EL (1993a) The reaction of ozone with MPAN,
    CH_2=$C(CH_3)C(O)OONO_2$. Environ Sci Technol 27: 2548-2552
Grosjean D, Williams II EL & Grosjean E (1993b) Atmospheric chemistry of isoprene and of
    its carbonyl products. Environ Sci Technol 27: 830-840
Grosjean D, Williams II EL & Grosjean E (1993c) Gas phase reaction of the hydroxyl radical
    with the unsaturated peroxyacyl nitrate CH_2=$(CH_3)C(O)OONO_2$. Int J Chem Kinet 25:
    921-929
Gu C-L, Rynard CM, Hendry DG & Mill T (1985) Hydroxyl radical oxidation of isoprene.
    Environ Sci Technol 19: 151-155
Guenther AB, Zimmermann PR, Harley PC, Monson RK & Fall R (1993) Isoprene and
    monoterpene emission rate variability: model evaluations and sensitivity analysis. J
    Geophys Res 98: 12609-12617
Guenther AB, Zimmermann PR & Wildermuth M (1994) Natural volatile organic compound
    emission rate estimates for U.S. woodland landscapes. Atmos Environ 28: 1197-1210
Haagen-Smit AJ (1952) Chemistry and physiology of Los Angeles smog. Indust Eng Chem
    44: 1342-1346
Haagen-Smit AJ, Bradley CE & Fox MM (1953) Ozone formation in photochemical
    oxidation of organic substances. Indust Eng Chem 45: 2086-2089
Haagen-Smit AJ, Darley EF, Zaitlin M, Hull H & Nobel WM (1952) Investigation of injury to
    plants from air pollution in the Los Angeles area. Plant Physiol 27: 18-34
Haagen-Smit AJ & Fox MM (1956) Ozone formation in photochemical oxidation of organic
    substances. Indust Eng Chem 48: 1484-1487

Hakola H, Arey J, Aschmann SM & Atkinson R (1994) Product formation from the gas-phase reactions of OH radicals and O_3 with a series of monoterpenes. J Atmos Chem 18: 75-102

Harley RA & Cass GR (1995) Modeling the atmospheric concentrations of individual volatile organic compounds. Atmos Environ 29: 905-922

Hough AM & Derwent RG (1990) Changes in the global concentration of tropospheric ozone due to human activities. Nature 344: 645-648

Isaksen ISA & Hov Ø (1987) Calculation of trends in the tropospheric concentration of O_3, OH, CO, CH_4 and N_2O. Tellus 39B: 271-285

Kim YP, Seinfeld JH & Saxena P (1993) Atmospheric gas-aerosol equilibrium I. thermodynamic model. Aerosol Sci Technol 19: 157-181

Koppmann R, Plass-Dülmer C, Ramacher B, Rudolph J, Kunz H, Melzer D & Speth P (1998) Measurements of carbon monoxide and nonmethane hydrocarbons during POPCORN, J Atmos Chem 31: 53-72

König G, Brunda M, Puxbaum H, Hewitt CN, Duckham SC & Rudolph J (1995) Relative contributions of oxygenated hydrocarbons to the total biogenic VOC emissions of selected mid-European agricultural and natural plant species. Atmos Environ 29: 861-874

Leone JA & Seinfeld JH (1985) Comparative analysis of chemical reaction mechanisms for photochemical smog. Atmos Environ 19: 437-464.

Lin X, Trainer M & Liu SC (1988) On the nonlinearity of the tropospheric ozone production. J Geophys Res 93: 15879-15888

Logan JA (1985) Tropospheric ozone: seasonal behavior, trends, and anthropogenic influence. J Geophys Res 90: 10463-10882

Makar PA, Fuentes JD, Wang D, Staebler RM % Wiebe HA (1999) Chemical processing of biogenic hydrocarbons within and above a temperate deciduous forest. J Geophys Res 104: 3581-3603

Marenco A, Gouget H, Nedelec P & Pages J-P (1994) Evidence of a long-term increase in tropospheric ozone from Pic du Midi data series: consequences: positive radiative forcing. J Geophys Res 99: 16617-16632

McKee DJ (ed) (1994) Tropospheric Ozone: Human Health and Agricultural Impacts. CRC Press, Boca Raton, Florida, U.S.A.

Meyers TP & Paw UKT (1987) Modeling the plant canopy micrometeorology with higher-order closure principles. Agric For Meteorol 41: 143-163

Middleton JT, Kendrick Jr. JB & Schwalm HW (1950) Injury to herbaceous plants by smog or air pollution. Plants Dis Rep 34: 245-252

Milford JB, Russell AG & McRae GJ (1989) Spatial patterns in photochemical pollutant response to NO_X and ROG reductions. Environ Sci Technol 23:1290-1301

Paulson SE & Seinfeld JH (1992) Development and evaluation of a photochemical mechanism for isoprene. J Geophys Res 97: 20703-20715

Plass-Dülmer PD, Brauers T & Rudolph J (1998) POPCORN: A field study of photochemistry in northeastern Germany. J Atmos Chem 31: 5-31

Saxena P, Hildemann LM, McMurry PH & Seinfeld JH (1995) Organics alter hygroscopic behavior of atmospheric particles. J Geophys Res 95: 1837-1851.

Saxena P & Hildemann LM (1996) Water-soluable in atmospheric particles: a critical review of the literature and application of thermodynamics to identify candidate compounds. J Atmos Chem 24: 57-109

Sillman S (1995) The use of NO_X, H_2O_2, and HNO_3 as indicators for ozone-NO_X-hydrocarbon sensitivity in urban indicators. J Geophys Res 100: 14175-14188

Skov H, Hjorth J, Jensen NR & Restelli G (1992) Products and mechanisms of the reactions of the nitrate radical (NO₃) with isoprene, 1,3-butadien and 2,3-dimethyl-1,3-butadien in air. Atmos Environ 26A: 2771-2783

Staudt M, Bertin N, Frenzel B & Seufert G (2000) Variations in the amount and composition of monoterpenes emitted by young Pinus pinea trees – implications for emissions modeling. J Atmos Chem 35: 77-99

Steinbrecher R, Hauff K, Hakola H & Rössler J (1999) A revised parametrisation for emission modelling of isoprenoids for boreal plants. In: Laurila Th & Lindfors V (eds) Biogenic VOC Emission and Photochemistry in the Boreal Regions of Europe, pp 29-43. European Commission, Brussels, Belgium

Steinbrecher R, Hauff K, Rabong R & Steinbrecher J (1997) The BEMA-project: isoprenoid emission of oak species typical for the Mediterranean area: source strength and controlling variables. Atmos Environ 31: 79-88

Stockwell WR, Kirchner F, Kuhn M & Seefeld S (1997) A new mechanism for regional atmospheric chemistry modeling. J Geophys Res 102: 25847-25879

Stockwell WR, Kramm G, Scheel H-E, Mohnen VA & Seiler W (1996) Ozone formation, destruction and exposure in Europe and the United States. In: Sandermann Jr. H, Wellburn AR & Heath RL (eds) Forest Decline and Ozone: A Comparison Of Controlled Chamber and Field Experiments, pp 1-38. Springer Verlag, New York, U.S.A.

Wesely ML (1989) Parameterization of surface resistances to gaseous dry deposition in regional-scale numerical models. Atmos Environ 23: 1293-1304

Volz A & Kley D (1988) Evaluation of the Montsouris series of ozone measurements made in the nineteenth century. Nature 332: 240-242

Walton S, Gallagher MW & Duyzer JH (1997) Use of a detailed model to study the exchange of NO$_X$ and O₃ above and below a deciduous canopy. Atmos Environ 31: 2915-2931

ENVIRONMENTAL FACTORS INFLUENCING TRACE GAS EXCHANGE

Chapter 6.1

Acid rain and N-deposition

Sharon J. Hall[1] and Pamela A. Matson[2]
[1]*Environmental Science Program, The Colorado College, Colorado Springs, CO 80903, U.S.A.*
[2]*Department of Geological and Environmental Sciences, Stanford University, Stanford, CA 94305, U.S.A.*

1.   INTRODUCTION

Atmospheric deposition of nitrogen and sulfur compounds to terrestrial ecosystems has increased significantly over the last several decades due to the emissions of reactive trace gases associated with human activity. Whereas the stable, long-lived greenhouse gases such as carbon dioxide (CO_2), methane (CH_4), and nitrous oxide (N_2O) operate on a global scale, the effects of reactive, short-lived gases such as nitric oxide and nitrogen dioxide (NO and NO_2, combined as NO_X), sulfur dioxide (SO_2), and ammonia (NH_3) are primarily regional. In the troposphere, SO_2, NO_X, and NH_3 combine with water and other atmospheric constituents and can enter into ecosystems as acids and particulates within a few to thousands of kilometers from their source.

The acidification of ecosystems from sulfur and nitrogen deposition has been well studied in northern latitudes where, historically, the problem has been most severe. Early studies primarily highlighted the detrimental and complex effects of acidic sulfur compounds on aquatic and terrestrial systems, including acute and chronic impacts on populations of plants and animals in the northern U.S. and Europe (Schindler et al. 1985; Ulrich 1989; Likens et al. 1996). Since the implementation of strict SO_2 emissions controls during the mid 1980's, emissions of SO_2 have decreased relative to emissions of NO_X in developed nations (Rodhe et al. 1995; Holland et al.

279

R. Gasche et al. (eds.), Trace Gas Exchange in Forest Ecosystems, 279–306.

1999). Today the focus of many studies has shifted to the synergistic relationship between SO_2 and NH_3 deposition (McLeod et al. 1990), and the indirect acidification potential of both reduced and oxidized reactive nitrogen compounds (van Breemen et al. 1982; Nihlgård 1985; Heij and Schneider 1991; Galloway 1995). Absolute emissions of sulfur still dominate over nitrogen in some parts of Europe and especially in developing countries in Asia, but the relative importance of nitrogen emissions and deposition will likely increase as global populations and rates of food production rise, and as ecosystems in the northern hemisphere become nitrogen saturated (Ågren and Bosatta 1988; Henriksen and Brakke 1988).

Future scenarios of SO_2, NO_X, and NH_3 emissions suggest that acid precipitation and nitrogen deposition will move from a predominantly northern hemisphere issue to a global issue as industrialization and agricultural practices intensify in the developing world during the 21st century (Busch et al. 2001). Primary production in many terrestrial ecosystems, especially in temperate forests, is limited by nitrogen (Vitousek and Howarth 1991). Thus, it is not surprising that anthropogenic nitrogen input in the temperate zone has had consequences for the structure and function of ecosystems beyond the effect of acidity alone. As atmospheric deposition of pollutants shifts to a more global distribution, however, it is unclear whether the pattern of responses in temperate forests will apply in tropical and sub-tropical forests as well. Because it is likely that primary production in many tropical forests is limited by some element other than nitrogen, we might expect the consequences of nitrogen deposition to be quite different (Matson et al. 1999).

In this chapter, we first review current estimates of sulfur and nitrogen emissions and deposition from anthropogenic sources. We then evaluate the consequences of increased deposition into both temperate and tropical forests, showing that the effects may be different between these systems depending on whether primary production is limited by nitrogen or by some other element, such as phosphorus. Finally, we will discuss the effect of anthropogenic nitrogen deposition on the emissions of the trace gases, N_2O, NO, and CH_4, from forest soils. In general, it is difficult to explore the effects of acidity on ecosystem function independent of the effects of nitrogen because of several factors. First, industrial emissions produce both sulfur and nitrogen compounds that are associated with acid precipitation, and nitrogen is becoming an increasingly important constituent of atmospheric deposition worldwide. Second, it is widely known that nitrogen deposition both directly and indirectly causes soil acidification. For these reasons, in the following discussion we will concentrate on the effects of nitrogen deposition in forest ecosystems with the understanding that these ecosystems are also being directly and indirectly acidified as well.

2. BACKGROUND – ANTHROPOGENIC EMISSIONS AND DEPOSITION OF NITROGEN AND SULFUR COMPOUNDS

SO_2, NO_X, and NH_3 are reactive compounds in the atmosphere with lifetimes of hours to days. In the atmosphere, they are rapidly transformed by chemical processes into a variety of nitrogen and sulfur species. Of these, the formation of sulfuric acid (H_2SO_4) and nitric acid (HNO_3) have received the most attention in the past several decades because of their direct detrimental effects on biological processes. However, anthropogenic emissions of nitrogen and sulfur can also be transformed into a number of other species that are not directly acidic but can significantly alter the cycling and transformation of nutrient elements within ecosystems. For nitrogen, these include various nitrogen oxides (HO_2NO_2, N_2O_5, PAN (peroxyacetyl nitrate), and HNO_2) and NH_3, which can adsorb onto the surfaces of particles or, in the presence of sulfur, be deposited to leaf and soil surfaces as $(NH_4)_2SO_4$ (van Breemen et al. 1982). Historically, most deposition was thought to occur in rainwater; now, with the establishment of extensive atmospheric deposition networks, at least in developed countries, it is estimated that at least half of all atmospheric inputs to ecosystems are in dry form, as ions or particulates (Lovett 1994).

Fossil fuel combustion is the primary anthropogenic source of atmospheric SO_2 and NO_X, whereas agricultural practices (fertilized agriculture and livestock wastes) are the dominant source of NH_3 and a secondary source of NO_X (Galloway 1996). Because of their different sources, the distribution of anthropogenic SO_2, NO_X, and NH_3 emissions and deposition differ from one another in space and time. Nearly 75 Tg of SO_2-S and 21 Tg of NO_Y-N were emitted to the atmosphere in 1990 from fossil fuel use, comprising an increase of over 300% and 50% of background sources of oxidized sulfur and nitrogen, respectively, in the 20th century (Table 1). Because of their source, deposition of sulfur and nitrogen oxides is concentrated where fossil fuel use is greatest, near urban and industrial areas in developed nations in North America, Europe, and Japan.

In contrast, global emissions of NH_3 (and to a lesser extent, NO_X) from intensive agriculture has increased significantly only in the last several decades, concurrent with increases in chemical fertilizer use and livestock production since 1950. Furthermore, because of its predominately agricultural source, NH_3 emissions have been historically concentrated in rural areas of both developed and developing countries, including those in Asia.

Table 1. 1990 Global anthropogenic emissions of reactive nitrogen and sulfur compounds (Tg N yr^{-1}). Modified from Galloway (1996), Holland et al. (1997), and Davidson and Kingerlee (1997).

	Fossil fuel	Burning	Agriculture Animal waste	Soils	Anthro-pogenic Total	Pre-indusrial Total	Fraction deposited to Terrestrial systems (%)
SO$_2$	75	--	--	--	75	20[b]	50
NO$_Y$[a]	21	8	--	5[c]	34	22[d]	71
NH$_3$	--	5	32	10	47	21	83

[a] NO$_Y$ = NO$_X$ + any other single nitrogen species with an oxygen atom; [b] Includes only SO$_2$ emissions from volcanoes; [c] From Davidson & Kingerlee (1997); [d] Pre-industrial values include 3 Tg NO$_X$-N from lightning, 0.8 Tg NO$_Y$-N from natural biomass burning, and 18 Tg NO$_X$-N/yr from soils, assuming that cultivation and fertilizer management increase NO$_X$ fluxes from ecosystems by at least 50% (From Davidson and Kingerlee (1997), assuming that cultivated land was derived equally from forest and grassland in both temperate and tropical systems).

Globally, approximately 47 Tg of nitrogen as NH$_3$ are emitted to the atmosphere each year from agricultural practices, which is more than twice the amount of NO$_X$ produced in fossil fuel combustion (Schlesinger and Hartley 1992) and twice as large as pre-industrial fluxes (Table 1). Nitric oxide is another product of agricultural practices: NO is produced by microbial processes in soils, and emissions can be large enough to contribute significantly to regional tropospheric ozone episodes, especially in rural areas outside of the influence of combustion activities (Hall et al. 1996). Fertilized agriculture is estimated to contribute over 5 Tg of nitrogen as NO to the atmosphere each year, constituting 25% of the total global emissions of NO from soils (Davidson and Kingerlee 1997). Biomass burning associated with land clearing for agriculture is also a source of NO$_X$ (8.0 Tg N yr^{-1}); together with soil emissions, agriculture currently releases a total of 13 Tg of N per year as NO$_X$, nearly half of that emitted by combustion activities (Table 1).

Using best estimates, it has been calculated that greater than 50% of the global anthropogenic emissions of each of these gases are deposited to terrestrial ecosystems each year (Table 1), with up to 10% of S emissions and more than 10% of N emissions falling on forests (Holland et al. 1997)[1].

---

[1] Holland et al. (1997) calculated that 10% of all fossil fuel NO$_Y$ emissions falls on forest ecosystems. NH$_3$ and SO$_2$ emissions were not included in the study; however, assuming that SO$_2$ deposition has an equal distribution as NO$_Y$ (because of the fossil fuel source) and assuming that even a fraction of the ammonia deposited to terrestrial systems falls on forests, we can estimate that >> 10% of total nitrogen deposition and 10% of S eposition is received by forest ecosystems yearly.

Because of the patchy distribution of reactive trace gases in the atmosphere, some forests receive more deposition than others: nitrogen and sulfur inputs have reached up to 100 kg ha^{-1} yr^{-1} downwind from some urban centers and livestock facilities in regions of Europe (Ineson et al. 1998), while estimates for the United States are somewhat lower (up to 30 kg N ha^{-1} yr^{-1}; Fenn et al. 1998; Lovett 1994). Nevertheless, compared to pre-industrial rates (2 kg N ha^{-1} yr^{-1}; Galloway 1996), and average global fertilizer application to agricultural systems (~50 kg N ha^{-1} yr^{-1}; Matthews 1994), atmospheric nitrogen deposition can represent a significant perturbation to temperate forest nitrogen cycling and associated ecosystem processes. In many cases, forests are receiving much more nitrogen from the atmosphere than their annual growth requires.

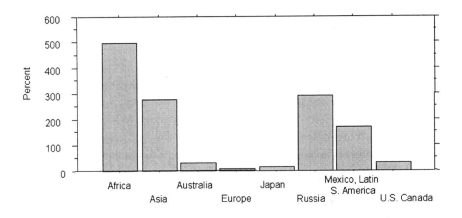

Figure 1. Projected increase in NO$_X$ emissions by year 2020. From Galloway (1995).

Not surprisingly, anthropogenic emissions of SO$_2$, NO$_X$, and NH$_3$ are expected to increase in the next century along with increases in the global population, *per capita* income, and food production. In addition, the distribution of nitrogen and sulfur deposition is expected to change significantly. Assuming the continuation of current trends, by year 2050, developing nations will increase their fossil fuel emissions of NO$_X$ and SO$_2$ by several hundred percent (Figure 1), and over half of the total emissions will be produced in tropical and subtropical areas. NH$_3$ emissions are also expected to increase with a shift in distribution toward developing countries, where nearly two thirds of all fertilizer use will occur (Matthews 1994; Galloway et al. 1995), and biomass burning will most likely continue to be a major regional source of oxidized nitrogen compounds in the tropics (Andreae et al. 1988).

Clearly, acid precipitation and nitrogen deposition will be important perturbations in the majority of the world's ecosystems in the near future. What is uncertain is how (and how quickly) different ecosystem types will respond to nitrogen deposition and how their responses will affect future atmospheric composition.

3.   CONSEQUENCES OF ENHANCED ACID PRECIPITATION AND NITROGEN DEPOSITION TO FOREST ECOSYSTEMS

3.1   Temperate Forests

The effects of chronic acid and nitrogen deposition have been studied extensively in temperate ecosystems in the northern latitudes, where the consequences have become so severe that many lakes, streams, and forests are limed as a part of routine management practices. Acid deposition has been shown to have complex effects at all levels of biological organization, including direct negative impacts on plant and animal populations and diversity, and indirect impacts due to aluminum toxicity, decreased phosphorus availability, base cation leaching from soils, and increased heavy metal mobility and uptake (Johnson et al. 1982; Schindler et al. 1985; Likens et al. 1996). Furthermore, the natural recovery of acidified ecosystems is slow even when acidic inputs are decreased, because many soils and waters have lost essential nutrients and species have changed in abundance within communities, in some cases permanently. The striking effects of acidity on forest ecosystem function has promoted successful mitigation strategies to decrease anthropogenic SO_2 emissions and increase the pH of soils and fresh waters through management (Rodhe et al. 1995). Controls on fossil fuel NO_X emissions have been less successful, and regulatory efforts to reduce agricultural emissions of NH_3 and NO_X are nascent or nonexistent (Hall et al. 1996; Holland et al. 1999).

In addition to its potential as an acidifying agent, nitrogen is also the element most limiting to primary production in most temperate forest ecosystems. Thus, atmospheric nitrogen deposition has been shown to cause complex changes in ecosystem properties and processes separate from the effects induced by acidity, especially when inputs are large and continuous. After following the responses of a range of temperate forests to experimental nitrogen inputs, researchers developed the conceptual model of "nitrogen saturation" (Ågren and Bosatta 1988; Aber et al. 1989), a state that occurs in forest ecosystems when nitrogen availability exceeds biological demand.

Central to this model is the assumption that primary production in forest ecosystems is nitrogen-limited before nitrogen inputs begin, such that most of the added nitrogen, at least initially, will be retained in the ecosystem in plant and microbial growth. Indeed, most forests in North America and Europe are known to grow in response to nitrogen additions until production is limited by some other factor (Kauppi et al. 1992; Kurz et al. 1995; Spiecker et al. 1996; Binkley and Högberg 1997).

Based on this principle, research and modeling activities during the last several years have suggested that atmospheric nitrogen deposition could be inadvertently fertilizing northern temperate forests, causing an apparent increase in forest production (Kauppi et al. 1992; Binkley and Högberg 1997) and potentially storing large amounts of atmospheric carbon in plant biomass (between 0.3 and 2.0 Pg C, Townsend et al. 1996; Holland et al. 1997). More recent experimental data have reduced this estimate considerably (0.25 Pg C yr^{-1}; Nadelhoffer et al. 1999), showing that trees are poor competitors for nitrogen compared to soils. Nevertheless, all studies indicate that nitrogen-limited temperate forest ecosystems (especially soils) are generally strong sinks for anthropogenic nitrogen: less than 10% of nitrogen additions are lost as NO_3^- through leaching or as N_2O, NO, or dinitrogen (N_2) emissions from soil in many temperate forests even after years of nitrogen additions (Magill et al. 1997; Nadelhoffer et al. 1999).

According to the nitrogen saturation model, after continued nitrogen deposition – perhaps up to decades or more (depending on how nitrogen-limited the system is to begin with) – nitrogen-limited forests are thought to become nitrogen saturated, where nitrogen availability in soil is in excess of plant and microbial growth requirements (Fig. 2). Although some of the added nitrogen soils may not be accessible to biota, rates of nitrogen cycling in many temperate forests have increased significantly following long-term nitrogen additions, including the microbially-mediated processes of nitrogen mineralization (conversion of organically-bound nitrogen to NH_4^+; Aber et al. 1998), nitrification (conversion of NH_4^+ to NO_2^- and NO_3^-; Fenn et al. 1998), denitrification (conversion of NO_3^- to N_2O and N_2; Gundersen 1991; Tietema et al. 1993), and nitrogen trace gas emissions (see below).

In some regions, excess nitrogen has also caused a shift in the competitive balance in plant communities. For example, research has shown that chronic nitrogen additions decrease plant species diversity by allowing weedy, fast-growing plants adapted to high nitrogen conditions to dominate over others (Berendse et al. 1993). In a few extreme cases in northern Europe and the northeastern United States, long-term N inputs have caused more severe changes in forest soil chemistry and nitrogen cycling: nearly all of the nitrate received in precipitation has been shown to move directly through the system and into stream water in some regions (Durka et al. 1994;

Dise and Wright 1995), causing leaching and removal of essential base cations that buffer soil pH (especially calcium and magnesium) and possibly leading to aluminum toxicity, plant nutrient deficiencies, and forest decline (Schulze 1989; McNulty et al. 1996).

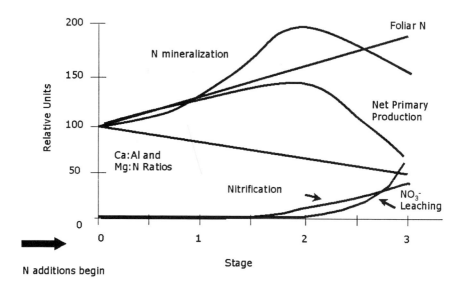

Figure 2. Conceptual model of nitrogen saturation in temperate forest ecosystems. From Aber et al. (1998).

3.2   Tropical Forests

After more than a decade of research, it is clear that acid and nitrogen deposition can cause serious and sometimes irreversible ecological change to plant and animal communities in nitrogen-limited, temperate forest ecosystems as well in those systems connected to these forests by streams and estuaries. However, despite projections of increasing nitrogen emissions and transport in the developing world, few studies have evaluated the response of tropical ecosystems to anthropogenic nitrogen inputs (Kaplan et al. 1988; Keller 1988; Bakwin et al. 1990) especially over the long-term (Hall and Matson 1999; Erickson et al. 2001).

Although field data are few, there are reasons to expect that tropical forests growing on highly weathered soils will respond very differently to nitrogen inputs than most nitrogen-limited temperate forests. First, many tropical forests occur on highly weathered soils such as Oxisols and Ultisols (Vitousek and Sanford 1986) that are naturally acidic, cation poor, and rich in biologically-available nitrogen relative to other elements such as phosphorus. Thus, primary production in many humid tropical forests is likely to be limited by phosphorus or some element other than nitrogen (Vitousek and Sanford 1986; Vitousek et al. 1997; Tanner et al. 1998; Martinelli et al. 1999). If this is true, nitrogen additions may not increase carbon storage in plant biomass or organic matter as in temperate systems; in fact, carbon storage may decrease if acid- and nitrate-induced nutrient deficiencies to plant growth (e.g. phosphorus or cations) or aluminum toxicity become severe (Matson et al. 1999; Asner et al. 2000). Second, rates of nitrogen cycling are higher in many tropical forests than in most temperate forests, with the notable exception of nitrogen saturated ones (Vitousek and Sanford 1986; Matson and Vitousek 1987; Keller 1988; Vitousek and Matson 1988; Tietema and Verstraten 1991). Typically, soil nitrate concentrations and leaching are high in humid tropical forests, where

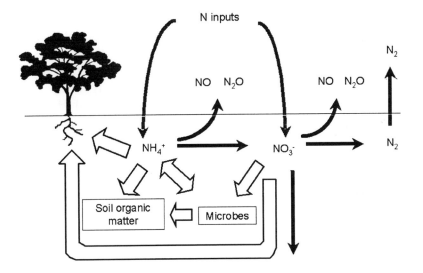

Figure 3. Pathways of anthropogenic nitrogen retention (white arrows) and nitrogen loss (black arrows) in forest ecosystems.

up to 100% of nitrogen that is mineralized is also nitrified (Vitousek and Matson 1988; Matson and Vitousek 1987; Neill et al. 1997; Hall and Matson 2002), and nitrogen trace gas losses per unit of nitrogen cycled are naturally large (Keller 1988; Matson 1989). For these reasons, anthropogenic nitrogen may not be retained in tropical forests to the extent of that in nitrogen-limited temperate forests (Fig. 3, white arrows); instead, nitrogen inputs may be immediately available to nitrifying and denitrifying microorganisms and thus a greater fracton of added nitrogen may be lost as N_2O, NO, or N_2 gases (Fig. 3, black arrows).

4.   IMPLICATIONS FOR TRACE GAS EMISSIONS FROM SOILS

As discussed in earlier chapters, microbiological activity in soils one of the most important natural sources and sinks of many atmospheric trace gases such as N_2O, NO, and CH_4. For example, CH_4 and N_2O together are responsible for 25% of the projected climate warming expected next century, and nearly 50% their global budgets are comprised of emissions and consumption from soil (Conrad 1996; IPCC 1996). As this book attests, microbial response to environmental change is the subject of current, active research programs. However, one of the most difficult challenges in this field lies in understanding how to link properties and processes that occur at the temporal and spatial scale of a microorganism to those that occur at the scale of an ecosystem, a region, or the globe. Our knowledge in this area is in part limited by technology: it has been estimated that we can culture less than 10% of the existing bacterial species, especially many of those responsible for nutrient transformations in soil (Conrad 1996). However, we are also limited by research focus: while we are beginning to gain understanding of the complexity of microbial community structure and function in the laboratory, fewer studies have taken the complementary step towards the identification of ecosystem-level variables (if they exist) that can accurately represent the range of microbial physiology, metabolism and function and which also can be easily incorporated into models that integrate across time, scales, and systems (Schimel and Gulledge 1998).

In the following section we review the current state of research on the effects of nitrogen deposition on N_2O, NO, and CH_4 emissions from soil, focusing on research that has identified integrative variables that can be modeled at the ecosystem scale. Fluxes of each of these gases from soil have been shown to be altered by nitrogen deposition, but the magnitude and global importance of their effects are uncertain in temperate forests and largely unexplored in other forest types.

4.1   Nitrous oxide and nitric oxide

N_2O is a stable compound in the troposphere with a long residence time and absorption properties that make it an extremely effective greenhouse gas with a global warming potential of nearly 300 times that of CO_2 over the next 20 years (IPCC 1996). When injected into the stratosphere, N_2O is transformed by high energy radiation into various nitrogen oxide molecules that can catalyze the destruction of stratospheric ozone. The concentration of N_2O in the atmosphere has risen significantly since the beginning of the century and is currently increasing by 0.25% yr^{-1}. In contrast, nitric oxide (NO) is a reactive gas in the troposphere, with a lifetime of hours to days, and it is one of the most important players in tropospheric ozone formation and regional air chemistry. Both gases are produced by industrial processes; however, while combustion is the largest anthropogenic source of NO (21 $Tg\ yr^{-1}$), indirect emissions from microbial activity in cultivated soils are the primary source of N_2O from human activity (3 $Tg\ yr^{-1}$). Total anthropogenic sources of N_2O and NO are half as large and equal to natural sources of these gases, respectively.

In general, N_2O and NO are produced and consumed by the processes of nitrification and denitrification in soil, but it is well understood that these two processes are performed by a suite of known (and unknown) microorganisms, each of which may have different growth requirements (Conrad 1996). As reviewed earlier in this book, environmental controls over soil nitrogen trace gas emissions have been historically difficult to define in the field because of the heterogeneity of microbial species and chemical transformations that are involved. However, recent work has identified a few integrative variables that have been successful in predicting nitrogen trace gas emissions on a regional scale. One such variable is nitrogen availability: N_2O and NO fluxes from soil are highly correlated with rates of nitrogen mineralization and especially nitrification in a number of laboratory and field studies in both wet and dry temperate and tropical ecosystems (Matson and Vitousek 1987; Castro et al. 1993; Fenn et al. 1996; Hall et al. 1996; Magill et al. 1997; Davidson et al. 2000)[2]. Firestone and Davidson (1989) suggested that while nitrogen trace gas emissions as a whole are controlled by the amount of nitrogen moving through the nitrification and denitrification processes, environmental variables such as water-filled pore space determine whether denitrification or nitrification dominates and thus whether N_2 and N_2O, or NO is the primary product at the soil surface (Firestone and Davidson 1989).

In support of these trends, nitrogen fertilization and irrigation associated with intensive agriculture have been shown to significantly increase the emissions of N_2O and NO from cultivated soils worldwide (Williams et al. 1992; Hall et al. 1996; Matson et al. 1998), especially when competition for

---

[2] However, most studies have measured only net rates of nitrogen mineralization and nitrification, which may seriously underestimate gross turnover of NH_4^+ and NO_3^- (Davidon 1992).

nitrogen between plants and microorganisms is low. Thus, in situations where nitrogen availability in soil is in excess of requirements by plants and heterotrophic microorganisms (i.e. if fertilizer is applied to bare soil or in excess of crop demand), rates of nitrification, denitrification, and associated nitrogen trace gases emissions may be high enough to make a significant difference to atmospheric processes. Agricultural emissions are the largest anthropogenic source of N_2O globally (IPCC 1996), they can produce large enough concentrations of NO to alter regional ozone formation (Hall et al. 1996), and they can represent an important economic loss of fertilizer nitrogen to the atmosphere (Matson et al. 1998).

4.1.1   Temperate Forests

Because agricultural and forest soils are different from one another in a number of ways, it can be expected that N_2O and NO emissions from temperate forests will respond differently than agricultural systems to intentional fertilization or inadvertent fertilization through atmospheric deposition. First, many relatively undisturbed temperate forest soils support low net and gross rates of nitrification relative to nitrogen mineralization (Hart et al. 1994; Magill et al. 1997; Tietema 1998), suggesting that immobilization of nitrogen into plant and microbial biomass is rapid, competition for nitrogen is strong, and relatively little nitrogen may be available to nitrifying and denitrifying bacteria. Second (and related to the first), under highly nitrogen-limited conditions where available NH_4^+ is scarce, population numbers of nitrifying microorganisms, especially chemolithotrophic bacteria, may be low enough initially that incoming nitrogen deposition (or increased nitrogen availability through increased nitrogen mineralization) is not completely utilized by existing numbers of active nitrifying microorganisms (Johnson 1992). It has been hypothesized that small, chronic nitrogen inputs to temperate forest ecosystems (e.g. atmospheric nitrogen deposition) favor the growth of nitrifying populations over time (Johnson 1992). Nitrogen losses following nitrogen additions may be proportional to the nitrifying population size: that is, in highly nitrogen-limited systems that support small populations of nitrifiers, NO_3^- leaching losses (and presumably nitrogen trace gas emissions) following nitrogen additions are relatively small. However, after repeated nitrogen inputs (perhaps over many years), nitrifying populations are large enough that they may be able to transform NH_4^+ into NO_3^- or nitrogen traces gases without delay.

Indeed, a number of studies have shown that N_2O and NO emissions from nitrogen-limited temperate forests are initially low but increase significantly following long-term experimental fertilization or years of high atmospheric inputs (Schmidt et al. 1988; Tietema et al. 1991; Brumme and Beese 1992; Matson et al. 1992; Castro et al. 1993, 1995; Papen et al. 1993; Nilsson 1995; MacDonald et al. 1997; Magill et al. 1997; Butterbach-Bahl et al. 1998; Fenn et al. 1998; Ineson et al. 1998; Skiba et al. 1999). Only a few of these studies have examined whether these elevated fluxes are of significance to atmospheric processes or total ecosystem nitrogen losses (Gundersen 1991; Davidson and Kingerlee 1997; Magill et al. 1997). However, the existing data suggest that, unlike in agricultural systems, rates of nitrogen trace gas emissions from temperate forest ecosystems are relatively small or non-existent compared to rates of nitrogen leaching and especially to rates of nitrogen inputs (Bowden et al. 2000). Although it has not been systematically evaluated across systems, it is possible that competition for nitrogen by soil processes other than heterotrophic nitrogen immobilization (such as abiotic sequestration by soil organic matter) is high in some forest ecosystems, even following years of chronic nitrogen inputs (Aber et al. 1998; Nadelhoffer et al. 1999), such that the size and activity of the nitrifying community remains relatively small.

4.1.2   Tropical forests

As discussed earlier, the extensive research in temperate forests of Europe and North America has suggested that forests respond differently to nitrogen deposition depending on their initial location along the continuum of nitrogen accumulation and saturation (e.g. Fig. 2). With this model in mind, there are a number of reasons to expect that nitrogen trace gas emissions may be a more important pathway of nitrogen loss in tropical systems that receive nitrogen deposition compared to temperate systems. First, tropical wet and dry forests are the largest biogenic source of N_2O and NO, respectively, on a global scale (IPCC 1996; Davidson and Kingerlee 1997). The combination of high temperatures and precipitation, high nitrogen availability, and oftentimes highly clayey soils in tropical wet forests creates ideal conditions for N_2O emissions via nitrification in surface soil horizons and denitrification in anaerobic microsites or deeper in the soil profile (Robertson 1989). On the other hand, high nitrogen availability combined with a warm and dry climate (punctuated by intense rain events) in tropical dry forests can create conditions that support high rates of NO emissions produced by nitrification and likely by abiotic reactions involving nitrite accumulation (also from nitrification) during the dry season (Johansson et al. 1988; Sanhueza et al. 1990; Davidson et al. 1991). Second, because nitrogen availability is naturally high relative to other nutrients in many highly-weathered tropical systems, these soils are likely to support large standing populations of nitrifying bacteria that may be able to process anthropogenic nitrogen additions without delay (Johnson 1992).

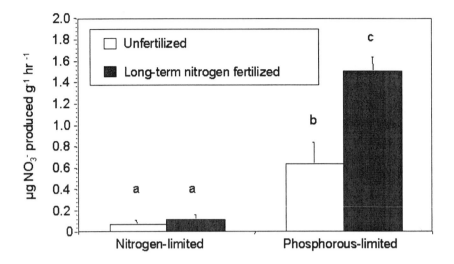

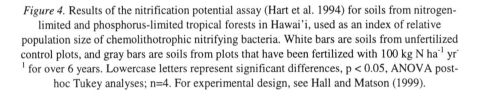

Figure 4. Results of the nitrification potential assay (Hart et al. 1994) for soils from nitrogen-limited and phosphorus-limited tropical forests in Hawai'i, used as an index of relative population size of chemolithotrophic nitrifying bacteria. White bars are soils from unfertilized control plots, and gray bars are soils from plots that have been fertilized with 100 kg N ha^{-1} yr^{-1} for over 6 years. Lowercase letters represent significant differences, $p < 0.05$, ANOVA post-hoc Tukey analyses; n=4. For experimental design, see Hall and Matson (1999).

Although several studies have shown that one-time nitrogen fertilization can significantly increase emissions of N_2O from tropical forest soils (Kaplan et al. 1988; Keller 1988; Bakwin et al. 1990), only a few studies have examined the effects of long-term nitrogen additions on nitrogen trace gas emissions in tropical ecosystems (Hall and Matson 1999; Erickson et al. 2001). Hall and Matson (1999) found that long-term nitrogen additions to a phosphorus-limited tropical forest caused significant increases in the population size of nitrifying bacteria (Fig. 4), and that this increase was highly correlated to the magnitude and duration of N_2O and NO emissions immediately following nitrogen fertilization (Fig. 5). These trends were not found in an otherwise similar nitrogen-limited tropical forest, where population sizes of nitrifying bacteria and nitrogen oxide fluxes remained small even after a decade of nitrogen additions. Notably, nitrifier populations were larger in unfertilized soil of the phosphorus-limited forest than in long-term fertilized soil in the nitrogen-limited forest, and this trend was also correlated with the magnitude of nitrogen oxide emissions immediately following fertilization. It is possible that the overall magnitude of peak nitrogen trace gas emissions in both the nitrogen-limited and phosphorus-limited forests may be reflective of the size of the nitrifying community in each of these systems.

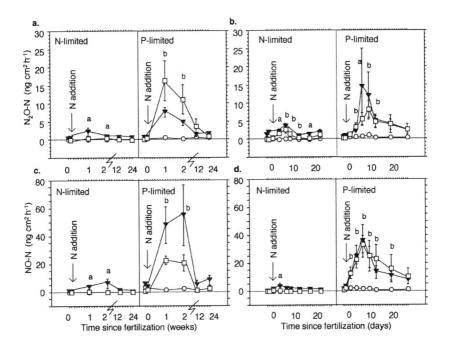

Figure 5. Nitrous oxide (a and c) and nitric oxide (b and d) emissions from soils in nitrogen (N) -limited and phosphorus (P) -limited tropical forests in Hawai'i before and immediately following N additions of 50 kg N ha^{-1}. ▼: Long Term fertilized (LT), □: First-time fertilized (FT), ○: Control (C). a = LT > C (p < 0.05), b = LT and FT > C (p < 0.05). Error bars are ± SE. All fluxes were log transformed prior to ANOVA and post-hoc Tukey analyses. Figures 5a and 5b: 1996 gas emissions, 1 and 2 weeks after the first N addition to the FT plots. Figures 5c and 5d: 1997 gas emissions following N additions; FT plots have been fertilized for 1 yr and the LT plots have been fertilized for 12 and 6 yrs in the N-limited and P-limited sites, respectively. From Hall and Matson (1999).

4.2   Methane

Methane is another greenhouse gas whose dynamics in soil are mediated primarily by microorganisms. Methane has a relatively short lifetime in the atmosphere compared to N_2O (~12 yrs), but its greenhouse warming potential over the next several decades is more than 50 times that of CO_2, and its concentration is increasing in the atmosphere by 0.6% yr^{-1}, faster than that of CO_2 and N_2O combined (IPCC 1996). Concentrations of CH_4 have more than doubled since the beginning of the century, primarily due to fossil fuel exploitation, landfills, biomass burning, and agricultural practices, including livestock rearing and rice cultivation. Currently, anthropogenic sources are more than twice as large as natural sources globally.

Despite more than 375 Tg of CH_4-C that is emitted annually due to human activity, atmospheric concentrations are increasing by only one tenth this rate due to important sinks in the atmosphere and soils. The most important CH_4 sink is atmospheric oxidation by the OH⁻ radical, whose concentrations have also increased recently due to enhanced stratospheric exposure to ultraviolet radiation from loss of ozone (Bekki et al. 1994). However, oxidation by soil microorganisms also contributes more than 5–10% of the total CH_4 destroyed, which is similar in size to the annual atmospheric increase ($\sim$37 Tg yr^{-1}; IPCC 1996).

Soil bacteria both produce and consume methane, although the types of organisms involved and their optimum environmental conditions differ for each process. Methanogens are a group of strictly anaerobic bacteria that reduce CO_2 to CH_4 when no other electron acceptors are available. Such environments occur primarily in freshwater-saturated soils found in wetlands and cultivated rice fields and rarely occur in upland forest ecosystems. On the other hand, both temperate and tropical forest ecosystems are known to be significant sinks for atmospheric CH_4 due to the activity of methanotrophic bacteria (Keller et al. 1986; Delmas et al. 1992; Mancinelli 1995). Methanotrophs are a diverse group of microorganisms which are aerobic, microaerophyllic (preferring oxygen levels lower than atmospheric levels), and which oxidize CH_4 to CO_2. All use the enzyme group, methane monooxygenase (MMO) to catalyze the first step in the oxidation reaction, although its function differs between species. MMO is relatively nonspecific: one type of this enzyme is evolutionarily related to ammonium monooxygenase (AMO; the enzyme present in nitrifying bacteria) and it has the ability to oxidize ammonium to nitrite without any energy gain. Thus, methanotrophic bacteria can oxidize ammonia and nitrifying bacteria can oxidize methane, although the affinity of each enzyme is strongest with its own particular substrate (e.g. MMO and CH_4; Mancinelli 1995).

Several groups of methanotrophs have been long recognized to be important regulators of methane flux from saturated soils and sediments, where CH_4 concentrations are extremely high, but the organisms that consume atmospheric methane in upland soils (at only 1.7 ppm) have only been recently identified using novel isotopic and molecular techniques (Bull et al. 2000; Jensen et al. 2000). In general, in order to use a substrate for growth, microorganisms must gain enough energy from an oxidation reaction to satisfy their maintenance requirements. Atmospheric concentrations of methane are so low that high enzymatic affinity is necessary for methanotrophs to gain the energy necessary for biosynthesis (Mancinelli 1995). The most well characterized methanotrophs (those from water saturated systems) have a low affinity for methane, so currently little is known about the physiology and metabolism of atmospheric methane consumers and how environmental conditions regulate their activity and distribution.

Nevertheless, in the field, oxidation of atmospheric CH_4 in soils by these novel microorganisms has been shown to be significantly altered by acid and nitrogen deposition in some systems. Because most of this work has been performed in temperate agricultural and forest systems (similar to the situation with nitrogen trace gases) little to nothing is known about how (or if) anthropogenic nitrogen inputs will alter methane consumption in other forest types, including those in tropical latitudes.

4.2.1 Temperate ecosystems

Several studies have shown that nitrogen availability can significantly inhibit CH_4 consumption in experimentally fertilized and nitrogen deposition-affected forests (Steudler et al. 1989; Willison et al. 1995; Paul and Clark 1996; Primé and Christensen 1997; Butterbach-Bahl et al. 1998; Cai and Mosier 2000; Hutsch 2001), at least temporarily (Borjesson and Nohrstedt 2000; Cai and Mosier 2000; Hutsch 2001), and in some cases permanently (Hütch et al. 1994). In other studies, CH_4 oxidation rates decline only following repeated fertilization over time (Mancinelli 1995; Gulledge and Schimel 2000). Furthermore, several studies have even shown that nitrogen additions have no effect on soil CH_4 oxidation rate, regardless of duration or amount (Hütch et al. 1993; Mosier and Schimel 1991; Dunfield et al. 1995; Hütsch 1996; Willison et al. 1996; Christensen et al. 1997; Gulledge et al. 1997; Ineson et al. 1998; Gulledge and Schimel 2000; Bradford et al. 2001; Steinkamp et al. 2001).

Given that the microorganisms responsible for CH_4 consumption in upland soils have not been well characterized physiologically, it is not surprising that the patterns and controlling processes of nitrogen-mediated inhibition are unclear. Unlike the processes of nitrification and denitrification, which require nitrogen as a substrate for N_2O and NO production, methane oxidation may be indirectly altered by nitrogen availability via numerous pathways, including enzyme inhibition, a change in community location within the soil profile, or a change in microbial community composition. Thus, currently it is not possible to predict the response of methane oxidation to nitrogen deposition through the use of ecosystem-level controlling variables. Before we can accomplish this task, we need to know more about how the physiologies of the individual species involved and how their communities respond to environmental change in general, and specifically, how they respond to changes in resource (i.e. nitrogen) availability. Understanding the range of possible controlling mechanisms at the physiological scale is a first step in the identification of variables that may integrate across these processes at the ecosystem scale.

In an excellent review of current literature, Gulledge and Schimel (1997) have identified three patterns of nitrogen inhibition of methane oxidation in field and laboratory studies. In some cases, nitrogen fertilization inhibits CH_4 oxidation *immediately*, within hours to days after application. In others, inhibition of methane oxidation is *delayed*, i.e. it happens only after long term nitrogen fertilization. Finally, some experiments have even shown *no inhibition* of methane oxidation due to nitrogen availability. Based on their work and many others, several controlling mechanisms are possible (Fig. 6).

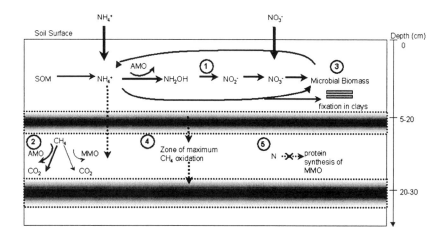

Figure 6. Mechanisms of nitrogen inhibition of CH_4-oxidation in aerobic, upland soils. 1: Direct toxicity by NH_2OH and NO_2- and potentially other organic compounds and salts, 2: Nitrifying bacteria outcompete methanotrophs for CH_4 3: Nitrogen immobilization by biotic and abiotic processes until nitrogen saturation, 4: Movement of CH_4-oxidation zone lower in soil profile results in decreased CH_4 (atmospheric) diffusion to oxidation site, 5: Nitrogen inhibition of MMO synthesis.

First, nitrogen availability may immediately inhibit methane oxidation because MMO is nonspecific, and NH_3 may competitively interfere with CH_4 oxidation for the active site of the MMO enzyme (Conrad 1996). Second, nitrite (NO_2^-) and hydroxylamine (NH_2OH) are the direct products of NH_3 oxidation by MMO, but NO_2^- can inhibit the synthesis of important reducing equivalents such as NADH + H^+, and NH_2OH reversibly inhibits MMO activity (King and Schnell 1994 Mancinelli 1995; Conrad 1996;

Primé and Christensen 1997). Third, although this idea has been less explored, it is also possible that soluble organic compounds other than NH_4^+ and NO_2^- may be responsible for inhibition of CH_4 oxidation (Amaral and Knowles 1997). Last, there is evidence to suggest that methanotrophic growth may be not be inhibited by nitrogen but may be negatively affected by high salt concentrations, including the salts of ammonium and nitrate associated with nitrogen fertilization and deposition (Adamsen and King 1993; Crill et al 1994; Dunfield et al. 1995; MacDonald et al. 1997; Whalen 2000).

Delayed inhibition of methane oxidation by nitrogen may be caused by a shift in community composition from methanotrophs to nitrifying bacteria, despite the low affinity of AMO for CH_4, as nitrogen additions favor the growth of nitrifying bacteria over methanotrophs in the surface soil (Mosier and Schimel 1991; Hütch et al. 1993; Willison et al. 1995, 1996; Steudler et al. 1996). Alternatively, because nitrogen may be immobilized immediately in nitrogen-limited systems, soil nitrogen concentrations may be initially too low following experimental nitrogen inputs to alter methanotrophic activity. After repeated fertilization, plant uptake, microbial immobilization, and abiotic fixation processes may become nitrogen saturated. Subsequently, availability may slowly increase over time and can be toxic to methanotrophic activity as described above (Mancinelli 1995; Gulledge et al. 1997). Another possibility is that high nitrogen inputs may lead to nitrogen toxicity in the upper soil layers, which gradually causes the zone of maximum oxidation to move further down the soil profile and effectively reduces the sink strength because of diffusional limitations (Conrad 1996). Finally, it is possible that high concentrations of inorganic nitrogen may directly inhibit synthesis of the MMO enzyme. As a result of this inhibition, ambient rates of CH_4 oxidation may decline as microbial turnover occurs, and following continued inputs methanotrophs may have a difficult time reestablishing (Gulledge et al. 1997; Fig. 6).

Despite the number of ways in which nitrogen can inhibit soil methane oxidation to some degree, in some cases nitrogen has no effect on methanotrophic activity. Based on physiological evidence, it is likely methanotrophic activity in these types of soils is dominated by microbes whose activity is insensitive to nitrogen availability and whose identify we have yet to determine (Gulledge et al 1997). However, other explanations are possible: for example, Swensen and Bakken (1999) found that methane oxidation in a Norwegian soil was masked by abiotic, fossil methane *release* from the mechanical weathering of sand particles during experimental incubations (Swensen and Bakken 1999). Alternatively, it has been suggested that some forest systems may be so N-limited that methanotrophs recover quickly from N fertilization (Steinkamp et al. 2001).

In sum, microbial physiology and community composition may be two of the most important factors to consider when examining the controls over methane oxidation in soil under elevated nitrogen deposition. Others may also be important, including the interaction of biological and geochemical processes (e.g. Swensen and Bakken 1999), but unifying principles across soils and systems remain elusive. Nevertheless, it is clear that elevated nitrogen deposition will likely inhibit methane oxidation at some places and at some times. What we need now is a quantitative picture of the identity of the microbial communities and how they are affected by environmental factors across a range of natural systems before we will be able to fully understand and model how an environmental perturbation (like increased nitrogen deposition) controls methane dynamics between soils and the atmosphere at regional and global scales.

5.   CONCLUSIONS AND RESEARCH NEEDS

Acid and nitrogen deposition to terrestrial ecosystems is increasing due to human activity and has been shown to alter the microbial production and consumption of the atmospherically important trace gases NO, N_2O, and CH_4 from soil. In general, the evidence from temperate forest systems suggests that N_2O and NO emissions from soil increase with nitrogen additions if the nitrifying and denitrifying microorganisms are no longer limited by nitrogen supply and all other conditions (e.g. temperature, moisture, pH, organic carbon availability) are satisfactory for their activity. However, heterotrophic demand and possibly abiotic reactions in temperate forest soils are often strong enough to retain large inputs of nitrogen over decades. In these situations, nitrogen trace gas emissions may be elevated compared to background values but may represent a small perturbation to total ecosystem nitrogen losses or to the regional and global flux of these gases. Furthermore, the limited available evidence suggests that nitrogen trace gas emissions following fertilization may be related to the size of the nitrifying population size and its activity. The effect of nitrogen additions in tropical forests have not been evaluated systematically; however, because nitrogen is often in excess of biological demand in these systems, nitrogen trace gas emissions may be larger and more important to regional atmospheric budgets than is predicted from models based on temperate forest responses alone.

Methane oxidation in forest soils has been shown to be inhibited by nitrogen availability, but the extent, the onset, and the duration of this effect is unclear. Several mechanisms are possible to explain the patterns of inhibition among soils and ecosystems. Microbial species composition and community structure are likely to be important controlling factors, but until we understand more about the organisms, their growth requirements and limitations, it will be difficult to evaluate the importance of anthropogenic nitrogen on inhibition of CH_4 oxidation in ecosystem and global models.

Current research programs are providing essential knowledge about how increased nitrogen availability alters microbial function and how microbial activity in turn alters trace gas dynamics, but future research is strongly limited by several factors. First, our understanding of how microorganisms will respond to global environmental change is limited by the identification and investigation of the suite of unknown species responsible for many biogeochemical transformations in soil. New technologies are emerging which will help elucidate the composition of microbial communities and how the activity of specific populations relate to their biogeochemical function. Future research will benefit from programs that effectively utilize this technology not only to identify the components and structure of the microbial community, but also to assess when and where it is important to predicting trace gas dynamics at the ecosystem and regional scales, and if and how changes therein are induced by human alteration of climate, land use or resource availability.

Second, even within the processes we do understand physiologically, difficulties still lie in relating what we know about soil microbial function to trace gas dynamics at the scale of ecosystems and regions. The identification of easily-measurable, ecosystem-level variables that integrate across spatial and temporal heterogeneity of microbial function will be necessary components in the development of successful regional and global scale models of biosphere-atmosphere interactions and their responses to anthropogenic environmental perturbations.

Finally, our knowledge of how soil microbial communities function under increased resource availability have been heavily biased towards temperate systems, where research in both agricultural systems and nitrogen-affected temperate systems is most concentrated. Nitrogen deposition is expected to be come a global problem in the next century as industrialization and agricultural intensification increase in developing countries. However, little is known about if and to what extent microbial communities differ between temperate and tropical forest ecosystems and how an increase in nitrogen deposition to tropical forests will alter trace gas exchange between soils and the atmosphere. To help us understand the numerous factors controlling the spatial and temporal heterogeneity of trace gas emissions from soils under human influence, future research is needed that links biological activity at the microbial scale to larger-scale ecosystem processes and that spans a range of temperate and tropical ecosystems across gradients of environmental change.

ACKNOWLEDGEMENTS

This review was supported by the Colorado College Gresham Riley Fellowship Program to SJH and the USDA Terrestrial Ecosystem and Global Change Program and the Andrew Mellon Foundation to PM.

REFERENCES

Aber J, McDowell W, Nadelhoffer K, Magill A, Berntson G, Kamakea M, McNulty S, Currie W, Rustad L & Fernandez I (1998) Nitrogen saturation in temperate forests: Hypotheses revisited. Bioscience 48: 921-934

Aber JD, Nadelhoffer KJ, Steudler P, & Melillo JM (1989) Nitrogen saturation in northern forest ecosystems. Bioscience 39: 378-386

Adamsen APS & King GM (1993) CH_4 consumption in temperate and subarctic forest soils: rates, vertical zonation, and responses to water and nitrogen. Appl Environ Microb 59: 485-490

Ågren GI & Bosatta E (1988) Nitrogen saturation of terrestrial ecosystems. Environ Poll 54: 185-197

Amaral JA & Knowles R (1997) Inhibition of methane consumption in forest soils and pure cultures of methanotrophs by aqueous forest soil extracts. Soil Biol Biochem 29: 1713-1720

Andreae MO, Browell EV, Garstang M, Gregory GL, Harriss RC, Hill GF, Jacob DJ, Pereira MC, Sachse GW, Setzer AW, Dias PLS, Talbot RW, Torres AL & Wofsy SC (1988) Biomass burning emissions and associated haze layers over Amazonia. J Geophys Res 93: 1509-1527

Asner G, Townsend A, Riley W, Matson P, Neff J & Cleveland C (2000) Physical and biogeochemical controls of terrestrial ecosystem responses to nitrogen deposition. Biogeochemistry 54: 1-39

Bakwin PS, Wofsy SC & Fan S (1990) Measurements of reactive nitrogen oxides (NO_Y) within and above a tropical forest canopy in the wet season. J Geophys Res 95: 16,765-72

Bekki S, Law KS & Pyle JA (1994) Effect of ozone depletion on atmospheric CH_4 and CO concentrations. Nature 371: 595-597

Berendse F, Aerts R & Bobbink R (1993) Atmospheric nitrogen deposition and its impact on terrestrial ecosystems. In: Vos CC & Opdam P (eds) Landscape Ecology of a Stressed Environment, pp 105-121. Chapman and Hall, London, U.K.

Binkley D & Högberg P (1997) Does atmospheric deposition of nitrogen threaten Swedish forests? For Ecol Manage 92: 119-152

Borjesson G & Nohrstedt HO (2000) Fast recovery of atmospheric methane consumption in a Swedish forest soil after single-shot N-fertilization. For Ecol Manage 134: 83-88

Bowden RD, Rullo G, Stevens GR & Steudler PA (2000) Soil fluxes of carbon dioxide, nitrous oxide, and methane at a productive temperate deciduous forest. J Environ Qual 29: 268-276

Bradford MA, Wookey PA, Ineson P & Lappin-Scott HM (2001) Controlling factors and effects of chronic nitrogen and sulfur deposition on methane oxidation in a temperate forest soil. Soil Biol Biochem 33: 93-102

Brumme R & Beese F (1992) Effects of liming and nitrogen fertilization on emissions of CO_2 and N_2O from a temperate forest. J Geophys Res 97: 12851-12858

Bull ID, Parekh NR, Hall GH, Ineson P & Evershed RP (2000) Detection and classification of atmospheric methane oxidizing bacteria in soil. Nature 405: 175-178

Busch G, Lammel G, Beese FO, Feichter J, Dentener FJ & Roelofs GJ (2001) Forest ecosystems and the changing pattern of nitrogen input and acid deposition today and in the future based on a scenario. Environ Sci Pollut R 8: 95-102

Butterbach-Bahl K, Gasche R, Huber CH, Kreutzer K & Papen H (1998) Impact of N-input by wet deposition on N-trace gas fluxes and CH_4-oxidation in Spruce forest ecosystems of the temperate zone in Europe. Atmos Environ 32: 559-564

Cai ZCC & Mosier AR (2000) Effect of NH_4Cl addition on methane oxidation by paddy soils. Soil Biol Biochem 32: 1537-1545

Castro MS, Steudler PA, Melillo JM, Aber JD & Millham S (1993) Exchange of N_2O and CH_4 between the atmosphere and soils in spruce-fir forests in the northeastern United States. Biogeochemistry 18: 119-135

Castro MS, Steudler PA, Melillo JM, Aber JD & Bowden RD (1995) Effects of nitrogen fertilization on the fluxes of N_2O, CH_4, and CO_2 in a Florida slash pine plantation. Can J Forest Res 24: 9-13

Christensen TR, Michelsen A, Jonasson S & Schmidt IK (1997) Carbon dioxide and methane exchange of a subarctic heath in response to climate change related environmental manipulations. Oikos 79: 34-44

Conrad R (1996) Soil microorganisms as controllers of atmospheric trace gases (H_2, CO, CH_4, OCS, N_2O, and NO). Microbiol Rev 60: 609-640

Crill PM, Martikainen PJ, Nykanen H & Silvola J (1994) Temperature and N fertilization effects on methane oxidation in a drained peatland soil. Soil Biol Biochem 26: 1331-1339

Davidson EA (1992) Sources of nitric oxide and nitrous oxide following wetting of dry soil. Soil Sci Soc Am J 56: 95-101

Davidson EA & Kingerlee W (1997) A global inventory of nitric oxide emissions from soils. Nut Cycl Agroecosys 48: 37-50

Davidson EA, Keller M, Erickson HE, Verchot LV & Veldkamp E (2000) Testing a conceptual model of soil emissions of nitrous and nitric oxides. BioScience 50: 667-680

Davidson EA, Vitousek PM, Matson PA, Riley R, Garcia-Mendez G & Maass JM (1991) Soil emissions of nitric oxide in a seasonally dry tropical forest of Mexico. J Geophys Res 96: 15439-15445

Delmas RA, Servant J & Tathy JP (1992) Sources and sinks of methane and carbon dioxide exchanges in mountain forest in equatorial Africa. J Geophys Res 97: 6169

Dise NB & Wright RF (1995) Nitrogen leaching from European forests in relation to nitrogen deposition. Forest Ecol Manag 71: 153-161

Dunfield P, Knowles R, Dumont R & Moore TR (1995) Effect of nitrogen fertilizers and moisture content on the CH_4 and N_2O fluxes in a humisol: measurements in the field and intact soil cores. Biogeochemistry 29: 199-222

Durka W, Schulze ED, Gebauer G & Voerkelius S (1994) Effects of forest decline on uptake and leaching of deposited nitrate determined from ^{15}N and ^{18}O measurements. Nature 372: 765-767

Erickson H, Keller M & Davidson EA (2001) Nitrogen oxide fluxes and nitrogen cycling during postagricultural succession and forest fertilization in the humid tropics. Ecosystems 4: 67-84

Fenn ME, Poth MA & Johnson DW (1996) Evidence for nitrogen saturation in the San Bernardino Mountains in Southern California. Forest Ecol Manag 82: 211-230

Fenn ME, Poth MA, Aber JD, Baron JS, Bormann BT, Johnson DW, Lemly AD, McNulty SG, Ryan DF & Stottlemyer R (1998) Nitrogen excess in North American ecosystems: Predisposing factors, ecosystem responses, and management strategies. Ecol Appl 8: 706-733

Firestone MK & Davidson EA (1989) Microbiological basis of NO and N_2O production and consumption in soil. In: Andreae MO & Schimel DS, (eds) Exchange of Trace Gases between Terrestrial Ecosystems and the Atmosphere, pp 7-21. John Wiley & Sons Ltd, New York, U.S.A.

Galloway JN (1995) Acid deposition: perspectives in time and space. Water Air Soil Poll 85: 15-24

Galloway JN (1996) Anthropogenic mobilization of sulfur and nitrogen: Immediate and delayed consequences. Annu Rev Energ Env 21: 261-292

Galloway JN, Schlesinger WH, Levy H, Michaels A & Schnoor JL (1995) Nitrogen fixation: Anthropogenic enhancement - environmental response. Global Biogeochem Cycl 9: 235-252

Gulledge J & Schimel JP (2000) Controls on soil carbon dioxide and methane fluxes in a variety of taiga forest stands in interior Alaska. Ecosystems 3: 269-282

Gulledge J, Doyle AP & Schimel JP (1997) Different NH_4^+-inhibitions patterns of soil CH_4 consumption: a result of distinct CH_4-oxidizer populations across sites? Soil Biol Biochem 29: 13-21

Gundersen P (1991) Nitrogen deposition and the forest nitrogen cycle: role of denitrification. Forest Ecol Manag 44: 15-28

Hall SJ & Matson PA (1999) Nitrogen oxide emissions after nitrogen additions in tropical forests. Nature 400: 152-155

Hall SJ & Matson PA (2002) Nutrient Status of Tropical Rain Forests Influences Soil N Dynamics After N Additions. Ecology, in press

Hall SJ, Matson PA & Roth PM (1996) NO_X emissions from soil: Implications for air quality modeling in agricultural regions. Annu Rev Energ Env 21: 311-346

Hart SC, Stark JM, Davidson EA & Firestone MK (1994) Nitrogen mineralization, immobilization, and nitrification. In: SSSA, (ed) Methods of Soil Analysis, Part 2, Microbiological and Biochemical Properties, Vol. SSSA Book Series No. 5, pp 985-1018. Soil Science Society of America, Madison, WI, U.S.A.

Heij GJ & Schneider T. 1991. Acidification research in the Netherlands. Final Report of the Dutch Priority Programme on Acidification, Elsevier, Amsterdam, The Netherlands

Henriksen A & Brakke DF (1988) Increasing contributions of nitrogen to the acidity of surface waters in Norway. Water Air Soil Poll 42: 183-201

Holland DM, Principe PP & JES II (1999) Trends in atmospheric sulfur and nitrogen species in the eastern United States for 1989-1995. Atmos Environ 33: 37-49

Holland EA, Braswell BH, Lamarque JF, Townsend A, Sulzman J, Müller JF, Dentener F, Brasseur G, II HL, Penner JE & Roelofs GJ (1997) Variations in the predicted spatial distribution of atmospheric nitrogen deposition and their impact on carbon uptake by terrestrial ecosystems. J Geophys Res 102: 15849-15866

Hütch BW (1996) Methane oxidation in soils of two long-term fertilization experiments in Germany. Soil Biol Biochem 28: 773-782

Hütch BW, Webster CP & Powlson DS (1993) Long-term effects of nitrogen fertilization on methane oxidation in soil of the Broadbalk wheat experiment. Soil Biol Biochem 25: 1307-1315

Hütch BW, Webster CP & Powlson DS (1994) CH_4 oxidation in soil as affected by landuse, soil pH, and N fertilization. Soil Biol Biochem 26: 1613-1622

Hutsch BW (2001) Methane oxidation, nitrification, and counts of methanotrophic bacteria in soils from a long-term fertilization experiment ("Ewiger Roggenbau" at Halle). J Plant Nutr Soil Sci 164: 21-28

Ineson P, Coward PA, Benham DG & Robertson SMC (1998) Coniferous forests as "secondary agricultural" sources of nitrous oxide. Atmos Environ 32: 3321-3330

IPCC. 1996. Climate change 1995: The science of climate change Cambridge University Press, Cambridge, U.K.

Jensen S, Holmes AJ, Olsen RA & Murrell JC (2000) Detection of methane oxidizing bacteria in forest soil by monooxygenase PCR amplification. Microbial Ecol 39: 282-289

Johansson C, Rodhe H & Sanhueza E (1988) Emission of NO in a tropical savanna and a cloud forest during the dry season. J Geophys Res 93: 7180-93

Johnson D (1992) Nitrogen Retention in Forest Soils. J Environ Qual 21: 1-12

Johnson DW, Turner J & Kelly JM (1982) The effects of acid precipitation on forest nutrient status. Water Resour Res 18: 449-461

Kaplan WA, Wofsy SC, Keller M & daCosta JM (1988) Emission of NO and deposition of O3 in a tropical forest ecosystem. J Geophys Res 93: 1389-95

Kauppi PE, Mielikainen K & Kuusela K (1992) Biomass and carbon budget of European Forests, 1971-1990. Science 256: 70-74

Keller M (1988) Emissions of N_2O from tropical forest soils: response to fertilization with NH_4^+, NO_3^-, and PO_4^-. J Geophys Res 93: 1600-1604

Keller M, Kaplan WA & Wofsy SC (1986) Emissions of N_2O, CH_4, and CO_2 from tropical soils. J Geophys Res 91: 11791-11801

King GM & Schnell S (1994) Effect of increasing atmospheric methane concentration on ammonium inhibition of soil methane consumption. Nature 370: 282-284

Kurz WA, Apps MJ, Beukema SJ & Lekstrum T (1995) 20th century carbon budget of Canadian forests. Tellus 47: 170-177

Likens GE, Driscoll CT & Buso DC (1996) Long-term effects of acid rain: Response and Recovery of a Forest Ecosystem. Science 272: 244-246

Lovett GM (1994) Atmospheric deposition of nutrients and pollutants in North America: An ecological perspective. Ecol Appl 4: 629-650

MacDonald JA, Skiba U, Sheppard LJ, Ball B, Roberts JD, Smith KA & Fowler D (1997) The effect of nitrogen deposition and seasonal variability on methane oxidation and nitrous oxide emission rates in an upland Spruce plantation and moorland. Atmos Environ 31: 3693-3706

Magill AH, Aber JD, Hendricks JJ, Bowden RD, Melillo JM & Steudler PA (1997) Biogeochemical response of forest ecosystems to simulated chronic nitrogen deposition. Ecol Appl 7: 402-415

Mancinelli RL (1995) The regulation of methane oxidation in soil. Annu Rev Microbiol 49: 581-605

Martinelli LA, Piccolo MC, Townsend AR, Vitousek PM, Cuevas E, McDowell W, Robertson GP, Santos OC & Treseder K (1999) Nitrogen stable isotope composition in leaves and soils: tropical versus temperate forests. Biogeochemistry 46: 45-65

Matson PA (1989) Regional extrapolation of trace gas flux based on soils and ecosystems. In: Andreae MO & Schimel DS (eds) Exchange of Trace Gases between Terrestrial Ecosystems and the Atmosphere, pp 97-108. John Wiley & Sons, New York, U.S.A.

Matson PA & Vitousek PM (1987) Cross-system comparisons of soil nitrogen transformations and nitrous oxide flux in tropical forest ecosystems. Global Biogeochem Cycl 1: 163-170

Matson PA, Naylor R & Ortiz-Monasterio I (1998) Integration of Environmental, Agronomic, and Economic Aspects of Fertilizer Management. Science 280: 112

Matson PA, McDowell WH, Townsend AR & Vitousek PM (1999) The globalization of N deposition: ecosystem consequences in tropical environments. Biogeochemistry 46: 152-155

Matson PA, Gower ST, Volkmann C, Billow C & Grier CC (1992) Soil nitrogen cycling and nitrous oxide flux in a Rocky Mountain Douglas-fir forest: effects of fertilization, irrigation and carbon additions. Biogeochemistry 18: 101-117

Matthews E (1994) Nitrogenous fertilizers: Global distribution of consumption and associated emissions of nitrous oxide and ammonia. Global Biogeochem Cycl 8: 411-39

McLeod AR, Holland MR, Shaw PJA, Sutherland PM, Darrall NM & Skeffington RA (1990) Enhancement of nitrogen deposition to forest trees exposed to SO_2. Nature 347: 277-279

McNulty SG, Aber JD & Newman SD (1996) Nitrogen saturation in a high elevation New England spruce-fir stand. Forest Ecol Manage 84: 109-121

Mosier AR & Schimel DS (1991) Influence of agricultural nitrogen on atmospheric methane and nitrous oxide. Chem Industry 23: 874-877

Nadelhoffer KJ, Emmett BA, Gundersen P, Kjønaas OJ, Koopmans CJ, Schleppi P, Tietema A & Wright RF (1999) Nitrogen deposition makes a minor contribution to carbon sequestration in temperate forests. Nature 398: 145-148

Neill C, Piccolo MC, Cerri CC, Steudler PA, Melillo JM & Brito M (1997) Net nitrogen mineralization and net nitrification rates in soils following deforestation for pasture across the southwestern Brazilian Amazon Basin landscape. Oecologia 110: 243-252

Nihlgård B (1985) The ammonium hypothesis - and additional explanation to the forest dieback in Europe. Ambio 14: 2-8

Nilsson LO (1995) Forest biogeochemistry interactions among greenhouse gases and N deposition. Water Air Soil Poll 85: 1557-1562

Papen H, Hellmann B, Papke H & Rennenberg H (1993) Emission of N-oxides from acid irrigated and limed soils of a coniferous forest in Bavaria. In: Oremland RS (ed) Biogeochemistry of Global Change: Radiatively Active Trace Gases, pp 245-59. Chapman & Hall, London, U.K.

Paul EA & Clark FE (1996) Soil Microbiology and Biochemistry Academic Press, San Diego.

Primé A & Christensen S (1997) Seasonal and spatial variation of methane oxidation in a Danish Spruce forest. Soil Biol Biochem 29: 1165-1172

Robertson GP (1989) Nitrification and denitrification in humid tropical ecosystems. In: Procter J (ed) Mineral Nutrients in Tropical Forest and Savanna Ecosystems, pp 55-69. Blackwell Scientific, Brookline Village, MA, U.S.A.

Rodhe H, Grennfelt P, Wisniewski J, Ågren C, Bengtsson G, Johansson K, Kauppi P, Kucera V, Rasmussen L, Rosseland B, Schotte L & Selldén G (1995) Acid Reign '95? Conference Summary Statement. Water Air Soil Poll 85: 1-14

Sanhueza E, Hao WM, Scharffe D, Donoso L & Crutzen PJ (1990) N_2O and NO emissions from soils of the northern part of the Guayana Shield, Venezuela. J Geophys Res 95: 22481-22488.

Schimel J & Gulledge J (1998) Microbial community structure and global trace gases. Glob Change Biol 4: 745-758

Schindler DW, Mills KH, Malley DF, Findlay DL, Shearer JA, Davies IJ, Turner MA, Linsey GA & Cruikshank DR (1985) Long-term ecosystem stress: The effects of years of experimental acidification on a small lake. Science 228: 1395-1401

Schlesinger WH & Hartley AE (1992) A global budget for atmospheric NH_3. Biogeochemistry 15: 191-211

Schmidt J, Seiler W & Conrad R (1988) Emission of nitrous oxide from temperate forest soils into the atmosphere. J Atmos Chem 6: 95-115

Schulze ED (1989) Air pollution and forest decline in a spruce (*Picea abies*) forest. Science 244: 776-782

Skiba U, Sheppard LJ, Pitcairn CER, Van Dijk S & Rossall MJ (1999) The effect of N deposition on nitrous oxide and nitric oxide emissions from temperate forest soils. Water Air Soil Poll 116: 89-98

Spiecker H, Mielikäinen K, Köhl M & Skovsgaard JP (eds.) 1996. Growth Trends in European Forests: Studies from 12 countries, pp 1-372

Steinkamp R, Butterbach-Bahl K & Papen H (2001) Methane oxidation by soils of an N limited and N fertilized spruce forest in the Black Forest, Germany. Soil Biol Biochem 33: 145-153

Steudler PA, Bowden RD, Melillo JM & Aber JD (1989) Influence of nitrogen fertilization on methane uptake in temperate forest soils. Nature 341: 314-316

Steudler PA, Jones RD, Castro MS, Melillo JM & Lewis DL (1996) Microbial controls of methane oxidation in temperate forest and agricultural soils. In: Murrell JC & Kelly DP (eds) Microbiology of Atmospheric Trace Gases, Springer Verlag, Berlin, Germany

Swensen B & Bakken LR (1999) Release of fossil methane from mineral soil particles and its implication for estimation of methane oxidation in a mineral subsoil. Biogeochemistry 47: 1-14

Tanner EVJ, Vitousek PM & Cuevas E (1998) Experimental investigation of nutrient limitation of forest growth on wet tropical mountains. Ecology 79: 10-22

Tietema A (1998) Microbial carbon and nitrogen dynamics in coniferous forest floor material along a European nitrogen deposition gradient. Forest Ecol Manage 101: 29-36

Tietema A & Verstraten JM (1991) Nitrogen cycling in an acid forest ecosystem in the Netherlands at increased atmospheric nitrogen input: the nitrogen budget and the effects of nitrogen transformations on the proton budget. Biogeochemistry 15: 21-46

Tietema A, Bouten W & Wartenbergh PE (1991) Nitrous oxide dynamics in an oak-beech forest ecosystem in the Netherlands. Forest Ecol Manage 44: 53-61

Tietema A, Riemer L, Verstraten JM, vanderMaas MP, vanWijk AJ & vanVoorthuyzen I (1993) Nitrogen cycling in acid forest soils subject to increased atmospheric nitrogen input. Forest Ecol Manage 57: 29-44

Townsend AR, Braswell BH, Holland EA & Penner JE (1996) Spatial and temporal patterns in terrestrial carbon storage due to deposition of fossil fuel nitrogen. Ecol Appl 6: 806-814

Ulrich B (1989) Effects of acidic precipitation on forest ecosystems in Europe. In: Adriano DC & Johnson AH (eds) Acidic Precipitation, Vol. 2, Biological and Ecological Effects, pp 189-272. Springer Verlag, New York, U.S.A.

van Breemen N, Burrough PA, Velthorst FJ, vanDobben HF, deWit T, Ridder TB & Reijnders HFR (1982) Soil acidification from atmospheric ammonium sulphate in forest canopy throughfall. Nature 299: 548-550

Vitousek PM & Sanford RL (1986) Nutrient cycling in moist tropical forest. Annu Rev Ecol Syst 17: 137-167

Vitousek PM & Matson PA (1988) Nitrogen transformations in a range of tropical forest soils. Soil Biol Biochem 20: 361-367

Vitousek PM & Howarth RW (1991) Nitrogen limitation on land and in the sea: How can it occur? Biogeochemistry 13: 87-115

Vitousek PM, Chadwick OA, Crews TE, Fownes JH, Hendricks DM & Herbert D (1997) Soil and ecosystem development across the Hawaiian Islands. GSA Today 7: 1-8

Whalen SC (2000) Influence of N and non-N salts on atmospheric methane oxidation by upland boreal forest and tundra soils. Biol Fert Soils 31: 279-287

Williams EA, Hutchinson GL & Fehsenfeld FC (1992) NO_X and N_2O emissions from soil. Global Biogeochem Cycl 6: 351-88

Willison TW, Webster CP, Goulding KWT & Fowler DS (1995) Methane oxidation in temperate forest soils: effects of land use and the chemical form of nitrogen fertilizer. Chemosphere 30: 539-546

Willison TW, Cook R, Müller A & Powlson DS (1996) CH_4 oxidation in soils fertilized with organic and inorganic-N; differential effects. Soil Biol Biochem 28: 135-136

Chapter 6.2

Tropospheric Ozone

Ozone induction of signal and defence pathways leading to the emission of plant volatiles

Christian Langebartels[1], Gabriele Thomas[1], Gerd Vogg[1], Jürgen Wildt[2], Dieter Ernst[1] and Heinrich Sandermann[1]

[1]*Institute of Biochemical Plant Pathology, GSF - National Research Center for Environment and Health, D-85764 Neuherberg, Germany*
[2]*Institut of Chemistry and Dynamics of the Geosphere (ICG-III Phytosphere), Forschungszentrum Jülich, D-52425 Jülich, Germany*

1.　　INTRODUCTION

The life style of land plants resembles that of ancient animal species that stayed in the sea all are in a major aspect filtrating organisms. This life form depends on the availability of finely dispersed nutrients in a transport medium, in the case of sessile animals, seawater. Similarly, plant roots take up mineral salts from soil water via their roots, and transport them in elaborated organs, xylem vessels. In the leaf, the trace gas CO_2 is filtered from the surrounding air, and enormous amounts of the transport medium air are necessary to obtain sufficient amounts of CO_2 for growth. Leaves are perfectly designed to fulfill their filtering function: those of coniferous trees show a very large outer surface. Deciduous trees and herbaceous plants, on the other hand, are characterized by high stomatal uptake rates as well as voluminous leaf gas spaces resulting in inner surfaces that are up to 15-fold higher than the outer leaf surface. As exemplified for the ethylene receptor (Solano and Ecker 1998), the recognition mechanisms for trace elements and trace gases in plants apparently comprise similar receptor proteins, so-called 'phosphorelais', as in bacteria and yeast. Along with trace elements in roots and mycorrhiza and with the trace gas CO_2 in leaves, contaminating organic and inorganic pollutants in water and air are also taken up by the filtrating

R. Gasche et al. (eds.), Trace Gas Exchange in Forest Ecosystems, 307–324.
© 2002 *Kluwer Academic Publishers. Printed in the Netherlands.*

of a necrotizing pathogen. The similarity of ozone and pathogen responses in plants have led to the concept that ozone (or ozone-derived reactive oxygen species, ROS) and pathogens trigger gene expression via common signaling pathways. The gaseous hormone ethylene, the signaling compound salicylic acid, as well as ROS and lipoxygenase pathway products (e.g. jasmonic acid) have been implicated in mediating ozone responses in plants (Sharma et al. 1996; Sandermann et al. 1998; Langebartels et al. 2002). Table 2 summarizes the current knowledge on their biosynthetic pathways, and on the ozone-sensitive steps.

Table 2. Ozone induction of components of signaling pathways in plants

Pathway	Level Transcript (T*), Enzyme (E)	Metabolite
Ethylene	ACC synthase (T, E)	ACC and conjugates,
	ACC oxidase (T)	Ethylene ↑
	SAM synthetase (T)	
Salicylic acid	Phenylalandine ammonia-lyase (T, E)	Salicylic acid and conjugates,
	Benzoic acid-2-hydroxylase (E)	Gentistic acid and conjugates
		Methyl salicylate ↑
Reactive oxygen species	NAD(P)H oxidase	Superoxide
	Oxalate oxidase	H_2O_2 (↑)
	Wall peroxidase	
Lipooxygenase products	Lipooxygenase (T)	Jasmonate and conjugates **, methyl jasmonate ** ↑
	Allene oxide synthase	C_6 aldehydes ↑ and alcohols ↑
		hexenyl acetate ↑

*T, E are indicated when ozone induction was demonstrated at the transcript or enzyme level, respectively; **: Exogenous application. No ozone induction demonstrated; ↑: volatile compound

2.1   Ethylene

Increased ethylene emission from ozone-exposed plants is an early, consistent marker for ozone sensitivity in herbaceous and tree species (Tingey et al. 1976; Sandermann 1996; Wellburn and Wellburn 1996). Daily exposure of European beech with twice-ambient ozone levels leads to re-curring peaks of ethylene emission. In contrast to herbaceous plants, the peak maxima for ethylene emission are delayed in trees and occur during the night, during ozone minima (Anegg and Kalisch unpublished results). It has been proposed that ozone exerts its toxicity through a chemical reaction with ethylene, yielding toxic products that initiate a self-propagating lipid peroxidation cycle (Elstner et al. 1985; Mehlhorn and Wellburn 1987). In trees, however, the peak episodes of ozone and ethylene emission can be separated, so that other, plant-based mechanisms have to be taken into accout to explain ozone toxicity.

In ozone-exposed plants ethylene biosynthesis is a result of the activation of specific isoforms of ACC synthase and ACC oxidase at the transcript level (Bae et al. 1996; Toumainen et al. 1997; Vahala et al. 1998). In addition, specific genes of SAM synthetase, an enzyme being involved in ethylene and polyamine biosynthesis as well as in methylation reactions, were ozone-responsive in tomato (Toumainen et al. 1997). ACC synthase is, in addition, rapidly induced by phosphorylation/dephosphorylation reactions at the protein level (Toumainen et al. 1997). Rodecaop and Tingey (1983) were the first to demonstrate that levels of the immediate precursor of ethylene, the cyclic compound 1-aminocyclopropane-1-carboxylate (ACC), are elevated along with ethylene emission. Conjugated forms of ACC (malonyl and glutamyl ACC) seem to be the major storage forms of ACC, but they can be hydrolyzed under certain conditions to release free ACC (Abeles et al. 1992).

Once activated, ethylene may affect stomatal uptake (Gunderson and Taylor 1991), leading to a transient reduction of ozone deposition, but also to decreased emission of volatiles from ozone-treated plants.

2.2   Salicylic Acid

Salicylic acid is a major phenolic compound involved in defence gene activation and initiation of cell death during oxidative stress (Durner et al. 1997; Reymond and Farmer 1998). It has been implicated in both local and systemic disease resistance towards pathogens. Salicylic acid was reported to interact with and to inhibit the H_2O_2-degrading enzymes catalase and ascorbate peroxidase which in turn could lead to higher cellular H_2O_2 levels (Wendehenne et al. 1998). In addition, a high-affinity binding protein for salicylic acid exists in tobacco, and its characterization is underway (Du and Klessig 1997). Salicylic acid can be formed from phenylalanine through o-coumaric acid or benzoic acid (Klessig et al. 1998). In tobacco, salicylic acid is primarily synthesized from benzoic acid via benzoic acid 2-hydroxylase activity which is TMV- and H_2O_2-inducible. A relatively late increase in benzoic acid 2-hydroxylase activity was found in tobacco (Yalpani et al. 1994), but no information on the induction processes is available so far for this enzyme. Free and conjugated salicylic acid was found to be induced by ozone in *Arabidopsis thaliana* and tobacco (Yalpani et al. 1994; Sharma et

312

al. 1996). In addition, a recent report describes the emission of the volatile metabolite, methyl salicylate, from ozone-sensitive tobacco, with low amounts being released in the tolerant counterpart (Heiden et al. 1999; see below).

2.3   Reactive Oxygen Species

Ozone readily transforms into reactive oxygen species upon decomposition in the apoplastic fluid surrounding leaf mesophyll cells. Thus, ozone-derived ROS have, for a long time, been postulated to trigger membrane deterioration and finally cell death in sensitive plant species and cultivars (reviewed by Heath and Taylor 1997). During recent years, however, this concept has been questioned, and ROS of cellular origin were recognized as amplifying factors in ozone damage. First, the ozone biomonitor plant tobacco Bel W3 was found to accumulate H_2O_2, a major ROS species, in the apoplastic fluid during the phase of post-cultivation in

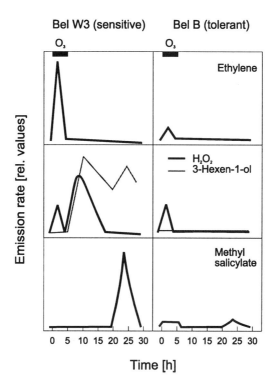

Figure 1. Time course of emission of plant volatiles from the ozone-sensitive tobacco cultivar Bel W3 and the ozone-tolerant cv. Bel B.

pollutant-free air (Fig. 1; Schraudner et al. 1998). It could subsequently be shown that H_2O_2 or superoxide accumulation preceded cell death in ozone-treated *A. thaliana* and birch, respectively (Rao and Davis 1999; Pellinen et al. 1999). ROS accumulation as well as lesion development were substantially reduced by an inhibitor of NAD(P)H oxidase and other flavin-containing oxidases. These results provide evidence that ozone triggers enzymatic processes that form ROS, but do not allow to assign (a) specific enzyme(s) for ROS production. NAD(P)H oxidases, but also diamine and polyamine oxidases, oxalate oxidase as well as wall peroxidases are currently considered as candidate catalysts for cellular ROS production (Scheel 1998; Bolwell 1999). It is conceivable that in different plant systems different enzymes are predominantly responsible for ROS production and transformation, e.g. by superoxide dismutase. Along this line, preliminary results of our group indicate that the predominant type of ROS species (H_2O_2 or superoxide) differs between ozone-exposed plant species and cultivars (Wohlgemuth et al. unpublished results).

2.4 Jasmonate

Finally, jasmonate is an important signal molecule in plant defence reactions. It is produced from linolenic acid via 13-lipoxygenase activity (Reymond and Farmer 1998), and occurs in free and conjugated form, as well as a volatile metabolite (methyl jasmonate). Ozone induction of lipoxygenase genes has been detected in soybean and *Lens culinaris*, but no substrate specificity has been reported (reviewed in Langebartels et al. 2002). Treatment with methyl jasmonate gas protected tobacco plants against ozone injury (Örvar et al. 1997), and ozone sensitivity of a poplar clone has been proposed to result from jasmonate insensitivity (Koch et al. 1998). In addition to jasmonate biosynthesis, 13-lipoxygenase together with hydroperoxide lyase activity ultimately converts 13-hydroperoxylinolenic acid to volatile C_6 aldehydes and alcohols, among which 3-hexen-1-ol and 3-hexen-1-al are major components of the 'green odour' in plants (Hatanaka 1993). It was recently demonstrated that these defined products of a 13-lipoxygenase pathway are released from ozone-treated tobacco during post-cultivation in pollutant-free air (Fig. 2; Heiden et al. 1999). The spectrum of the major emitted compounds clearly points to a specific activation of biosynthetic pathways by the air pollutant ozone rather than to unspecific lipid peroxidation processes which would yield other products such as *trans*-2,*cis*-6-nonadienal.

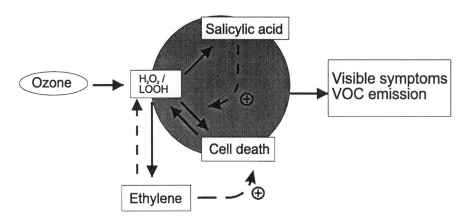

Figure 2. Model of the ozone-induced oxidative cell death cycle leading to visible symptoms and elevated emissions of volatile organic compounds (VOC). LOOH, lipid hydroperoxides and products of the lipoxygenase pathway.

3. RELEASE OF SIGNAL AND DEFENCE MOLECULES INTO THE LEAF APOPLAST

Primary reactions of ozone with plant constituents do occur in the thin (about 2 μm) layer of apoplastic fluid surrounding the protoplast of leaf mesophyll cells. Constituents of this apoplastic fluid, in particular antioxidant compounds and enzymes, have been studied intensively, and are considered responsible for the detoxification of ozone and ozone-derived ROS (Heath and Taylor 1997). Ascorbic acid, in particular, was found in up to millimolar concentrations in the apoplastic fluid, and it was calculated that these levels are sufficient to prevent injury from low ozone concentrations (Polle 1997). However, the role of ascorbate in the apoplastic fluid has also been questioned as oxidation of fluorescent marker dyes could not be reduced by simultaneous infiltration with ascorbate (Jakob and Heber 1998). An ascorbate-deficient mutant of *A. thaliana* was hypersensitive to ozone (Conklin et al. 1997), demonstrating an important function of ascorbate in ozone tolerance.

The release of signal molecules from mesophyll cells into the apoplastic fluid has recently been studied (Moeder et al. 1999). Studies with ozone-sensitive tobacco and *Vicia faba* cultivars revealed a controlled release (between 10 and 30% of the total content) of the ethylene precursor ACC and of salicylic acid into the apoplastic fluid. This extracellular location may be of interest for several reasons: first, ACC may be used as a substrate of an apoplastic-localized ACC oxidase. Apoplastic ACC oxidase was found in apple and tomato fruits (Ramassamy et al. 1998) but no data are currently available for leaves. Secondly, ACC and salicylic acid may be involved in short-distance signalling of defence responses. ACC accumulation has been found to occur in the vicinity of pathogen-induced lesions (Spanu and Boller 1989), and salicylic acid injected into the apoplastic space had rapid access to the vascular tissue (Moeder and Langebartels unpublished results). The apoplastic fluid could therefore function as a link between the local and systemic transport of signals.

H_2O_2 accumulation in pathogen-infected and ozone-treated cells occurs in the apoplastic fluid, outside the plasmalemma (Lamb and Dixon 1997; Schraudner et al. 1998). H_2O_2 accumulation in ozone-treated birch was first observed in the apoplast of palisade and spongy mesophyll cells, while intracellular accumulation, especially in mitochondria, was a late event and obviously occurred together with cell death (Pellinen et al. 1999). The concentrations for H_2O_2, ACC and salicylic acid in the apoplastic fluid are in the range of 0.1 to 1 mM, and it seems possible that the controlled release of signal and defence compounds into this thin fluid layer is effective in impairing pathogen growth. When diluted and transported to neighboring cells, the concentrations are still in the range to mediate local defence reactions.

When the signaling pathways summarized in Table 2 are studied in parallel, it is remarkable to note that all of them can form volatile metabolites. This is, of course, evident for the ethylene pathway where the hormone itself is the volatile compound. Along this line, the receptor proteins for ethylene in *A. thaliana* and tomato are localized in the plasmalemma (Solano and Ecker 1998). They show sequence similarity to 'phosphorelais' proteins in the bacterial periplasmic space which deliver extracellular signals, e.g. redox status, nutrient availability, xenobiotics and heavy metals, to the DNA (Parkinson and Kofoid 1992). Methyl salicylate, which occurs as a volatile salicylic acid metabolite in ozone-treated and TMV-infected tobacco, was effective in mediating defence reactions from the infected to a control plant, via the air path (Shulaev et al. 1997). It has not yet been determined whether H_2O_2 is emitted from plants. On the other hand, H_2O_2 as well as organic peroxides are common components of oxidative air pollution, albeit at small absolute concentrations (usually <1 nL L^{-1}; Cape 1997). Defined 13-lipoxygenase products, C_6 alcohols and aldehydes, are known to occur in wounded and herbivore-infested plants (Paré and Tumlinson 1999). Bean plants infected with *Pseudomonas syringae* pv *phaseolicola* showed a similar spectrum of C_6 volatiles (Croft et al. 1993). Methyl jasmonate, finally, the volatile form of the growth regulator jasmonate, is released from pathogen-infected plants. It is effective in triggering a sub-set of defense-related genes, e.g. thionin *THI2.1*, defensin *PDF1.2* and hevein-like protein *HEL* (Reymond and Farmer 1998).

The emission kinetics of plant volatiles were analyzed in ozone-sensitive and -tolerant tobacco (Fig. 1). Only small quantities were emitted from both cultivars under control conditions. Ethylene was the first compound to be released into the gas phase. Its emission peaked during the exposure period, and was high in the ozone-sensitive cultivar, but not in the tolerant one (Langebartels et al. 1991). An all-or-nothing response was found for C_6 aldehydes and alcohols, with cis-3-hexen-1-ol as lead substance (Heiden et al. 1999). This volatile peaked after 10 h in Bel W3, and was below the detection level in Bel B. The peak occurred together with peak II of H_2O_2 accumulation in the apoplastic fluid of cv. Bel W3 (Fig. 1; Schraudner et al. 1998), and well before visible injury took place. It is probable that these cytotoxic C_6 aldehydes and alcohols play a role in lesion formation. Methyl salicylate occurred very late in ozone-treated tobacco, at a time when cell death was already visible in Bel W3 (Fig. 1). The emitted dose in Bel W3 was by a factor of 15 higher than in Bel B. It is postulated that methyl salicylate is emitted only when cells are undergoing cell death, and that the molecule is present in the liquid phase in an earlier phase. Small amounts of methyl salicylate were emitted from cv. Bel B during the light period also in pollutant-free air (Heiden et al. 1999). As salicylic acid is known to potentiate defence reactions even at low concentrations (Draper 1997), this daily accumulation may predispose the tolerant cultivar towards oxidative stress as well as pathogen infection.

When the three episodes of ozone induction are considered *in toto*, it is evident that following ozone exposure, and probably oxidative stress in general, intracellular defence reactions are activated first, often at the level of transcription (episode I). Thus, the responses are directed, in the case of pathogen attack, to protect the invaded cell. In episode II, local cell-to-cell signaling leads to an induction of antioxidant and antimicrobial defence responses in the vicinity of the attacked cell. Episode III comprises systemic movement of signal molecules in the vascular tissue (salicylic acid and additional compounds; Klessig et al. 1998) and, in parallel, emission of volatile compounds which may transmit the message through the gas phase to distal tissues of the leaf. As shown for methyl salicylate, airborne signals potentially activate defence responses even in non-infected neighboring plants (Shulaev et al. 1997). Volatile methyl jasmonate was also active in defence gene activation (Reymond and Farmer 1998). In addition to volatile metabolites, conjugates of ACC, salicylic and gentistic acid, jasmonate and cis-3-hexen-1-ol (the volatile hexenyl acetate) are formed in episode III. Both processes help to limit the effective time period for these highly active signal molecules, by releasing them into the surrounding air (volatile compounds) or by storing them (in an inactive conjugate form) in the vacuole for rapid release of the signal in subsequent stress situations.

4. REACTIVE OXYGEN SPECIES, ETHYLENE AND SALICYLATE CO-OPERATE IN AN OXIDATIVE CELL DEATH CYCLE

Superoxide radicals are necessary and sufficient to initiate 'run-away' cell death in *Arabidopsis* mutants while H_2O_2 alone does not initiate HR-like lesions (Jabs et al. 1996; Dorey et al. 1999). Ozone exposure activated the accumulation of H_2O_2, not superoxide, in tobacco (Schraudner et al. 1998), birch (Pellinen et al. 1999) and tomato (Mittelstraß et al. unpublished) suggesting that H_2O_2 is not the sole signal involved in lesion formation. It is now commonly accepted that ROS signalling interacts with at least three other pathways, i.e. salicylic acid as well as ethylene and jasmonate signaling (Dong 1998; McDowell and Dangl 2000). An oxidative cell death (OCD) cycle was postulated for plant-pathogen interactions. In this model, ROS, salicylate and cell death are implicated in a self-amplifying cycle ultimately promoting lesion development and pathogen killing (Van Camp et al. 1998). Thus, H_2O_2 functions both upstream and downstream of salicylic acid in the pathway. Salicylic acid potentiates ROS toxicity and thereby contributes to cell death. This oxidative cell death cycle could also operate under abiotic stress, e.g. ozone exposure, as summarized in Figure 2. Ozone leads to an initial (peak I) increase of H_2O_2 or other ROS which in turn increase salicylate levels (Rao and Davis 1999), this signal promoting a second burst of H_2O_2 (peak II) in sensitive tobacco (Schraudner et al. 1998). Recent studies indicate that ethylene is not only an indicator of damage but an active component enhancing the damage process. When ethylene biosynthesis, perception or signalling was blocked, by inhibitors in ozone-treated tobacco and tomato or by mutation in the *Arabidopsis* mutant *ein2* (Tuomainen et al. 1997; Overmyer et al. unpublished results), ROS accumulation and subsequent cell death were both repressed. These results suggest that ethylene promotes ROS accumulation and plays a role in the initiation and propagation phases of lesion development.

Leaves normally release small quantities of volatile compounds, but when a plant is damaged by abiotic or biotic factors, the quantity and the number of volatiles increase dramatically (Croft et al. 1993; Heiden et al. 1999). Under various stress conditions the interaction between ROS, ethylene, salicylate (and other signals, e.g. jasmonate and NO; Durner and Klessig 1999; McDowell and Dangl 2000) is thought to determine the degree and extent of lesion formation. It is this common mechanism that forms the basis of the principal structural uniformity of the emitted compounds in various plant families, albeit with a specific 'bouquet' for individual species. When the OCD cycle is operative in ozone-sensitive plants, it leads to highly increased VOC emission by two to three orders of magnitude (Fig. 2). This is not only found for the above potential signal molecules (e.g. 50x, ethylene; 130x, hexenyl acetate; 600x, methyl salicylate) but also for several sesquiterpenes that are emitted on the day following the treatment (e.g. >100x for valencene; Heiden et al. 1999). The ozone-triggered emission of sesquiterpenes and other compounds that possess ozone generation potential in turn leads to back-coupling effects, and potentially elevated ozone levels in combination with nitrogen oxides. Depending on the nature of the compounds, ozone reduction by chemical reactions between ozone and VOC in the gas phase is also possible. When plants in the field are considered, highest ozone generation (or reduction) potential is expected when the OCD cycle is triggered, and visible symptoms occur, either caused by abiotic stress (ozone, wounding) or biotic interactions.

5. CONCLUSIONS AND PERSPECTIVES

Research during recent years has shifted from the analysis of ozone as a direct oxidant to its potential role as a trigger of plant responses that reduce or amplify ozone effects in ozone-tolerant and -sensitive plants, respectively. Various studies have demonstrated that plant responses to ozone are highly controlled, e.g. only specific isoforms of antioxidant or ethylene biosynthetic enzymes are induced. It can therefore be concluded that unspecific oxidative reactions play no major role in ozone toxicity (Langebartels et al. 2002). It is proposed that the newly discovered damage amplification reactions are to be incorporated into the currently discussed 'critical levels' concept (level II approach) for ozone (Kärenlampi and Skärby 1996).

Various signal pathways are activated in parallel in ozone-sensitive tobacco and other species. It can therefore be speculated that ozone or ozone-derived ROS affect processes that occur 'upstream' of the above signal pathways. The earliest detectable cellular events after pathogen invasion are ion fluxes across the plasma membrane as well as mechanisms involving protein phosphorylation and GTP-binding proteins (McDowell and Dangl 2000). Protein phosphorylation has been implicated in the transduction of ROS, ethylene and salicylic acid signals. In the case of salicylic acid, two phosphoproteins acting 'upstream' and 'downstream' of salicylic acid have been characterized (Conrath et al. 1997; Zhang and Klessig 1997). The latter, a salicylic acid induced protein kinase (SIPK) belongs to the mitogen-activated protein (MAP) kinase family (Zhang and Klessig 1997). Ethylene signaling involves a MAP kinase cascade 'downstream' of recognition by the ethylene receptor (Solano and Ecker 1998). In addition, the ethylene biosynthetic enzyme, ACC synthase, is activated postranslationally by phosphorylation events, and inhibition of protein kinases led to reduced ozone induction (Tuomainen et al. 1997). H_2O_2 can also activate MAP kinase cascades as recently shown in tobacco and *A. thaliana* (Desikan et al. 1999; Kovtun et al. 2000). It is therefore possible that ozone is recognized by (a) redox sensitive protein(s), an 'ozone receptor' according to Sandermann (1996), and the signal is initially mediated by induction of (MAP) kinase activity and/or reduction of protein phosphorylases. Ozone may be used as a tool to unravel these oxidative stress signaling mechanisms.

The plethora of effects triggered by ozone influence the disposition state of the plant in subsequent stress situations. This has been demonstrated in plant-pathogen as well as -herbivore interactions where the same or overlapping pathways are used (Paré and Tumlinson 1999; McDowell and Dangl 2000). For example, the pattern of volatile compounds renders a leaf attractive or disagreeable to herbivores, but also allows insect parasitoids and predators to locate hosts on infested plants (Paré and Tumlinson 1999). Periods of high ambient ozone essentially do not co-incide with infection periods of most pathogens. However, as various ozone-induced metabolic changes are maintained in plants over days to months, ozone episodes in the field may through these 'memory' effects also be connected to a later infection period (Sandermann 2000). The changed metabolite status of ozone-exposed plants may lead to either enhanced or decreased likelihood of disease, depending on the individual herbivore species or viral, bacterial or fungal pathogen. As ozone responses have now been described at the transcript, protein and metabolite level, and the similarity to pathogen and herbivore responses is evident, the underlying mechanisms of predisposition or enhanced resistance can be analyzed in the coming years.

Recent findings of NO as an important signaling molecule in plants (Delledonne et al. 1998, Klessig et al. 1998)), suggest that its effects as a phytotoxic air pollutant have to be re-analyzed. It is, for example, interesting to note that Mehlhorn and Wellburn, (1987) found stress ethylene emission in pea after NO (or NO_2) exposure, as well as increased sensitivity towards ozone by pre-treatment with NO. NO is known to work synergistically with salicylic acid and ROS to activate defined defence responses, but also antagonistic effects with salicylic acid and H_2O_2 have been reported (Durner and Klessig, 1999). Up to now, most of the above data were obtained with model plants, in particular tobacco and *A. thaliana*. Information on forest trees and crop plants is not readily available, but can build on the progress achieved with model plants.

ACKNOWLEDGEMENTS

The excellent contributions of our coworkers as well as cooperation with M. Schraudner, Forschungszentrum Jülich, and J. Kangasjärvi, Helsinki, are gratefully acknowledged. This work was supported by grants from BStMLU, DFG (SFB 607), BMBF and EU-FAIR.

REFERENCES

Abeles FB, Morgan PW & Saltveit ME (1992) Ethylene in Plant Biology. Academic Press, New York, U.S.A.

Bae GY, Nakajima N, Ishizuka K & Kondo N (1996) The role in ozone phytotoxicity of the evolution of ethylene upon induction of 1-aminocyclopropane-1-carboxylic acid synthase by ozone fumigation in tomato plants. Plant Cell Physiol 37: 129-134

Bauer S, Galliano H, Pfeiffer F, Messner B, Sandermann H & Ernst D (1993) Isolation and characterization of a cDNA clone encoding a novel short-chain alcohol dehydrogenase from Norway spruce (*Picea abies* L. Karst). Plant Physiol 103: 1479-1480

Bolwell GP (1999) Role of active oxygen species and NO in plant defence responses. Curr Opin Plant Biol 2: 287-294

Buschmann K, Etscheid M, Riesner D & Scholz F (1998) Accumulation of a porin-like mRNA and a metallothionein-like mRNA in various clones of Norway spruce upon long-term treatment with ozone. Eur J For Pathol 28: 307-322

Cape JN (1997) Photochemical oxidants - What else is in the atmosphere besides ozone? Phyton 37: 45-58

Conklin P, Pallanca J, Last R & Smirnoff N (1997) L-Ascorbic acid metabolism in the ascorbate-deficient *Arabidopsis* mutant vtc1. Plant Physiol 115: 1277-1285

Conrath U, Silva H & Klessig DF (1997) Protein dephosphorylation mediates salicylic acid-induced expression of PR-1 genes in tobacco. Plant J 11: 747-757

Croft KPC, Jüttner F & Slusarenko AJ (1993) Volatile products of the lipoxygenase pathway evolved from *Phaseolus vulgaris* (L.) leaves inoculated with *Pseudomonas syringae* pv *phaseolicola*. Plant Physiol 101: 13-24

Delledonne M, Xia Y, Dixon RA & Lamb C (1998) Nitric oxide functions as a signal in plant disease resistance. Nature 394: 585-588

Desikan R, Clarke A, Hancock JT & Neill SJ (1999) H_2O_2 activates a MAP kinase-like enzyme in *Arabidopsis thaliana* suspension cultures. J Exp Bot 50: 1863-1866

Dong X (1998) SA, JA, ethylene, and disease resistance in plants. Curr Opin Plant Biol 1: 316-323

Dorey S, Kopp M, Geoffroy P, Fritig B & Kauffmann S (1999) Hydrogen peroxide from the oxidative burst is neither necessary nor sufficient for hypersensitive cell death induction, phenylalanine ammonia lyase stimulation, salicylic acid accumulation, or scopoletin consumption in cultured tobacco cells treated with elicitin. Plant Physiol 121: 163-172

Draper J (1997) Salicylate, superoxide synthesis and cell suicide in plant defence. Trends Plant Sci 2: 162-166

Du H & Klessig DF (1997) Identification of a soluble, high affinity salicylic acid-binding protein from tobacco. Plant Physiol 113: 1319-1327

Durner J & Klessig DF (1999) Nitric oxide as a signal in plants. Curr Opin Plant Biol 2: 369-374

Durner J, Shah J & Klessig DF (1997) Salicylic acid and disease resistance in plants. Trends Plant Sci 2: 266-274

Durner J, Wendehenne D & Klessig D (1998) Defense gene induction in tobacco by nitric oxide, cyclic GMP, and cyclic ADP-ribose. P Natl Acad Sci USA 95: 10328-10333

Elstner EF, Osswald W & Youngman RJ (1985) Basic mechanisms of pigment bleaching and loss of structural resistance in spruce (*Picea abies*) needles: advances in phytomedical diagnostics. Experientia 41: 591-597

Galliano H, Cabané M, Eckerskorn C, Lottspeich F, Sandermann H & Ernst D (1993) Molecular cloning, sequence analysis and elicitor-/ozone-induced accumulation of cinnamyl alcohol dehydrogenase from Norway spruce (*Picea abies* L.). Plant Mol Biol 23: 145-156

Gunderson CA & Taylor GE (1991) Ethylene directly inhibits foliar gas exchange in *Glycine max*. Plant Physiol 95: 337-339

Hatanaka A (1993) The biogeneration of green odour by green leaves. Phytochemistry 34: 1201-1218

Heath RL & Taylor GE (1997) Physiological processes and plant responses to ozone exposure. In: Sandermann H, Wellburn AR & Heath RL (eds) Forest decline and ozone: A comparison of controlled chamber and field experiments. Ecol. Studies Vol. 127, pp 317-368. Springer Verlag, Berlin, Germany

Heiden AC, Hoffmann T, Kahl J, Kley D, Klockow D, Langebartels C, Mehlhorn H, Sandermann H, Schraudner M, Schuh G & Wildt J (1999) Emission of volatile organic compounds from ozone-exposed plants. Ecol Appl 9: 1160-1167

Jabs T, Dietrich RA & Dangl JL (1996) Initiation of runaway cell death in an *Arabidopsis* mutant by extracellular superoxide. Science 273: 1853-1856

Jakob B & Heber U (1998) Apoplastic ascorbate does not prevent the oxidation of fluorescent amphiphilic dyes by ambient and elevated concentrations of ozone in leaves. Plant Cell Physiol 39: 313-322

Kärenlampi L & Skärby L (1996) Critical levels for ozone in Europe: Testing and finalizing the concepts. UN-ECE workshop report. University of Kuopio, Kuopio

Kiiskinen M, Korhonen M & Kangasjärvi J (1997) Isolation and characterization of cDNA for a plant mitochondrial phosphate translocator (*Mpt1*). Ozone stress induces Mpt1 mRNA accumulation in birch (*Betula pendula* Roth). Plant Mol Biol 35: 271-279

Klessig DF, Durner J, Shah J & Yang Y (1998) Salicylic acid-mediated signal transduction in plant disease resistance. In: Romeo JT, Downum KR & Verpoorte R (eds) Phytochemical signals and plant-microbe interactions, Vol 32, pp 119-137. Plenum Press, New York, U.S.A.

Kley D, Kleinman M, Sandermann H & Krupa S (1999) Photochemical oxidants: State of the science. Environ Pollut 100: 19-42

Koch JR, Scherzer AJ, Eshita SM & Davis KR (1998) Ozone sensitivity in hybrid poplar is correlated with a lack of defense-gene activation. Plant Physiol 118: 1243-1252

Kovtun Y, Chiu WL, Tena G & Sheen J (2000) Functional analysis of oxidative stress-activated mitogen-activated protein kinase cascades in plants. P Natl Acad Sci USA 97: 2940-2945

Lamb C & Dixon RA (1997) The oxidative burst in plant disease resistance. Annu Rev Plant Phys 48: 251-275

Langebartels C, Kerner K, Leonardi S, Schraudner M, Trost M, Heller W & Sandermann H (1991) Biochemical plant responses to ozone. I. Differential induction of polyamine and ethylene biosynthesis in tobacco. Plant Physiol 95: 882-889

Langebartels C, Schraudner M, Heller W, Ernst D & Sandermann H (2002) Oxidative stress and defense reactions in plants exposed to air pollutants and UV-B radiation. In: Inzé D & Van Montagu M (eds) Oxidative stress in plants. pp. 105-135. Taylor and Francis, London, U.K.

McDowell JM & Dangl JL (2000) Signal transduction in the plant immune response. Trends Biochem Sci 25: 79-82

Mehlhorn H & Wellburn AR (1987) Stress ethylene formation determines plant sensitivity to ozone. Nature 327: 417-418

Miller JD, Arteca RN & Pell EJ (1999) Senescence-associated gene expression during ozone-induced leaf senescence in *Arabidopsis*. Plant Physiol 120: 1015-1023

Moeder W, Anegg S, Thomas G, Langebartels C & Sandermann H (1999) Signal molecules in ozone activation of stress proteins in plants. In: Smallwood MF, Calvert CM & Bowles DJ (eds) Plant responses to environmental stress, pp 43-49. BIOS, Oxford, U.K.

Örvar BL, McPherson J & Ellis BE (1997) Pre-activating wounding response in tobacco prior to high-level ozone exposure prevents necrotic injury. Plant J 11: 203-212

Pääkkönen E, Seppänen S, Holopainen T, Kokko H, Kärenlampi S, Kärenlampi L & Kangasjärvi J (1998) Induction of genes for the stress proteins PR-10 and PAL in relation to growth, visible injuries and stomatal conductance in birch (*Betula pendula*) clones exposed to ozone and/or drought. New Phytol 138: 295-305

Paré PW & Tumlinson JH (1999) Plant volatiles as a defense against insect herbivores. Plant Physiol 121: 325-331

Parkinson JS & Kofoid EC (1992) Communication modules in bacterial signaling proteins. Annu Rev Genet 26: 71-112

Pellinen R, Palva T & Kangasjärvi J (1999) Subcellular localization of ozone-induced hydrogen peroxide production in birch (*Betula pendula*) leaf cells. Plant J 20: 349-356

Polle A (1997) Defense against photooxidative damage in plants. In: Scandalios JG (ed) Oxidative stress and the molecular biology of antioxidant defenses, pp 623-666. Cold Spring Harbor Laboratory Press, Plainview, U.S.A.

Ramassamy S, Olmos E, Bouzayen M, Pech JC & Latché A (1998) 1-Aminocyclopropane-1-carboxylate oxidase of apple fruit is periplasmic. J Exp Bot 49: 1909-1915

Rao MV & Davis KR (1999) Ozone-induced cell death occurs via two distinct mechanisms in *Arabidopsis*: The role of salicylic acid. Plant J 17: 603-614

Reymond P & Farmer EE (1998) Jasmonate and salicylate as global signals for defense gene expression. Curr Opin Plant Biol 1: 404-411

Rodecap KD & Tingey DT (1983) The influence of light on ozone-induced 1-aminocyclopropane-1-carboxylic acid and ethylene production from intact plants. Z Pflanzenphysiol 110: 419-427

Sandermann H (1994) Higher plant metabolism of xenobiotics: The 'green liver' concept. Pharmacogenetics 4: 225-241

Sandermann H (1996) Ozone and plant health. Annu Rev Phytopathol 34: 347-366

Sandermann H (2000) Ozone / biotic disease interactions: Molecular biomarkers as a new experimental tool. Environ Pollut 108: 327-332

Sandermann H, Ernst D, Heller W & Langebartels C (1998) Ozone: An abiotic elicitor of plant defense reactions. Trends Plant Sci 3: 47-50

Scheel D (1998) Resistance response physiology and signal transduction. Curr Opin Plant Biol 1: 305-310

Schneiderbauer A, Back E, Sandermann H & Ernst D (1995) Ozone induction of extensin mRNA in Scots pine, Norway spruce and European beech. New Phytol 130: 225-230

Schraudner M, Möder W, Wiese C, Van Camp W, Inzé D, Langebartels C & Sandermann H (1998) Ozone-induced oxidative burst in the ozone biomonitor plant, tobacco Bel W3. Plant J 16: 235-245

Schubert R, Fischer R, Hain R, Schreier PH, Bahnweg G, Ernst D & Sandermann H (1997) An ozone-responsive region of the grapevine resveratrol synthase promoter differs from the basal pathogen-responsive sequence. Plant Mol Biol 34: 417-426

Sharma YK & Davis KR (1997) The effects of ozone on antioxidant responses in plants. Free Radical Biol Med 23: 480-488

Shulaev V, Silverman P & Raskin I (1997) Airborne signalling by methyl salicylate in plant pathogen interactions. Nature 385: 718-721

Solano R & Ecker JR (1998) Ethylene gas: perception, signaling and response. Curr Opin Plant Biol 1: 393-398

Spanu P & Boller T (1989) Ethylene biosynthesis in tomato plants infected by *Phytophthora infestans*. J Plant Physiol 134: 533-537

Tingey DT, Standley C & Field RW (1976) Stress ethylene evolution: A measure of ozone effects on plants. Atmos Environ 10: 969-974

Tuomainen J, Betz C, Kangasjärvi J, Ernst D, Yin ZH, Langebartels C & Sandermann H (1997) Ozone induction of ethylene emission in tomato plants: Regulation by differential accumulation of transcripts for the biosynthetic enzymes. Plant J 12: 1151-1162

Tuomainen J, Pellinen R, Roy S, Kiiskinen M, Eloranta T, Karjalainen R & Kangasjärvi J (1996) Ozone affects birch (*Betula pendula* Roth) phenylpropanoid, polyamine and active oxygen detoxifying pathways at biochemical and gene expression level. J Plant Physiol 148: 179-188

Vahala J, Schlagnhaufer CD & Pell EJ (1998) Induction of an ACC synthase cDNA by ozone in light-grown *Arabidopsis thaliana* leaves. Physiol Plant 103: 45-50

Van Camp W, Van Montagu M & Inzé D (1998) H_2O_2 and NO: redox signals in disease resistance. Trends Plant Sci 3: 330-334

Wegener A, Gimbel W, Werner T, Hani J, Ernst D & Sandermann H (1997a) Molecular cloning of ozone-inducible protein from *Pinus sylvestris* L. with high sequence similarity to vertebrate 3-hydroxy-3-methylglutaryl-CoA synthase. Biochim Biophys Acta 1350: 247-252

Wegener A, Gimbel W, Werner T, Hani T, Ernst D & Sandermann H (1997b) Sequence analysis and ozone-induced accumulation of polyubiquitin mRNA in *Pinus sylvestris*. Can J For Res 27: 945-948

Wellburn FAM & Wellburn AR (1996) Variable patterns of antioxidant protection but similar ethene emission differences in several ozone-sensitive and ozone-tolerant plant selections. Plant Cell Environ 19: 754-760

Wendehenne D, Durner J, Chen Z & Klessig DF (1998) Benzothiadiazole, an inducer of plant defenses, inhibits catalase and ascorbate peroxidase. Phytochemistry 47: 651-657

Yalpani N, Enyedi AJ, León J & Raskin I (1994) Ultraviolet light and ozone stimulate accumulation of salicylic acid, pathogenesis-related proteins and virus resistance in tobacco. Planta 193: 372-376

Zhang S & Klessig DF (1997) Salicylic acid activates a 48 kD MAP kinase in tobacco. Plant Cell 9: 809-824

Zinser C, Ernst D & Sandermann H (1998) Induction of stilbene synthase and cinnamyl alcohol dehydrogenase mRNAs in Scots pine (*Pinus sylvestris* L.) seedlings. Planta 204: 169-176

SUBJECT INDEX

325

Tree Physiology

1. F.J. Bigras and S.J. Colombo (eds.): *Conifer Cold Hardiness*. 2001
    ISBN 0-7923-6636-0
2. S. Huttunen, H. Heikkilä, J. Bucher, B. Sundberg, P. Jarvis and R. Matyssek (eds.):
    Trends in European Forest Tree Physiology Research. Cost Action E6: EUROSILVA.
    2001 ISBN 1-4020-0023-5
3. R. Gasche, H. Papen and H. Rennenberg (eds.): *Trace Gas Exchange in Forest
    Ecosystems*. 2002 ISBN 1-4020-1113-X

KLUWER ACADEMIC PUBLISHERS – DORDRECHT / BOSTON / LONDON

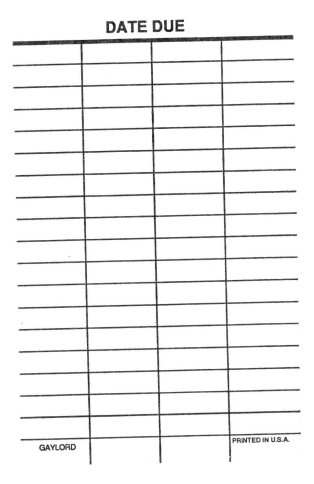